I0043592

Randal Charles John Nixon

Elementary plane trigonometry,

that is, plane trigonometry without imaginaries

Randal Charles John Nixon

Elementary plane trigonometry,
that is, plane trigonometry without imaginaries

ISBN/EAN: 9783744738941

Printed in Europe, USA, Canada, Australia, Japan

Cover: Foto ©berggeist007 / pixelio.de

More available books at **www.hansebooks.com**

Clarendon Press Series

ELEMENTARY PLANE TRIGONOMETRY

NIXON

London
HENRY FROWDE
OXFORD UNIVERSITY PRESS WAREHOUSE
AMEN CORNER, E.C.

New York
112 FOURTH AVENUE

Clarendon Press Series

ELEMENTARY PLANE TRIGONOMETRY

that is

Plane Trigonometry without Imaginaries

by

R. C. J. NIXON, M.A.

AUTHOR OF 'EUCLID REVISED,' 'GEOMETRY IN SPACE, ETC.

Oxford

AT THE CLARENDON PRESS

1892

Oxford

PRINTED AT THE CLARENDON PRESS

BY HORACE HART, PRINTER TO THE UNIVERSITY

PREFACE

It is customary to define an elementary course of trigonometry as *Trigonometry to the end of solution of Triangles.* But this definition is obviously very vague, and settles neither what of the subject is included nor what is excluded. Casting about for some natural line of demarcation between the regions of elementary and higher trigonometry, it occurred to me that such a line is given by the use or non-use of the symbol $\sqrt{-1}$. Thus Elementary Trigonometry may be appropriately defined as that part of the subject which can be done *without* the use of imaginaries; while the remainder, which requires imaginaries, will then constitute Higher Trigonometry.

The present book treats of Elementary Trigonometry, defined as above; and the points proposed in its composition are—

1°, to include only what specially concerns the Trigonometrical Functions—excluding all that seems more properly to appertain to Arithmetic, Algebra, Mensuration, or Pure Geometry; hence, for example, the theory of logarithms is omitted, as strictly belonging to treatises on Algebra: the only exceptions to this rule are that a few extraneous matters, which seemed indispensable to the subject, yet could not be well disposed of by reference to text-books, are treated of in a Preliminary Chapter; and that, for the same reasons, a Note on Convergence precedes the treatment of series.

2°, to utilize the gain of space thus acquired for a more thorough and complete amplification than is usual of what does specially concern the Trigonometrical Functions.

3°, to give all definitions and proofs in their fullest generality, and with the strictest accuracy *from the first*; my experience being entirely against provisional and limited definitions or proofs, followed afterwards by more complete extensions.

4°, to work out in full a large number of specimen Examples—150 of these are given—and to add hints to many of the Exercises.

In the arrangement of the order of treatment, and in the details of the proofs, I have not followed, or even been guided by, any previous treatise; but have relied on my own judgment, based on the experience gained in twenty-five years' labour as a teacher of the subject. Exact logical accuracy of sequence and method has been everywhere aimed at; and the rule followed has been to introduce everything when, and not until, it is needed, and then to treat it completely. Of course a subject of this kind cannot be *learned* without studying it in books; but, in *writing* this work, I have had very little direct help from any books, unless it be from the two Trigonometries of the late Professor De Morgan. For the unique accuracy and thoroughness of all De Morgan's writings I feel unbounded admiration, and to them I am largely indebted. There have lately appeared some four or five Trigonometries which, judging from their advertisements, are of an elaborate character. To avoid the possibility of being influenced by them in any way, I have avoided even looking at these; nor have I any idea what they contain or do not contain. Indeed any other course with regard to contemporaneous works seems to be quite unjustifiable. For an author to make himself acquainted with the modern books with which he is, by the nature of the case, in direct competition, may be a very ingenious way of surpassing them; but the proceeding is, to say the least, questionable.

The whole book was read, both in MS. and 'proof,' by pupils of mine at the Belfast Institution; and I relied on them, almost exclusively, to aid me in correcting press errors. They also worked a large number—from a half to three-quarters—of the 1200 Exercises. I have thus, as the book was in progress, been in constant touch with the exact class for whom it is intended—the upper boys at schools, who are preparing to compete for University Scholarships.

My grateful acknowledgements are due to—

1°, the Rev. J. Milne (of Dulwich) for leave to take what I liked from his *Companion*: this kind liberality was very useful to me in the chapter wherein the Lemoine and Brocard circles are briefly treated of.

2°, Professor Purser (of Queen's College, Belfast) for his ingenious elementary proof of *Gregorie's Series*, printed on pages 316 to 318: this proof has been just recently invented by him, and is now here first published.

3°, Mr. R. Chartres (of Manchester) for his original Theorem, printed on page 330, of which he has given me the copyright: by means of this very useful Theorem, the summation of numbers of series, hitherto considered to require the use of $\sqrt{-1}$, can be effected by elementary trigonometry.

Some other new proofs sent to me are acknowledged in their respective places.

The late Professor Wolstenholme not only gave me permission to make any use I pleased of his *Problems*, but also, when I sent him a list of those which I proposed to extract, most kindly returned me answers and corrections in all cases where such were needed: every *Problem* taken is indicated in its place.

Of the other Exercises many are taken from the *Educational Times' Reprints*—references to the volumes being added—a few are private, and the rest come from various public sources, frequently indicated. Those to which the letters Q. C. B. are appended were set by Professor Purser.

In the subsidiary matters of type, diagrams and pagination,

neither expense nor trouble has been spared to make the work perspicuous. In regard to the settlement of the details connected with these, I have to thank the authorities of the *Clarendon Press* for full permission to have my own ideas carried out in their integrity.

As far as practicable the pages have been arranged so that a diagram and its corresponding text may be on view together, without turning a page: see, for example, pages 48 to 63. Attention to such details adds greatly to a Student's ease and comfort in studying.

I hope some day to supplement this Elementary Trigonometry by a small companion volume to be called *De Moivre's Theorem and its Consequences*.

R. C. J. NIXON

ROYAL ACADEMICAL INSTITUTION, BELFAST
May, 1892

CONTENTS

including formulæ which should be known

Abbreviated references, used throughout the volume

Ox'	*Oxford*
Camb' . . .	*Cambridge*
Math' Tri' . . .	*Mathematical Tripos*
T. C. D. . .	*Trinity College, Dublin*
L. U. . .	*London University*
R. U. . . .	*Royal University*
Q. C. B. . . .	*Queen's College, Belfast*
I. I.	*Irish Intermediate*
Q. J. . .	*Quarterly Journal*
E. T. . .	*Educational Times*
E. R.	*Euclid Revised*

Particular Colleges at Cambridge are indicated thus—
Pet' for *Peterhouse*: *Joh'* for *John's*: &c

Corrigenda

Page	Line	For	Put
27	7	at end of book	in Contents
29	18	P (in 1st den'r)	N
56	6 from end	P (in 2nd den'r)	Q
103	last but one	$\ngtr 1$	is pos'
133	22	$4 x_1 x_2 x_3 \Sigma x_1$	$2 x_1 x_2 x_3 \Sigma x_1$
136	12	$\tan(\pi/7 \pm \phi)$, &c.	$\tan(\pi/7 + \phi)$, $-\tan(\pi/7 - \phi)$, &c
,,	6 from end	$+$	$-$
149	17	$2(R-r)$	$2(R+r)$
169	14	64800	648000
,,	last but one	π	α
183	7	cot C	cos C
193	Example 1	19″	41″
,,	Exercise 2	9·590951	9·5909351
,,	Exercise 4	9·657056	9·9657056
211	last but one	Hyparchus	Hipparchus
220	last	c′	C′
221	last	$\sec(\alpha-\phi)$	$\cos(\alpha-\phi)$
222	last but one	a^2	a
223	6 and 18	$\log b + \log c$	$\frac{1}{2}(\log b + \log c) + \log 2$
225	Exercise 8	$+px$	$-px$
231	5 from end	A	B
236	Exercise 8	9·717937	9·717973
,,	,,	·8729345	·8929345
,,	,,	9·8746794	9·8946794
289	6 from end	$S_1 A$	$S_2 A$

PRELIMINARY

Angles—Functions—Graphs

TRIGONOMETRY, in its modern developments, is the Science of the mathematical representation of Periodic Magnitude—that is of magnitude which alternately increases to a *maximum*, then decreases to a *minimum*; and continues this process in such a way that the same values of the magnitude are continually reproduced, in the same order, and at equal intervals whether of time or space *.

This representation is effected by the aid of angles; and, in making it, we find that the Euclidian definition of an angle is too limited for our purposes. It may indeed be noted that, though Euclid gives a definition of an angle which limits its magnitude to less than the sum of two right angles, he does tacitly assume the existence of greater angles—

1°, in iii. 20, where if the angle at the circumference is greater than a right angle, then that at the centre will be greater than two right angles; and,

2°, in vi. 33, where the statement, that *any* equimultiples whatever are to be taken of the angles, implies that there is *no limit at all* to the possible magnitude of an angle.

Our enlarged conception of an angle is this—

Def'—If a straight line is supposed to rotate, in one plane, about an extremity which remains fixed, then, in passing from any one position to any other position, it is said to **generate an angle**; the rotating line is called the **generator** of the angle;

---

* Defining *undulating magnitude* as "magnitude which becomes alternately greater and less, without any termination to succession of increase and decrease," De Morgan calls Trigonometry the "Science of continually undulating magnitude."

and the amount of turning, which carries the generator round from the one position to the other, is the measure of the angle between the two positions.

Examples

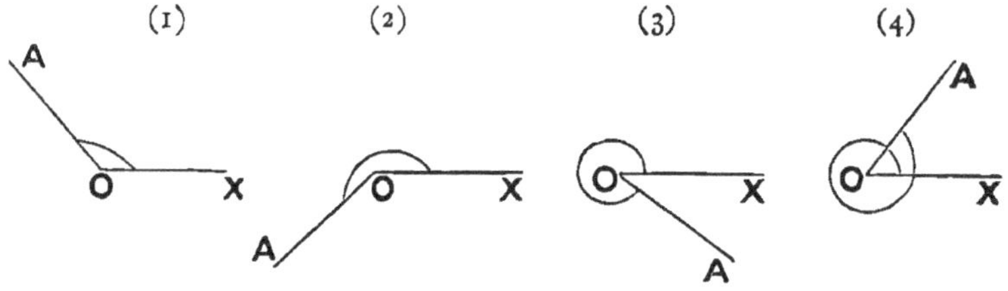

In these diagrams, OX represents the initial, and OA the final, position of the rotating line—the rotation being supposed to have taken place in a manner contrary to that in which the hands of a clock move.

The rotation represented is—

in (1)	through	more	than	one	and	less than	two	right angles
,, (2)	,,	,,	,,	two	,,	,,	three	,,
,, (3)	,,	,,	,,	three	,,	,,	four	,,
,, (4)	,,	,,	,,	four	,,	,,	five	,,

Def'—Angles conceived of as formed by the rotation of a line through more than two right angles have been sometimes distinguished as *__goniometrical angles.__

Note—To every goniometrical angle corresponds a geometrical angle whose arms are the initial and final position of the rotating line. For example, in figure (3), to the goniometrical angle, formed by rotation anti-clock-wise from OX to OA, corresponds the geometrical angle AOX, representing the inclination of OA to OX in the ordinary (i. e. Euclidian) sense. To prevent confusion between the graphic representation of a goniometrical, and the corresponding geometrical angle, it is necessary to indicate the amount and direction of turning in some such manner as is done in figures (2), (3), (4).

Def'—When two magnitudes are so related that any change in the one is invariably followed by a corresponding change in the other, then either is said to be a **function** of the other.

---

* *Peacock's Algebra*; vol. II, p. 147.

Examples

If x, y are variable; and a, b are constant; then x and y will be *functions*, each of the other, if any such relation holds between them as—

$$(1)\quad x^3 + ax^2y + bxy^2 + y^3 = 0$$

$$\text{or}\quad (2)\quad y^2 = a^{bx}$$

$$\text{or}\quad (3)\quad y = b\log_a x$$

Again, all such expressions as $ax^2 + b$, or a^{bx}, or $b\log_a x$, are called *functions* of x.

Any function of x is commonly denoted by such notation as $f(x)$, or $F(x)$, or $\phi(x)$; and, when so denoted, we indicate the special value which the function has for a special value of x, by inserting that value in place of x: thus $f(a)$ would mean the value which $f(x)$ has when $x = a$; so also $\phi(0)$ would mean the value which $\phi(x)$ has when $x = 0$.

Now if an angle is unlimited in size, so that it is the expression of the amount of rotation, in one plane, of a straight line about one extremity, from an initial position to some other position; then *all* periodic magnitudes are expressible in terms of the ratios of certain lines constructed with respect to these two positions. It will be seen that these ratios are functions of the angle: they are called **the trigonometrical functions**; and it is the business of trigonometry to investigate their mutual relations and other properties.

Before defining these functions, it will be necessary (as they are functions of angles) to lay down some mode by which the magnitude of an angle may be estimated.

All measurement of angles is (and indeed, by the nature of the case, *must be*; for any other is inconceivable) made by reference to a circle whose centre is the vertex of the angle—that is must be *circular measure*; so that to give this name to any special mode of measurement (as is *generally done) is to indicate a speciality which has no existence.

---

* De Morgan—most accurate of writers—never uses it except to indicate that *all* measure of angles is *circular measure*: thus he says without qualification—"Angles are measured by arcs of given circles."

The units of measurement employed are two—

1°, the *practical unit* called **a degree**; which is the angle subtended at the centre of any circle, by the one-three-hundred and sixtieth part of its circumference:

2°, the *theoretical unit*, called a * **radian**; which is the angle subtended at the centre of any circle by an arc whose length is the same as the length of its radius.

These definitions involve the theorems that a *degree* and a *radian* (as defined) are the same for all circles. These will now be proved.

Lemma—*In any two circles any two arcs which subtend equal angles, each at the centre of its own circle, are proportional to their respective radii.*

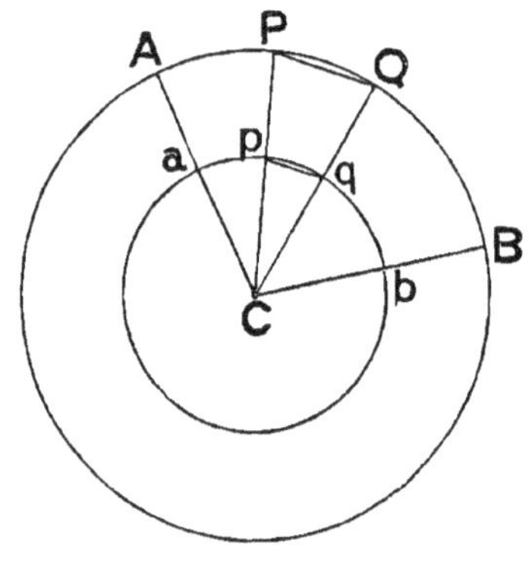

Place the ⊙s so as to have a common centre C.

Let AB, ab be arcs subtending the common angle ACB, or aCb. Divide AB into any number of equal parts, of which PQ is one; and let CP, CQ cut arc ab in p, q respectively: join PQ, pq.

Then, if CA = R, Ca = r, we have

ch'd PQ : ch'd pq = R : r

∴, *addendo*, Σ (ch'd PQ) : Σ (ch'd pq) = R : r

Now let the p'ts of division be indefinitely increased in number:

then Lim' Σ (ch'd PQ) = arc AB

and Lim' Σ (ch'd pq) = arc ab

∴, ultimately, arc AB : arc ab = R : r

* This word was invented by Professor James Thomson, F.R.S.

Cor′ (1)—Circumf′ ⊙ R : circumf′ ⊙ r = R : r

∴ circumf′ of a ⊙ $\propto$ its radius;

and ∴ = C × its radius,

where C is some constant.

It is the universal custom to denote this constant by 2π, so that we have

circumf′ of a ⊙ = π × its diameter.

Cor′ (2)—Since the area of a regular polygon

= $\frac{1}{2}$ rect′ under its perim′ and * apothem;

and that, when the number of its sides is increased indefinitely,

Limit of its area = area of its circum-⊙

„ perim′ = $2\pi r$

„ apothem = r

∴, ultimately, area ⊙ = πr^2

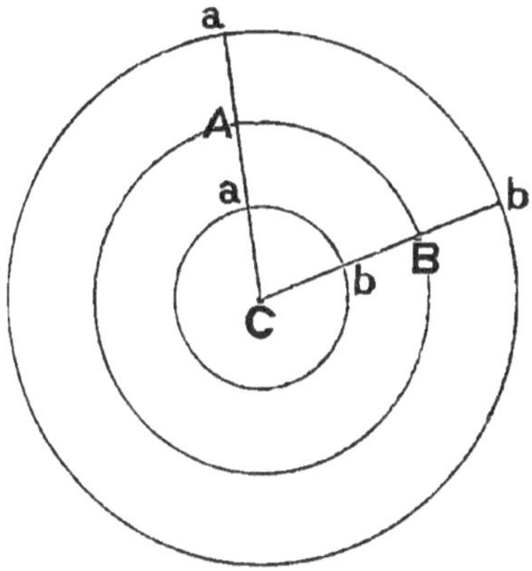

Now let AB be an arc of a ⊙, centre C; and let CA, CB (produced if necessary) intercept an arc ab on any concentric ⊙.

Then, by the *Lemma* and its *Cor′* (1),

arc AB : arc ab = CA : Ca

= whole circumf′ of ⊙ AB : whole circumf′ of ⊙ ab

* *Def′*—The *apothem* of a regular polygon is the perpendicular from its in-centre on any one of its sides.

∴, 1°, if arc AB = /360th of its own circumf′,*
then arc ab = /360th of its own circumf′:
i. e. the *def′* of a *degree* gives an $\wedge$ which is the same for all $\odot$s.
And, 2°, if arc AB = CA, then arc ab = Ca:
i. e. the *def′* of a *radian* gives an $\wedge$ which is the same for all $\odot$s.

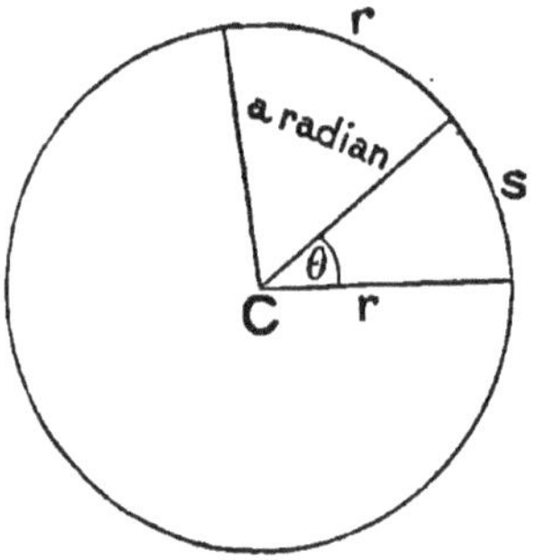

If θ is the radian measure of an $\wedge$ subtended by an arc s of a $\odot$ at its centre C; and r is the radius of the $\odot$; then, since an arc of length r subtends an $\wedge$ of one radian at C, we have, by *Euc′* vi. 33,

$$\theta : 1 = s : r$$
$$\therefore \quad s = r\theta$$
$$\text{or} \quad \theta = s/r$$

Hence—*any arc of a circle is equal to its radius multiplied by the radian measure of the angle it subtends at the centre.*

Otherwise—*if, with the vertex of any angle as centre, we describe any circle, then the radian measure of the angle is the ratio of the arc it subtends to the radius of the circle.*

* In accordance with the suggestion made by Dr. Mac Farlane in the *Educational Times* (vol′ XLVI) we shall use this / (Professor Stokes') symbol of division (sometimes called the *solidus*) as a symbol of operation, just as √ is used: thus /a means $\frac{1}{a}$. The symbol should be used with brackets, when the divisor consists of a product of factors: e. g. $\frac{K}{Rr}$ should be written K/(Rr).

Such a fraction as $\frac{x+y}{x-y}$ may be conveniently written $\overline{x+y}/\overline{x-y}$.

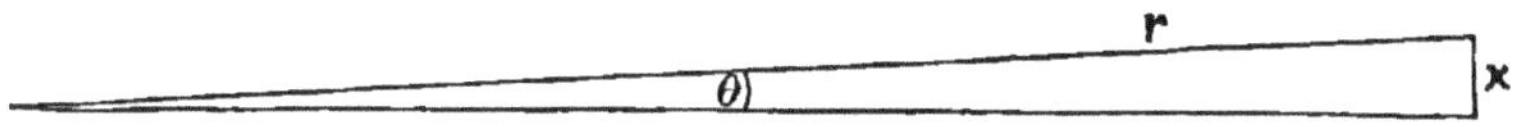

Note—If θ is very small and r very great, then s will be very nearly equal to its chord x: this enables us to approximate to the value of any one of the three x, θ, r, when the other two are given; for, in that case, approximately,

$$x = r\theta$$

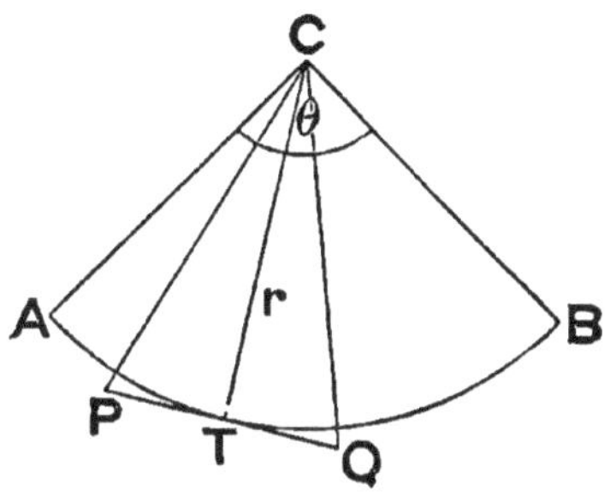

Deduction—Let ACB be a sector of a ⊙, whose arc AB subtends at C the ∧ whose radian measure is θ; so that arc AB $= r\theta$.

Let θ be divided into any number of equal parts; and let tang′ be drawn at the mid p′ts of the arcs subtended by each of these parts: this will give △s such as CPTQ, where PTQ touches arc AB at T.

$$\text{Now area } \triangle\, CPQ = \tfrac{1}{2}\, r \,.\, PQ$$

$$\therefore \Sigma\,(\triangle\, CPQ) = \tfrac{1}{2}\, r\, \Sigma\,(PQ)$$

But, when the number of these △s is indefinitely increased,

$$\text{Lim}'\, \Sigma\,(\triangle\, CPQ) = \text{area sector } ACB$$

$$\text{and Lim}'\, \Sigma\,(PQ) = \text{arc } AB = r\theta$$

∴, ultimately, area sector ACB $= \tfrac{1}{2}\, r^2\theta$

Cor′—The area of a ⊙ (as before) $= \tfrac{1}{2}\, r^2 . \, 2\pi = \pi r^2$

For practical purposes the degree is subdivided into 60 equal parts, called *minutes*; and the minute into 60 equal parts, called *seconds*; and the notation by which these are indicated is °, ′, ″: thus $10^\circ\, 52'\, 17''$ is read 10 degrees, 52 minutes, 17 seconds of *angle*.

Note (1)—To prevent confusion always use *m*, *s* for minutes and seconds of *time**: thus $10^{h}\,52^{m}\,17^{s}$ means 10 hours, 52 minutes, 17 seconds of *time*.

Note (2)—It is clear that the definition of a degree is quite arbitrary, and that the angle subtended by any other fraction of the circumference of a circle at its centre, might have been taken as the practical unit angle. As a matter of fact, at the time of the adoption of decimal notation in France, another unit, viz. that subtended at the centre of a circle by the one-four-hundredth of its circumference, was proposed; and, for a short time, brought into use. This Unit was called *a grade*; and was subdivided into 100 equal parts, called *minutes*; each minute again being subdivided into 100 equal parts, called *seconds*—indicated by the notation ^{g}, $'$, $''$: thus $10^{g}\,52'\,17''$ was read 10 grades, 52 minutes, 17 seconds.

But this *centesimal* angular division speedily passed out of use; and is now quite obsolete, and need engage the Learner's attention no further. For practical purposes the *degree*, and its sexagesimal subdivisions, are universally † and solely used at the present time.

An angle expressed in terms of the radian generally involves the constant π; how this comes about will readily appear by an example. Suppose we wish to express a right angle in terms of a radian. Let r be the radius of the circle of reference.

Then, a right $\wedge$: a radian = a quarter circumf': a radius

$$= 2\pi r/4 : r$$

$$= \pi/2 : 1$$

$$\therefore \text{ a right } \wedge = \pi/2 \text{ of a radian}$$

For brevity this is usually expressed

$$\text{a right } \wedge = \pi/2$$

But it is to be recollected that the unit of measure is the radian, and that $\pi/2$ means $\pi/2$ radians.

* Invariably used and recommended by the late Sir John Herschel: see his *Outlines of Astronomy*: 10th ed', p. 60.

† "Uniformity in nomenclature and modes of reckoning, in all matters relating to time, space, weight, measure, &c, is of such vast and paramount importance, in every relation of life, as to outweigh every consideration of technical convenience or custom."

(*Herschel's Outlines of Astronomy*: 10th ed', p. 88)

If the value 3·1416 (see p. 12) is substituted for π, we get that a right angle is 1·5708 of a radian, i. e. is rather more than a radian and a half; so that a radian is rather less than two-thirds of a right angle. This will serve as a rough guide, when the radian has to be represented graphically. The accurate relations between the two units of measure, viz.

$$\text{a radian} = 57{\cdot}2957795 \text{ \&c. degrees,}$$

$$\text{and, a degree} = {\cdot}017455329 \text{ \&c. of a radian,}$$

will be found hereafter: see p. 13.

Unless the contrary is expressly stated, whenever an angle is defined by means of π, the radian unit is understood.

When it is necessary to specially indicate that an angle is measured as a part or parts of the radian, we shall put a small r in connection with the letter denoting that angle: thus by ${}^r\theta$ we mean that θ is the value of the angle in terms of a radian as unit.

Note—Solely as a matter of printing convenience, x degrees is usually printed x°; though, of course, the algebraic meaning of x^0 is x to the power zero. The context prevents confusion.

As the ratio of any angle to two right angles must be the same whatever unit of measure is used, we get that if any (the same) angle contains x degrees, or θ radians, or y units, each of which subtends $/$mth of a circumference, then

$$x/360 = \theta/2\pi = y/m$$

Note (1)—It is advisable to learn to think of angles with equal readiness in terms of either unit: thus half a right angle should as readily suggest $\pi/4$ radians as 45 degrees.

Note (2)—Recollect that though, e. g., the third of a right angle is expressible either by 30 degrees, or by $\pi/6$ radians, it is as wrong to say $30 = \pi/6$, as it would be to say $3 = 36$, because 3 feet = 36 inches. To symbolise the fact that x degrees is equal to θ radians we put $x^\circ = {}^r\theta$.

Note (3)—The Learner should bear in mind that it is the general custom to speak of the measurement of an angle by the radian as *the circular measure* of that angle. The phrase will not be used in the following pages, for the reason given on p. 3.

To get an approximate value for the constant π.

If I_n, C_n are the respective areas of regular in- and circum-polygons of n sides, with respect to the same ⊙, then it is axiomatic that the area of the ⊙ lies between I_n and C_n.

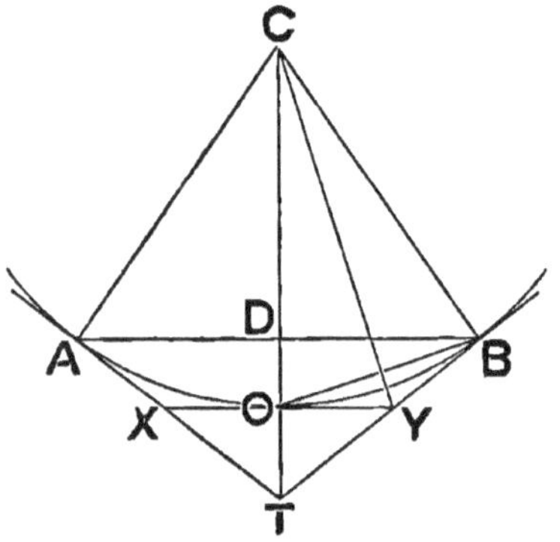

Now let C be the centre of a ⊙, AB a side of an in-polygon of n sides; let the tang's at A, B meet in T; and let CT cut ch'd AB in D, and arc AB in O.

At O draw tang' XOY, meeting TA, TB in X, Y respectively. Join CY, BO. Then it is easily seen that

$$\frac{I_n}{\Delta CDB} = \frac{C_n}{\Delta CTB} = \frac{I_{2n}}{\Delta COB} = \frac{C_{2n}}{2\Delta COY}$$

But CD : CO = CO : CT, ∵ T is pole of ADB

∴ Δ CDB : Δ COB = Δ COB : Δ CTB

$$\therefore \quad I_n : I_{2n} = I_{2n} : C_n \quad . \quad . \quad . \quad . \quad . \quad . \quad . \quad . \quad (1)$$

Again TY : YB = TC : CB = TC : CO

∴ TY : TB = TC : TC + CO

∴ YO : BD = TC : TC + CO

∴ 2 ΔCOY : 2 ΔCOB = Δ CTB : Δ CTB + Δ COB

$$\therefore \ C_{2n} : 2I_{2n} = C_n : C_n + I_{2n} \quad . \quad . \quad . \quad . \quad . \quad . \quad (2)$$

From results (1) and (2) we get the algebraic formulæ

$$I_{2n} = \sqrt{I_n . C_n}$$

$$\text{and} \quad C_{2n} = \frac{2I_{2n}^2}{I_n + I_{2n}}$$

Now taking the radius of the ⊙ as the unit of length, its square will be the unit of area.

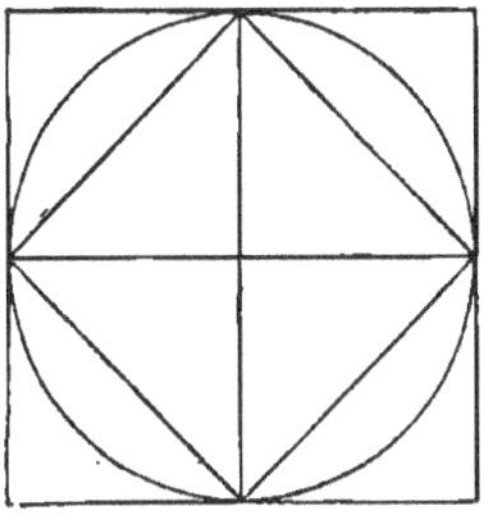

In the case when the pol's are the in- and circum- sq's, we see, by inspection, that

$$I_4 = 2 \quad C_4 = 4$$

Hence, applying the above formulæ, we get

$$I_8 = 2\sqrt{2} \qquad C_8 = \frac{16}{2 + 2\sqrt{2}} = 8(\sqrt{2} - 1)$$

$$\therefore \quad I_8 = 2{\cdot}8284271 \qquad C_8 = 3{\cdot}3137085$$

The calculations of the successive areas can now be proceeded with: they are rather laborious; but the Student whose curiosity is sufficient, and is equalled by his industry, will find that their successive values are—

$$\begin{array}{ll}
I_{16} = 3{\cdot}0614674 & C_{16} = 3{\cdot}1825979 \\
I_{32} = 3{\cdot}1214451 & C_{32} = 3{\cdot}1517249 \\
I_{64} = 3{\cdot}1365485 & C_{64} = 3{\cdot}1441184 \\
I_{128} = 3{\cdot}1403311 & C_{128} = 3{\cdot}1422236 \\
I_{256} = 3{\cdot}1412772 & C_{256} = 3{\cdot}1417504 \\
I_{512} = 3{\cdot}1415138 & C_{512} = 3{\cdot}1416321 \\
I_{1024} = 3{\cdot}1415729 & C_{1024} = 3{\cdot}1416025 \\
I_{2048} = 3{\cdot}1415877 & C_{2048} = 3{\cdot}1415951 \\
I_{4096} = 3{\cdot}1415914 & C_{4096} = 3{\cdot}1415933
\end{array}$$

These agree to 5 decimal places; and, as the ⊙ is intermediate in area to the two pol's, we see that the formula—

$$\text{area of a } \odot = 3{\cdot}14159 \times (\text{its radius})^2$$

is a close approximation.

$$\text{Hence } \pi = 3{\cdot}14159 \text{ very nearly.}$$

We shall habitually use 3·1416 as a good and sufficient approximation *.

Note (1)—It is often useful to remember that $31416 = 3 \times 7 \times 8 \times 11 \times 17$.

Note (2)—If the calculations be continued further, the areas of the polygons will be found to agree up to the figures 3·14159265; so that in a circle whose radius is a mile, the error in area, by taking these figures for π, is less than the ten-millionth of a square mile; that is less than a square foot.

The value of π to 8 decimal places may be recollected by the following jingle—

"A π receipt may thus be given;
"From 3 millions take 8 thousand seven;
"Increase remainder 5 per cent;
"Result to 8 decimals represent."

$$\begin{array}{r} 3000000 \\ 8007 \\ \hline 2991993 \\ 14959965 \\ \hline 3\cdot 14159265 \end{array} = \pi \text{ to 8 places.}$$

Note (3)—The following practical † rules are worth noting.

$$\begin{array}{lrll} \text{Since} & 22/7 = & 3\cdot\dot{1}4285\dot{7} & = A_1 \text{ say} \\ \therefore & A_1 \times \cdot 0004 = & \cdot 0012571\dot{4}2857\dot{} & = C_1 \text{ ,,} \\ \therefore & A_1 - C_1 = & 3\cdot 1416 \text{ exactly} & = A_2 \text{ ,,} \\ \text{Again} & C_1 \times \cdot 006 = & \cdot 00000754 & = C_2 \text{ ,,} \\ \therefore & A_2 - C_2 = & 3\cdot 14159245 & = A_3 \text{ ,,} \end{array}$$

A_3 is correct to 6 decimal places, i. e. as correct as 355/113.

Hence we get the following rules to multiply by π:

As a 1st approx$'$ use 22/7 for π.

As a 1st correction subtract ·0004 of the product from itself.

---

* The fraction 22/7 (originally found by Archimedes, and correct to *two* decimal places) is often used as a rough approximation for π. The fraction 355/113 (originally found by Metius, and correct to *six* decimal places) is a very close approximation to π. To recollect this last fraction put down each of the first three odd numbers twice, giving 113355; then the three right hand figures are the numerator, and the three left hand are the denominator. For an account of these approximations see an article by Dr. Glaisher in the *Messenger of Mathematics*: New Series, II. 123.

† Taken from *Milne's* '*Companion*,' by permission of the Editor.

This gives a 2nd approx′ as true as by using 3·1416 for π.

As a 2nd correction subtract ·006 of the 1st correction from the 2nd approx′.

This gives a 3rd approx′ true to 6 places; and is ∴ as true as by using 355/113 for π.

Similarly may be verified the following rules for approximately *dividing* by π:

1°, multiply by 7/22;
2°, add ·0004 of this result to itself;
3°, add ·006 of this last result to itself.

The final result is true for *six* places.

All questions of conversion, or reduction, of angles from one measure to another, evidently belong to the domain of simple Arithmetic. So also questions concerning areas and arcs of circles, or sectors of circles, belong to the province of Mensuration. Hence all such questions are foreign to Trigonometry properly so-called: however, to render this Preliminary Chapter more complete, we add here some Examples and Exercises of this character.

Examples

1. *To express a radian in terms of degrees, minutes, and seconds.*

$$\text{If} \quad x^\circ = \text{one radian}$$

$$\text{then} \quad \frac{x}{180} = \frac{1}{\pi}$$

$$\therefore \quad x = \frac{180}{3{\cdot}1416} = 57{\cdot}29578, \quad \text{nearly}$$

$$\text{i.e.} \quad \text{a radian} = 57^\circ{\cdot}29578$$

$$= 57^\circ\ 17'\ 44''{\cdot}8, \quad \text{nearly.}$$

NOTE—*It will be found that this* $= 57^\circ{\cdot}\tfrac{3}{10} - (\tfrac{1}{4})' - (\tfrac{1}{5})''$ *to within* $(\tfrac{1}{100})''$.

2. *To express a degree in terms of a radian.*

$$\text{If} \quad {}^r\theta = \text{one degree}$$

$$\text{then} \quad \frac{\theta}{\pi} = \frac{1}{180}$$

$$\therefore \quad \theta = \frac{3{\cdot}1416}{180} = {\cdot}01745, \quad \text{nearly}$$

$$\text{i.e.} \quad \text{a degree} = {\cdot}01745 \text{ of a radian, nearly.}$$

3. *To express* $23^\circ\ 14'\ 57''$ *in centesimal units.*

$$\begin{aligned} 23^\circ\ 14'\ 57'' &= 23^\circ\ 14'\cdot95 \\ &= 23^\circ\cdot2491\dot{6} \end{aligned} \quad \text{(dividing by 60 mentally in each case)}$$

$$\begin{aligned} &= (\tfrac{10}{9} \times 23\cdot2491\dot{6})^g \\ &= 25^g\cdot832407\dot{3} \\ &= 25^g\ 83^{\backprime}\ 24^{\backprime\backprime}\cdot07\dot{3} \end{aligned}$$

4. *To express* $23^g\ 14^{\backprime}\ 57^{\backprime\backprime}$ *in degrees and sexagesimal units.*

$$\begin{aligned} 23^g\ 14^{\backprime}\ 57^{\backprime\backprime} &= 23^g\cdot1457 \\ &= (\tfrac{9}{10} \times 23\cdot1457)^\circ \\ &= 20^\circ\cdot8313 \end{aligned}$$

$$\begin{aligned} &= 20^\circ\ 49'\cdot8678 \\ &= 20^\circ\ 49'\ 52''\cdot068 \end{aligned} \quad \text{(multiplying by 60 mentally in each case)}$$

5. *If a centesimal minute is taken as the unit of angular measure, what represents a sexagesimal minute?*

$$1' = (\tfrac{1}{60})^\circ = (\tfrac{10}{9} \times \tfrac{1}{60})^g = (\tfrac{100}{54})^{\backprime} = (\tfrac{50}{27})^{\backprime}$$

i.e. a sexagesimal minute is represented by $\frac{50}{27}$.

6. *What length of rope will enable an animal, tethered by it, to graze over an acre of ground?*

In a question like this, where rough approximation is evidently sufficient, we may take $\frac{22}{7}$ for π.

Let x yds. be the length of rope.

Then $\frac{22}{7}x^2 = 4840$

$\therefore\ x^2 = 1540$

whence $x = 39\cdot24$

i.e. $39\frac{1}{4}$ yds. is very nearly the length req'd.

7. *What is the length of a column which, at the distance of a mile, subtends an angle of* 1° *at the eye?*

Owing to the smallness of the $\wedge$, and the great distance of the column as compared to its length, we may consider the column as an arc of a $\odot$ of radius 1760 yds.

$$\therefore\ \text{length of column} = 1760 \times \frac{31416}{180 \times 10^4}\ \text{yds.}$$

$$= 30\cdot7178$$

i.e. the column is nearly 31 yds. high.

8. *Show that if the circumference of a circle whose radius is* 100 *miles is calculated by using* $355/113$ *for* π, *the error is less than* 4 *inches.*

Show also that the areal error, on the same supposition, is less than 2 *acres.*

By division $355/113 = 3{\cdot}14159292$ &c

But $\pi = 3{\cdot}14159265$ &c

$\therefore$ the error, by taking $355/113$ for π, $\not> {\cdot}0000003$

$\therefore$ error in circumf′ (which $= 200\,\pi$) $\not> {\cdot}00006$ of a mile

i.e. $\not> 3{\cdot}8016$ inches

and $\therefore$ < 4 inches.

Again areal error $\not> {\cdot}0000003 \times 10000$ sq′ miles

$\not> {\cdot}003 \times 640$ acres

$\not> 1{\cdot}92$ acres

and $\therefore$ < 2 acres.

Exercises

1. Express $\pi/13$ radians in degrees, and decimals of a degree.

2. Express 13° as a decimal of a radian.

3. Find approximately the radian measure of the angle which contains as many degrees as its supplement contains grades. *Joh′ Camb′*: ′80.

4. If the unit angle subtends an arc equal to the diameter, how will one-third of a right angle be expressed?

5. A train is going α miles an hour on an arc of a circle of β miles radius: how many seconds of angle will it turn through in n seconds of time?

6. What is the length of an object, which at the distance of a mile, subtends an angle of one minute?

7. Find (in degrees, minutes, and seconds) the angle which at the centre of a circle of 8 feet diameter, subtends an arc of 10 feet length.

8. At what distance would a man 6 feet high subtend an angle of one minute?

9. If the radius of a circle is 20 inches, find the radii of three concentric circles, by which the original circle is quadrisected.

10. If the measures of the angles of a triangle, referred to 1°, $100'$, $10\,000''$, respectively, as units, are in the proportion of 2 to 1 to 3, find the angles.

Math′ Tri′: ′73.

11. Find the length of an arc on the sea which subtends an angle of one minute at the centre of the Earth, supposing the Earth a sphere of diameter 7920 miles. *L. U.* ′84.

12. If the distance between the centres of the Earth and the Moon is sixty times the radius of the Earth, find the angle which the radius of the Earth subtends at the Moon's centre. *Trin' Camb'*: '44.

13. The apparent diameter of the Moon is 30′: find how far from the eye a circular plate of 6 inches diameter must be placed so as just to hide the Moon. *Trin' Camb'*: '49.

14. If two plumb lines, suspended from points at a fixed distance apart, are inclined to each other at a small angle of m'' when at the Earth's surface; and at an angle of n'' when at an elevation h; show that the Earth's radius is $nh/(m-n)$ nearly. *Pet' Camb'*: '50.

15. If the number of units of measure in any angle is equal to the number of degrees in that angle less the number of radians in it; how many degrees are there in the unit of measurement?

16. Calculate π to two decimal places on the assumption that an angle of 200° subtends at the centre of a circle an arc which is $3\frac{1}{2}$ times the radius. *Ox' Jun' Locals*: '86.

17. Calculate in inches, correct to four decimal places, the length of an arc of a circle which with a radius of a mile subtends an angle of one second at the centre. *Ox' Sen' Locals*: '86.

18. Show that with every unit the numbers expressing an angle of an equilateral triangle, and an angle of a regular hexagon, would together double the number expressing an angle of a square. *Joh' Camb'*: '76.

19. Assuming that the Earth is a sphere whose diameter is 7912 miles, show that (to three decimal places) the length of an arc on its surface, subtending a degree at its centre, is 69·045 miles; and that this result will be obtained whether we use 3·1416 or 3·1415926 for π.

20. Calculate π (to four decimal places) from the following data—In a circle, whose radius is 10 feet, an arc of 3 feet $11\frac{1}{8}$ inches subtends an angle of $22^\circ\ 30'$ at the centre.

21. If the diameter of a bicycle is 50 inches, show that the number of revolutions it makes in 9 seconds is very nearly the same as the number of miles per hour at which it is going. (Take $22/7$ for π)

22. Assuming that the Earth moves in a circle whose radius is 95 000 000 miles, and that a year is 365 days exactly; show that its rate of motion is nearly 19 miles per second.

23. Three equal circles (each of one foot radius) are so placed that each touches the other two; find (to three decimal places of square feet) the curvilinear area enclosed between the three circles.

24. Find (to three decimal places of square inches) the area of a segment of a circle (whose radius is one inch) cut off by a chord equal to the radius.

Q. C. B. '74

25. Calculate the velocity of light from the following *data*—The Earth's diameter is 7900 miles; the angle this diameter subtends at the Sun is $17''.8$; the time the Sun's light takes to reach the Earth is $8^m\ 13^s.3$. *Q. C. B.* '71.

26. Calculate the rate of the Moon's motion, in miles per hour, from the following *data*—The distance between the centres of the Earth and Moon is 59·964 times the Earth's radius; the time of the Moon's revolution round the Earth is 27 days, 7 hours, 43 minutes, 11 seconds; the Earth's radius is 3963 miles. *Q. C. B.* '81.

27. Calculate, so that the error shall be less than the /1000th part of the answer, the weight in grains of a globe of standard gold, whose radius is one metre; given that—Volume of a * sphere is $\frac{4}{3}\pi$ (radius)^3; specific gravity of standard gold is 17·157; a cubic inch of water weighs 252·46 grains; a metre is 39·37 inches. *Q. C. B.* '83.

28. If the Earth's orbit is assumed to be a circle whose diameter is 184 000 000 miles, find the area enclosed by this orbit.

What error is caused by taking 3·1416 instead of 3·1415926 for π?

29. It is known of a pair of regular polygons that the ratio of the number of grades in an angle of one to the number of radians in an angle of the other is 160 to π: show that the pair must be one of four. *Joh' Camb'*: '78.

30. When the unit of angular measure is the angle whose arc is π times the radius, the angles of a triangle are such that the number of degrees in one, the number of grades in another, and the number of units of angular measure in the third are all equal: find each angle in radian measure.

Pet' Camb': 66.

31. Of the angles of a triangle one contains as many grades as another contains degrees, and the third contains as many centesimal seconds as there are sexagesimal seconds in the sum of the other two: find the radian measure of each angle. *Joh' Camb'*: '72.

32. The apparent angular diameter of the Sun is half a degree: a planet is seen to cross its disc in a straight line at a distance from the centre equal to three-fifths of the radius: prove that the angle subtended at the Earth by the part of the planet's path projected on the Sun is $\pi/450$.

Math' Tri': '74.

---

* See *Geometry in Space*: p. 69.

33. ABC is a triangle such that, if each of its angles in succession is taken as the unit of measurement, and the measures formed of the sums of the other two, these measures are in arithmetical progression: show that the angles of the triangle are in harmonical progression. *Math' Tri'*: '79.

34. Show that there are eleven, and only eleven, pairs of regular polygons such that the number of degrees in an angle of the one is equal to the number of grades in an angle of the other; and that there are only four pairs when these angles are expressed in integers.

Find also the number of sides of each. *Math' Tri'*: '66 *and* '81.

NOTE—*If* x, y *are the numbers of sides, cond'ns of question give the indeterminate eq'n* $xy + 18x - 20y = 0$, or $(x - 20)(y + 18) + 360 = 0$.

35. For what pairs of regular polygons is it true that the number of degrees in each angle of the one is to the number of radians in each angle of the other as 144 to π?

36. Given, in two regular polygons, the ratio of the number of sides in one to the number of sides in the other; and given also the ratio of the number of degrees in an angle of one to the number of grades in an angle of the other; find the number of sides in each.

37. One angle of a triangle is x degrees, another is x grades, and the third is x radians; find, approximately, the magnitude of each angle in degrees, minutes, and seconds.

38. Two chords of a circle (radius 3 feet) intercept arcs $\pi/3$ and $\pi/5$: find (in degrees) the angle between the chords. *Ox' Jun' Math' Schol'*: '74.

39. Two circles cross each other orthogonally: if d is the distance between their centres; and if their radii are as $\sqrt{3}$ to 1; find (in terms of d) the area common to the circles.

As we have already, to some extent, seen in Algebra and Geometry, it is necessary and in the highest degree convenient, for purposes of mathematical generalization, to consider that if any process is conventionally taken as positive, then (what may be loosely called) the *reverse* process is to be taken as negative.

Examples

The positive processes—receipt of money; motion to the right; motion upwards; rotation like the hands of a clock—have as their respective negative analogues—payment of money; motion to the left; motion downwards; rotation opposite to that of the hands of a clock.

Def'—If two straight lines of unlimited length are taken at right angles to each other, they divide the plane (indefinitely extended) which contains them, into four parts; and each of these parts is called a **quadrant.**

Note—Of course, strictly speaking, a quadrant is the fourth part of a circle: the foregoing definition may be brought within this usual meaning by supposing a circle whose centre is the cross of the lines, and whose radius is of indefinite length.

In Trigonometry—and thence in mathematics generally—the following conventions are made.

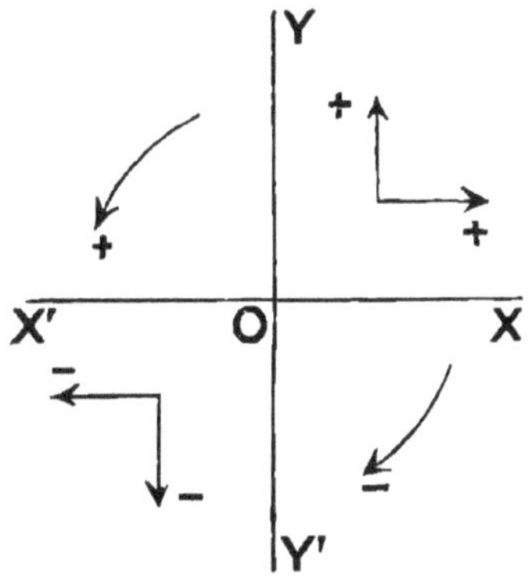

Def'—Let XOX', YOY' be two straight lines at right angles to each other, and indefinitely extended—where XOX' is drawn in direction from the right hand towards the left; and YOY' is drawn in direction from the top of the paper towards the bottom: then—

1°, the area separated off by OX, OY is called *the first quadrant;*
,, ,, ,, OY, OX' ,, *second* ,,
,, ,, ,, OX', OY' ,, *third* ,,
,, ,, ,, OY', OX ,, *fourth* ,,

2°, lines measured in direction X'OX are called *positive;*
and therefore ,, ,, ,, XOX' ,, *negative.*
,, ,, ,, Y'OY ,, *positive;*
and therefore ,, ,, ,, YOY' ,, *negative.*

3°, angles measured by rotation about O in an anti-clock-wise direction (that is following the cycle XYX'Y') are called *positive;* and therefore those measured by a clock-wise rotation (that is following the cycle XY'X'Y) are called *negative.*

Examples

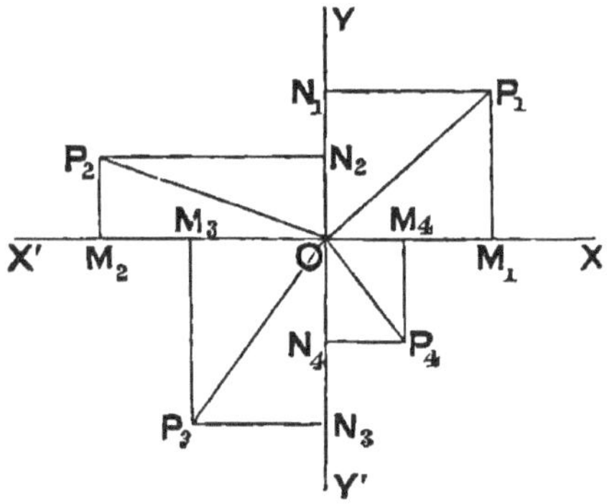

If OP_1, OP_2, OP_3, OP_4 are respectively in the 1st, 2nd, 3rd, and 4th quadrants; and P_1M_1, P_2M_2, &c. ⊥s on XOX′; P_1N_1, P_3N_3, &c. ⊥s on YOY′;

then OM_1, OM_4, ON_1, ON_2 are considered *positive*;

and ∴ OM_2, OM_3, ON_3, ON_4 ,, ,, *negative.*

Also, if OP_1, OP_2, &c. are imagined to have revolved, from coincidence with OX, in an anti-clock-wise direction, the ∧s XOP_1, XOP_2, XOP_3, XOP_4 are considered *positive*; and ∴ if they are imagined to have revolved in a clock-wise direction, the ∧s XOP_4, XOP_3, XOP_2, XOP_1 are considered *negative.*

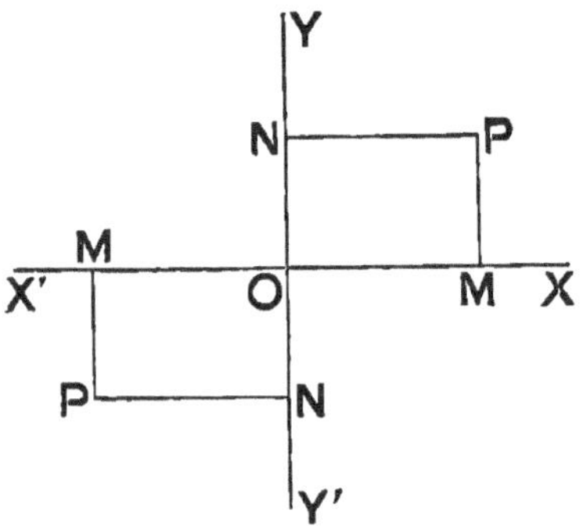

Def′—If XOX′, YOY′ are two fixed lines at right angles, and PM, PN the respective distances of any point P, in their plane, from them; then PM, PN are called the **rectangular coordinates** of P with respect to the *rectangular axes* XOX′, YOY′, and the *origin* O: also OM (to which PN is equal) is called the **abscissa** of P, and PM is called the **ordinate** of P.

Note—It is usual to denote OM, PM respectively by such pairs of letters as (x, y) or (h, k) or (a, b). Evidently the position of P, with respect to the axes, is completely determined when its coordinates are known.

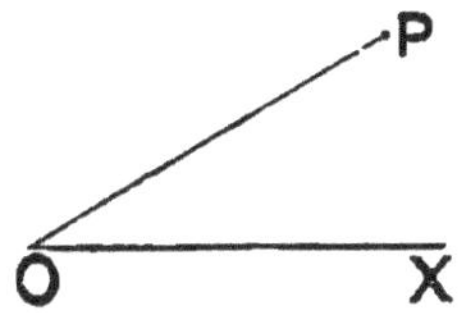

Def'—If O is a fixed point and OX a fixed direction, and if P is any point, then OP and the angle POX are called the **polar coordinates** of P with respect to the *pole* O, and *initial line* OX; also OP is called the **radius vector** of P, and POX the **vectorial angle** of P.

Note—It is usual to denote OP, and POX, respectively, by such pairs of letters as (r, θ) or (ρ, ϕ) or (a, α). Evidently the position of P, with respect to the pole and initial line, is completely determined when its polar coordinates are known.

Examples

1. The point $(-1, 2)$ is situated in the second quadrant, at a distance from OX, which is double its distance from OY.

2. The point $(0, -3)$ is in OY′.

3. If $x^2 + y^2 = 0$, then $x = 0$ and $y = 0$, simultaneously; so that $x^2 + y^2 = 0$ represents the point O.

4. So also $(x - \alpha)^2 + (y - \beta)^2 = 0$ represents the point (α, β).

5. In polar coordinates the point $(2, 30^\circ)$ is that got by letting a line revolve from its initial position OX, through 30°, and then measuring two units along it from O.

6. If (x', y') are the rectangular coordinates of a point P, and (x'', y'') of a point Q, then (as the Student will easily be able to prove, by dropping perpendiculars)

$$PQ^2 = (x' - x'')^2 + (y' - y'')^2.$$

7. We shall anticipate so much of what follows as to notice that if (x, y) are the rectangular, and (r, θ) the polar coordinates of the same point,

$$\text{then} \quad \left.\begin{aligned} x &= r\cos\theta \\ y &= r\sin\theta \end{aligned}\right\}$$

$$\text{and} \quad \left.\begin{aligned} r^2 &= x^2 + y^2 \\ \tan\theta &= y/x \end{aligned}\right\}$$

where the origin and axis of x are respectively the same as the pole and initial line.

After reading Chapter I the Student will easily see the truth of these statements.

From the relationships between the coordinates of points a complete branch of mathematics—termed *Coordinate Geometry*—has been developed; but, though the above definitions will be useful to us in Trigonometry, it would be irrelevant to our present purpose to enter further on that subject.

By means of rectangular coordinates we can give a graphic exhibition of successive changes of value of a function.

Let $\phi(x)$ be any function of x.

Then, if $y = \phi(x)$, and if (x, y) are taken as the rect'r coord's of a p't P, to every value of x there is a corresponding value of y.

Now if a set of values are assigned to x, and the corresponding values of y are found, then each simultaneous pair of values gives a position of P; and, if enough of such positions are determined, the curve which is the Locus of P can be approximately drawn.

Def'—The curve whose ordinates are the values of a function $\phi(x)$ corresponding to assigned values of x, is called the **graph** of $\phi(x)$.

Graphs may be (and frequently are with much advantage) drawn to represent any varying magnitude. Thus we might draw graphs to represent—

(*a*) the price of consols on successive days in a month;
or (*b*) the mortality of a town during successive weeks in a year;
or (*c*) the velocity of a train during successive minutes of an hour.

Example

It can be shown* that Boyle's Law, connecting the varying volume and pressure of a given quantity of gas at a constant temperature, has for its graph a curve known as the *rectangular hyperbola.*

Some graphs remain always at a finite distance: thus

$$y = \sqrt{a^2 - x^2}$$

* See *Magnus' Hydrostatics*: 3rd ed', p. 120.

is represented by a circle. But again other graphs are found to go off to an infinitely removed distance: such e.g. is the case with the graph of

$$y = \sqrt{x^2 - a^2}$$

For the satisfactory tracing of some of these, certain subsidiary lines called *asymptotes* are requisite.

Def'—If as a curve is indefinitely extended it continuously approaches to (without attaining within any finite distance) coincidence with a fixed straight line, so that it may be considered to touch that line at an infinitely removed distance along the line, then the straight line is called an **asymptote** to the curve.

For further information on the subject of **Graphs** the Student is referred to Chapter XV of Professor Chrystal's *Algebra*, where he will find diagrams and discussions of several graphs.

A large number of excellent Examples of the application of graph-drawing to practical questions in Political Economy will be found in Professor Marshall's *Principles of Economics*.

Algebraic Note

Sarrus' practical rule to write down the expansion of the three column determinant

$$\begin{vmatrix} \alpha & \beta & \gamma \\ \alpha' & \beta' & \gamma' \\ \alpha'' & \beta'' & \gamma'' \end{vmatrix}$$

Repeat the 1st two col's, and connect by diag's thus—

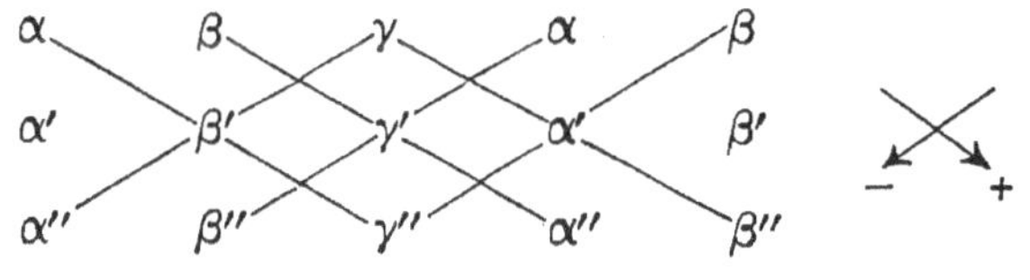

Put down the product of each set of 3 constituents lying along a diag' line; prefixing + if the line goes from the top towards the right, and − if from the top towards the left: this gives as result

$$\alpha\beta'\gamma'' + \beta\gamma'\alpha'' + \gamma\alpha'\beta'' - \beta\alpha'\gamma'' - \alpha\gamma'\beta'' - \gamma\beta'\alpha''$$

The above very useful rule is not usually given in Algebras; and ∴ has been printed here for reference.

CHAPTER I

The Trigonometrical Functions

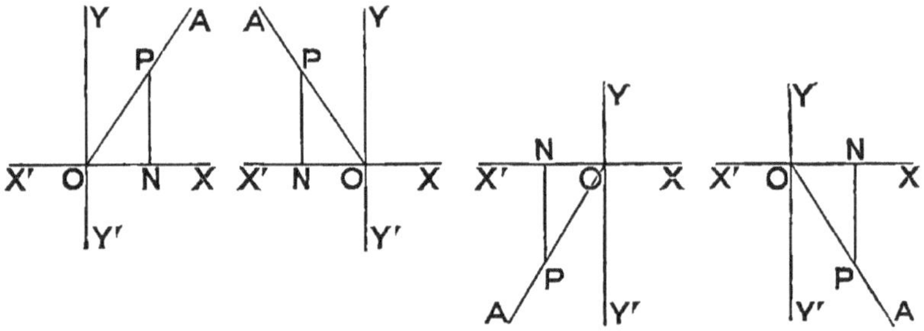

§ **1.** *Def's*—Let XOX′, YOY′ be ⊥ lines, in which OX, OY are pos′ direc′s.

Let OA, initially along OX, revolve about O either way round, in the plane of XOY, thro′ an angle α.

Then OA must be in one of the 4 quadrants into which the pl′ is divided by XOX′, YOY′; and ∴ must assume a position such as one of the 4 indicated in the fig′s.

In OA take any p′t P; and drop PN ⊥ on XOX′.

The sides of the △ PON will afford three ratios and their reciprocals: of these—

the ratio $\frac{NP}{OP}$ is called 'the sine of α'—usually written sin α

„ $\frac{ON}{OP}$ „ „ cosine „ „ „ cos α

„ $\frac{NP}{ON}$ „ „ tangent „ „ „ tan α

„ $\frac{ON}{NP}$ „ „ cotangent „ „ „ cot α

the ratio $\frac{OP}{ON}$ is called 'the **secant** of α'—usually written $\sec\alpha$

" $\frac{OP}{NP}$ " " **cosecant** " " " $\operatorname{cosec}\alpha$

The expression $1 - \cos\alpha$ is sometimes called the **versed-sine** of α, and is then written $\operatorname{vers}\alpha$; but this function is not much used.

Def'—The triangle PON, constructed as above for the angle α, is called the *auxiliary triangle* of α; and the ratios of the sides of the auxiliary triangle, when indicated by the names given above, are called * **the trigonometrical functions** *of the angle* α.

Note (1)—Of the lines which are the terms of the foregoing ratios, the revolving line OP is considered to remain invariably positive in all positions; while the other lines ON, NP follow the usual conventions, as to sign, explained in the Preliminary Chapter.

Note (2)—We shall use the contraction T. F. for the words 'trigonometrical function.'

Note (3)—From the mode of its construction, the auxiliary $\triangle$ is (for any definite $\wedge$) always of the same species, no matter where in OA the p't P is taken; and $\therefore$ the T. F.s are invariable for any (the same) $\wedge$.

Note (4)—The ratios defined as T. F.s depend only on the relative positions of the initial and revolving lines, and $\therefore$ will be the same for all $\wedge^s$ in which these lines occupy the same relative positions. Now the addition (positive or negative) of any multiple of 4 right $\wedge^s$ to an $\wedge$ will make no change in the relative positions of the lines forming that $\wedge$: hence—

If any $\wedge = \alpha^\circ = {}^r\theta$,

and $f(\alpha)$ or $f(\theta)$ denotes any T. F. of that $\wedge$,

then $f(\alpha) = f(m.\ 360^\circ + \alpha)$

and $f(\theta) = f(m.\ 2\pi + \theta)$

where m is any integer pos', neg', or zero.

Hence we see that the T. F.s are *periodic*; and that the extent of one of their periods is 2π. It will be seen hereafter that the tangent and cotangent have a shorter period, viz. π.

---

* Sometimes also *the circular functions*, and sometimes *the goniometrical ratios* of the angle.

§ **2.** The sine has been defined as NP/OP (*not* PN/OP) and hence from *def′* we see that—

for $\wedge$s between 0° and 180° the sine is *pos′*;
while „ „ 180° „ 360° „ „ *neg′*

Sim′ly for the other functions; so that the following table is at once apparent from the def′s.

$\wedge$ between	sin	cos	tan	cot	sec	cosec
0° and 90°	+	+	+	+	+	+
90° and 180°	+	−	−	−	−	+
180° and 270°	−	−	+	+	−	−
270° and 360°	−	+	−	−	+	−

Note—The auxiliary △ PON, with respect to an $\hat{\alpha}$, having been constructed (as above) then, calling PN its *perp′* (perpendicular) ON its *base*, and OP its *hyp′* (hypotenuse) the T. F.s are sometimes defined thus—

$\sin\alpha = \text{perp}'/\text{hyp}'$, $\cos\alpha = \text{base}/\text{hyp}'$, $\tan\alpha = \text{perp}'/\text{base}$

We do not recommend this mode of definition ∵ —

1°, it takes no account of the sign of the lines NP, ON; and,

2°, it is apt to give an inadequate (or even inaccurate) conception of the T. F.s for $\wedge$s greater than a r′t $\wedge$.

§ **3.** By def′, cot α, sec α, cosec α are respectively reciprocals of tan α, cos α, sin α, so that

$$\sin\alpha \,.\, \operatorname{cosec}\alpha = 1$$
$$\cos\alpha \,.\, \sec\alpha = 1$$
$$\tan\alpha \,.\, \cot\alpha = 1$$

Def'—Here we first come on what are called **trigonometrical identities**: that is relations between the functions which are true for *all* values of the angle (or angles) involved.

There is a large number of trigonometrical identities which must be recollected, if any success is to be attained in the applications of trigonometry. Those which are imperatively necessary to be committed to memory will be printed for reference in a list at the end of the book.

§ **4.** *To express any one* T. F. (*call it the first*) *in terms of any other one* (*call it the second*) *follow this plan*—

1°, construct fig's as in § 1:

2°, write down the def' of the 1st:

3°, if of the three sides of the auxil' △ one occurs in the 1st but not in the 2nd, eliminate that side by means of the relationship between the sides given by *Euc'* i. 47:

4°, divide each term of the fraction thus obtained by the den'r of the 2nd:

5°, replace the ratios by their T. F. equivalents.

The process will then be complete.

Example

To express cos α *in terms of* cosec α.

1°, take the fig's of § 1

2°, $\cos\alpha = \dfrac{ON}{OP}$

3°, $\therefore \;= \dfrac{\sqrt{OP^2 - NP^2}}{OP}$

4°, $\therefore \;= \dfrac{\sqrt{\left(\dfrac{OP}{NP}\right)^2 - 1}}{\dfrac{OP}{NP}}$

5°, $\therefore \;= \dfrac{\sqrt{\operatorname{cosec}^2\alpha - 1}}{\operatorname{cosec}\alpha}$

	$\sin\alpha$	$\cos\alpha$	$\tan\alpha$	$\cot\alpha$	$\sec\alpha$	$\operatorname{cosec}\alpha$
$\sin\alpha$	$\sin\alpha$	$\sqrt{1-\cos^2\alpha}$	$\frac{\tan\alpha}{\sqrt{1+\tan^2\alpha}}$	$\frac{1}{\sqrt{\cot^2\alpha+1}}$	$\frac{\sqrt{\sec^2\alpha-1}}{\sec\alpha}$	$\frac{1}{\operatorname{cosec}\alpha}$
$\cos\alpha$	$\sqrt{1-\sin^2\alpha}$	$\cos\alpha$	$\frac{1}{\sqrt{1+\tan^2\alpha}}$	$\frac{\cot\alpha}{\sqrt{\cot^2\alpha+1}}$	$\frac{1}{\sec\alpha}$	$\frac{\sqrt{\operatorname{cosec}^2\alpha-1}}{\operatorname{cosec}\alpha}$
$\tan\alpha$	$\frac{\sin\alpha}{\sqrt{1-\sin^2\alpha}}$	$\frac{\sqrt{1-\cos^2\alpha}}{\cos\alpha}$	$\tan\alpha$	$\frac{1}{\cot\alpha}$	$\sqrt{\sec^2\alpha-1}$	$\frac{1}{\sqrt{\operatorname{cosec}^2\alpha-1}}$
$\cot\alpha$	$\frac{\sqrt{1-\sin^2\alpha}}{\sin\alpha}$	$\frac{\cos\alpha}{\sqrt{1-\cos^2\alpha}}$	$\frac{1}{\tan\alpha}$	$\cot\alpha$	$\frac{1}{\sqrt{\sec^2\alpha-1}}$	$\sqrt{\operatorname{cosec}^2\alpha-1}$
$\sec\alpha$	$\frac{1}{\sqrt{1-\sin^2\alpha}}$	$\frac{1}{\cos\alpha}$	$\sqrt{1+\tan^2\alpha}$	$\frac{\sqrt{\cot^2\alpha+1}}{\cot\alpha}$	$\sec\alpha$	$\frac{\operatorname{cosec}\alpha}{\sqrt{\operatorname{cosec}^2\alpha-1}}$
$\operatorname{cosec}\alpha$	$\frac{1}{\sin\alpha}$	$\frac{1}{\sqrt{1-\cos^2\alpha}}$	$\frac{\sqrt{1+\tan^2\alpha}}{\tan\alpha}$	$\sqrt{\cot^2\alpha+1}$	$\frac{\sec\alpha}{\sqrt{\sec^2\alpha-1}}$	$\operatorname{cosec}\alpha$

The table opposite gives each T. F. in terms of each other. The Student is recommended to prove the truth of this table by going through the process of § 4 in each case.

§ **5.** There are modified forms of three of the connections between the T. F.s given opposite, which are so important as to be worth separate consideration.

Taking the fig's of § 1, we have, by *Euc'* i. 47,

$$NP^2 + ON^2 = OP^2$$

Divide each term of this eq'n successively by OP^2, ON^2, NP^2, and we get

$$\left(\frac{NP}{OP}\right)^2 + \left(\frac{ON}{OP}\right)^2 = 1$$

$$\left(\frac{NP}{ON}\right)^2 + 1 = \left(\frac{OP}{ON}\right)^2$$

and $$1 + \left(\frac{ON}{NP}\right)^2 = \left(\frac{OP}{NP}\right)^2$$

i. e. $$\sin^2\alpha + \cos^2\alpha = 1$$

$$\tan^2\alpha + 1 = \sec^2\alpha$$

and $$1 + \cot^2\alpha = \operatorname{cosec}^2\alpha$$

The following are also very important.

Since $$\frac{NP}{ON} = \frac{\frac{NP}{OP}}{\frac{ON}{OP}}$$

$$\therefore \quad \tan\alpha = \frac{\sin\alpha}{\cos\alpha}$$

Sim'ly $$\cot\alpha = \frac{\cos\alpha}{\sin\alpha}$$

By means of these connections between the T. F.s we are able, when the numerical value of any one of them is given, to find that of each of the rest.

Examples

1. *If* $\sin\alpha = \frac{3}{5}$, *find the value of each of the other* T. F.s.

$$\cos\alpha = \sqrt{1-\sin^2\alpha} = \sqrt{1-\tfrac{9}{25}} = \pm\tfrac{4}{5}$$

$$\tan\alpha = \frac{\sin\alpha}{\sqrt{1-\sin^2\alpha}} = \frac{\frac{3}{5}}{\sqrt{1-\frac{9}{25}}} = \pm\tfrac{3}{4}$$

And sim'ly for the rest.

Or we might proceed (after finding $\cos\alpha$) thus—

$$\tan\alpha = \frac{\sin\alpha}{\cos\alpha} = \frac{\frac{3}{5}}{\pm\frac{4}{5}} = \pm\tfrac{3}{4}$$

$$\cot\alpha = \frac{1}{\tan\alpha} = \pm\tfrac{4}{3}$$

And sim'ly for the rest.

2. *If* $\tan\alpha = \frac{m}{n}$, *find* $\sin\alpha$ *and* $\cos\alpha$.

$$\sin\alpha = \frac{\tan\alpha}{\sqrt{1+\tan^2\alpha}} = \frac{\frac{m}{n}}{\sqrt{1+\frac{m^2}{n^2}}} = \frac{m}{\sqrt{m^2+n^2}}$$

$$\cos\alpha = \frac{1}{\sqrt{1+\tan^2\alpha}} = \frac{1}{\sqrt{1+\frac{m^2}{n^2}}} = \frac{n}{\sqrt{m^2+n^2}}$$

These very useful results may be got otherwise.

For since $\frac{m}{n} = \tan\alpha = \frac{\sin\alpha}{\cos\alpha}$

$$\left.\begin{aligned}\therefore\quad \lambda m &= \sin\alpha\\ \text{and}\quad \lambda n &= \cos\alpha\end{aligned}\right\}\text{ where } \lambda \text{ is some const'.}$$

Square and add; then

$$\lambda^2(m^2+n^2) = \sin^2\alpha + \cos^2\alpha = 1$$

$$\therefore\quad \lambda = \frac{1}{\sqrt{m^2+n^2}}$$

$$\left.\begin{aligned}\therefore\quad \sin\alpha &= \frac{m}{\sqrt{m^2+n^2}}\\ \text{and}\quad \cos\alpha &= \frac{n}{\sqrt{m^2+n^2}}\end{aligned}\right\}\text{ as before.}$$

Exercises

1. If $\tan\alpha = \dfrac{2x(x+1)}{2x+1}$, find $\sin\alpha$ and $\cos\alpha$.

2. Given that $\tan\alpha = 1$, find each of the other T. F.s.

3. If $\cot\alpha = \dfrac{p}{q}$, find $\sin\alpha$ and $\cos\alpha$.

4. If $\sin\alpha = 1$, find the value of $\cos\alpha + \cot\alpha + \operatorname{cosec}\alpha$.

5. Given that $\sin\alpha = \cdot 012$, find each of the other T. F.s to three decimal places.

6. If $\operatorname{vers}\alpha = \dfrac{\sqrt{2}-1}{\sqrt{2}}$, find the value of

$$\sin\alpha + \cos\alpha + \tan\alpha + \cot\alpha + \sec\alpha + \operatorname{cosec}\alpha.$$

7. If $\tan^3\phi = \dfrac{\alpha}{\beta}$, show that $\alpha\operatorname{cosec}\phi + \beta\sec\phi = \left(\alpha^{\frac{2}{3}} + \beta^{\frac{2}{3}}\right)^{\frac{3}{2}}$.

8. Express each of the T. F.s of α in terms of $\operatorname{vers}\alpha$.

9. If $\cos\phi = n\sin\alpha$, and $\cot\phi = \sin\alpha/\tan\beta$,

prove that $\cos\beta = n/\sqrt{1 + n^2\cos^2\alpha}$.

NOTE—*Notice that this is the result of eliminating* ϕ *between the given eq'ns.*

10. In a triangle ABC the angle C is right: if AE, BD, drawn perpendicular to AB, meet BC, AC produced, in E, D, respectively; prove that $\tan CED = \tan^3 BAC$.

11. ABCD is a rectangle: AP perpendicular to BD: PX, PY respectively perpendicular to BC, CD: prove that

$$PX^{\frac{2}{3}} + PY^{\frac{2}{3}} = AC^{\frac{2}{3}}.$$

T. C. D. '33 and *Math' Tri'*: '41.

NOTE—$P\hat{B}X = P\hat{A}B = A\hat{D}B = \theta$ *say: put down the ratios which are the* sines *of these* $\wedge$s, *and multiply them together: this gives* $\sin^3\theta = PX/AC$: *form* $\cos^3\theta$ *sim'ly: then add the squares of the cube roots of these.*

§ 6. From the fundamental identities already given, other identities may be deduced: any such being proposed for demonstration, there are *four* ways of effecting this.

1°, start with a fundamental identity, and reduce it to the required identity.

Example

To prove that $\sin^4\alpha + \cos^4\alpha \equiv 1 - 2\sin^2\alpha\cos^2\alpha$.

We know that $\sin^2\alpha + \cos^2\alpha = 1$

∴, squaring each side, we get

$$\sin^4\alpha + \cos^4\alpha + 2\sin^2\alpha\cos^2\alpha = 1$$

$$\therefore\quad \sin^4\alpha + \cos^4\alpha \equiv 1 - 2\sin^2\alpha\cos^2\alpha$$

Or, 2°, take one side (as a rule the more complex) of the identity; and, modifying it by means of fundamental identities, reduce it, through successive changes, to the other side.

Example

To prove that

$$\sin^2\alpha\tan\alpha + \cos^2\alpha\cot\alpha + 2\sin\alpha\cos\alpha \equiv \tan\alpha + \cot\alpha.$$

Left side

$$= (1 - \cos^2\alpha)\tan\alpha + (1 - \sin^2\alpha)\cot\alpha + 2\sin\alpha\cos\alpha$$

$$= \tan\alpha - \cos^2\alpha\frac{\sin\alpha}{\cos\alpha} + \cot\alpha - \sin^2\alpha\frac{\cos\alpha}{\sin\alpha} + 2\sin\alpha\cos\alpha$$

$$= \tan\alpha - \cos\alpha\sin\alpha + \cot\alpha - \sin\alpha\cos\alpha + 2\sin\alpha\cos\alpha$$

$$= \tan\alpha + \cot\alpha$$

Or, 3°, reduce one side to as simple a form as you can; and then try to reduce the other side to the same form.

Example

To prove that

$$\operatorname{cosec}\alpha(\sec\alpha - 1) + \sin\alpha \equiv \cot\alpha(1 - \cos\alpha) + \tan\alpha.$$

$$\text{Left side} = \frac{1}{\sin\alpha\cos\alpha} - \frac{1}{\sin\alpha} + \sin\alpha$$

$$\text{Right side} = \frac{\cos\alpha}{\sin\alpha} - \frac{1 - \sin^2\alpha}{\sin\alpha} + \frac{\sin\alpha}{\cos\alpha}$$

$$= \frac{\cos^2\alpha + \sin^2\alpha}{\sin\alpha\cos\alpha} - \frac{1}{\sin\alpha} + \sin\alpha$$

$$= \frac{1}{\sin\alpha\cos\alpha} - \frac{1}{\sin\alpha} + \sin\alpha$$

∴ the identity is true.

Or, 4°, consider whether the supposition of the truth of the identity leads to a known result.

Example

To prove that $\dfrac{\text{cosec}\,\alpha + \cot\alpha}{\sec\alpha + \tan\alpha} \equiv \dfrac{\sec\alpha - \tan\alpha}{\text{cosec}\,\alpha - \cot\alpha}$.

The identity is true, *if* $\text{cosec}^2\alpha - \cot^2\alpha = \sec^2\alpha - \tan^2\alpha$

$$\text{i.e. } if\ \frac{1 - \cos^2\alpha}{\sin^2\alpha} = \frac{1 - \sin^2\alpha}{\cos^2\alpha}$$

But this *is* true, for each side $= 1$.

NOTE—The mode of arrangement of the argument in 4° should be observed: recollect *not* to start with the identity as *unconditionally* true, and then reduce both sides simultaneously; for this will simply lead to the truism $\therefore\ 0 = 0$

Exercises

Prove the truth of the following sixteen identities—

1. $\cos\alpha\tan\alpha + \sin\alpha\cot\alpha \equiv \sin\alpha + \cos\alpha$

2. $\tan\alpha + \cot\alpha \equiv \sec\alpha\,\text{cosec}\,\alpha$

3. $\sec^2\alpha + \text{cosec}^2\alpha \equiv \sec^2\alpha\,\text{cosec}^2\alpha$

4. $\tan^2\alpha - \sin^2\alpha \equiv \tan^2\alpha\sin^2\alpha$

5. $\cot^2\alpha - \cos^2\alpha \equiv \cot^2\alpha\cos^2\alpha$

6. $(\sin\alpha + \cos\alpha)^2 + (\sin\alpha - \cos\alpha)^2 \equiv 2$

7. $\sin^4\alpha - \cos^4\alpha \equiv \sin^2\alpha - \cos^2\alpha$

8. $\sin^2\alpha\cos^2\beta - \cos^2\alpha\sin^2\beta \equiv \sin^2\alpha - \sin^2\beta$

9. $\cos^2\alpha\cos^2\beta - \sin^2\alpha\sin^2\beta \equiv \cos^2\alpha - \sin^2\beta$

10. $\tan^2\alpha + \cot^2\alpha + 2 \equiv \sec^2\alpha\,\text{cosec}^2\alpha$

11. $\tan\alpha/(\tan\alpha - \tan\beta) \equiv \cot\beta/(\cot\beta - \cot\alpha)$

12. $\sin\alpha(1 + \tan\alpha) + \cos\alpha(1 + \cot\alpha) \equiv \text{cosec}\,\alpha + \sec\alpha$

13. $(\sec\alpha\sec\beta + \tan\alpha\tan\beta)^2 - (\tan\alpha\sec\beta + \sec\alpha\tan\beta)^2 \equiv 1$

14. $2(\sin^6\alpha + \cos^6\alpha) + 1 \equiv 3(\sin^4\alpha + \cos^4\alpha)$

15. $\sin\alpha\cos\alpha \equiv \sqrt{(\sin\alpha - \sin^3\alpha)^2 + (\cos\alpha - \cos^3\alpha)^2}$

16. $(\sin\alpha - \text{cosec}\,\alpha)^2 - (\tan\alpha - \cot\alpha)^2 + (\cos\alpha - \sec\alpha)^2 \equiv 1$

17. Show that the equation

$$\frac{\cos^3\theta}{\cos\alpha} + \frac{\sin^3\theta}{\sin\alpha} = 1$$

can be reduced to the form

$$\left(\frac{\cos\alpha}{\cos\theta} - \frac{\sin\alpha}{\sin\theta}\right)\left(\frac{\cos\alpha}{\cos\theta} + \frac{\sin\alpha}{\sin\theta} + 1\right) = 0$$

NOTE—*In given eq'n put* $\cos^2\alpha + \sin^2\alpha$ *and* $\cos^2\theta + \sin^2\theta$ *successively for* 1 ; *and, after rearrangement, divide the 1st result by the 2nd.*

18. If c stands for $\cos\alpha$, and s for $\sin\alpha$, prove that

$$c^{12} + 4c^{10}s^2 + 5c^8s^4 - 5c^4s^8 - 4c^2s^{10} - s^{12} \equiv c^2 - s^2$$

19. Prove that

$$\begin{vmatrix} 1 & \cos^4\theta & \sin^4\theta \\ 1 & (1+\sin^2\theta)^2 & \sin^4\theta \\ 1 & \cos^4\theta & (1+\cos^2\theta)^2 \end{vmatrix} \equiv 16\sin^2\theta\cos^2\theta$$

Ox' Jun' Math' Schol': '77

NOTE—*Write* s *for* $\sin\theta$, *and* c *for* $\cos\theta$, *and use Sarrus' rule given on p.* 23.

CHAPTER II

The possible values of the Trigonometrical Functions

§ 7. If the generator of an ∧ stops when it has gone thro′ 90°, or 180°, or 270°, or 360°—so that it is then on one of the lines bounding the quadrants—the def′s given of the T. F.s apparently cease to have a meaning. Thus when (in § 1) $\alpha = 0$, the base of the auxil′ △ vanishes, while its perp′ and hyp′ coalesce.

We proceed to examine whether a meaning can, in these cases, be assigned to the T. F.s.

Def′—Infinite magnitude is denoted by the symbol ∞.

Note—It is to be carefully noted that ∞ does not denote *definite* magnitude, but *endless increase* of magnitude. The relationships between finite, infinitely small, and infinitely great magnitude are

finite/infinitely small = infinitely great
finite/infinitely great = infinitely small

Or, if m denotes *finite* magnitude, and o denotes *endless decrease* of mag′,

$$\frac{m}{0} = \infty \quad \text{and} \quad \frac{m}{\infty} = 0$$

Def′—If $f(\theta)$ is a function of θ, and if, as θ approaches indefinitely near to α, $f(\theta)$ tends to become equal to a finite quantity (k say) and can, by bringing θ near enough to α, be made to differ from k by less than any assignable magnitude, then we say that $f(\alpha) = k$. But if as θ approaches indefinitely near to the value α, $f(\theta)$ diminishes so as to become less than any assignable magnitude, then we say that $f(\alpha) = 0$; and if as θ approaches indefinitely near to the value α, $f(\theta)$ increases so as to become greater than any assignable magnitude, then we say that $f(\alpha) = \infty$.

Note—In the two former cases the values k and o may be properly called the respective *Limits* of $f(\theta)$. But the value ∞ cannot properly be called a Limit: in this last case $f(\theta)$ is said to *increase without Limit.*

§ **8.** *To find meanings for the* T. F.*s of* 90°.

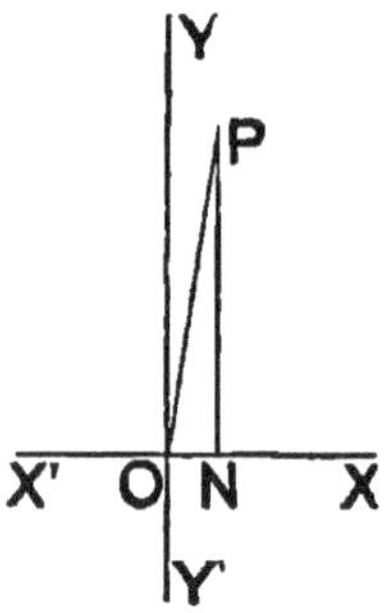

Suppose the generator OP of an ∧ to have turned thro′ $\alpha°$, where α very nearly $= 90°$; so that (XOX′, YOY′ being ⊥ lines, and OX the initial direction of OP)

$$X\hat{O}P \text{ very nearly } = 90°$$

Then PN (the ⊥ on OX) will very nearly coincide with OP.

∴, as α approaches 90°, $\sin\alpha$ (which $= NP/OP$) approaches unity.

And this approximation can be made nearer and nearer, by taking α sufficiently near 90°, until there is no assignable difference between $\sin\alpha$ and 1.

$$\therefore \quad \sin 90° = 1$$

$$\text{Sim'ly} \quad \operatorname{cosec} 90° = 1$$

Again, as α approaches 90°, ON becomes smaller, and can be made less than any assignable magnitude, by taking α sufficiently near 90°.

∴, as α approaches 90°, $\cos\alpha$ (which $= ON/OP$) becomes less than any assignable mag′.

$$\therefore \quad \cos 90° = 0$$

$$\text{Sim'ly} \quad \cot 90° = 0$$

Lastly, as α approaches 90°, $\tan\alpha$ (which $= NP/ON$) becomes greater than any assignable mag′.

$$\therefore \quad \tan 90° = \infty$$

$$\text{Sim'ly} \quad \sec 90° = \infty$$

By sim′r reasoning we could find values for the T. F.s of 0°, 180°, 270°, 360°.

The Student should investigate each case; and he will find that the following table is correct.

α	0°	90°	180°	270°	360°
$\sin\alpha$	0	1	0	−1	0
$\cos\alpha$	1	0	−1	0	1
$\tan\alpha$	0	∞	0	∞	0
$\cot\alpha$	∞	0	∞	0	∞
$\sec\alpha$	1	∞	−1	∞	1
$\operatorname{cosec}\alpha$	∞	1	∞	−1	∞

Note—In passing thro' the values 0 or ∞ there is a change of algebraic sign. We do not give any sign to the actual 0 or ∞; but say that the function is $\pm$ *just before* such value, and correspondingly $\mp$ *just after* it.

§ **9.** *Limits of the values of the* T. F.s

Taking the fig's of § 1,

since always $OP > ON$, unless they coincide,
and $OP > NP$, „ „ „ ;

but ON, NP may have any relative values;

we see that $\sin\alpha \not> 1$, and $\cos\alpha \not> 1$;
also that $\sec\alpha \not< 1$, and $\operatorname{cosec}\alpha \not< 1$;

but that $\tan\alpha$ and $\cot\alpha$ are unlimited in magnitude.

Examples

1. *The equation* $\sin\alpha = x + \frac{1}{x}$ *is impossible, if* x *is real.*

$$\text{For if} \quad y = x + \frac{1}{x}$$

$$\text{then} \quad x^2 - xy + 1 = 0$$

$$\therefore \quad 2x = y \pm \sqrt{y^2 - 4}$$

$\therefore$, for real values of x, the least value of y is 2

$$\text{i.e.} \quad x + \frac{1}{x} \nless 2$$

$$\therefore \quad \sin\alpha \neq x + \frac{1}{x}, \text{ for any real value of } x.$$

2. *The equation* $\sec\alpha = \frac{4xy}{(x+y)^2}$ *is only possible when* $x = y$.

$$\text{For since} \quad (x - y)^2 > 0, \text{ unless } x = y$$

$$\therefore \quad x^2 + y^2 > 2xy, \quad \text{,,} \quad \text{,,}$$

$$\therefore \quad (x + y)^2 > 4xy, \quad \text{,,} \quad \text{,,}$$

$$\frac{4xy}{(x+y)^2} < 1, \quad \text{,,} \quad \text{,,}$$

$$\therefore \quad \sec\alpha \neq \frac{4xy}{(x+y)^2} \quad \text{,,} \quad \text{,,}$$

§ **10.** We are now in a position to trace the gradual changes in the values of each T. F., as the angle changes from zero to four right angles.

To trace the changes of magnitude and sign in $\sin\alpha$, *as* α *changes continuously from* 0° *to* 360°.

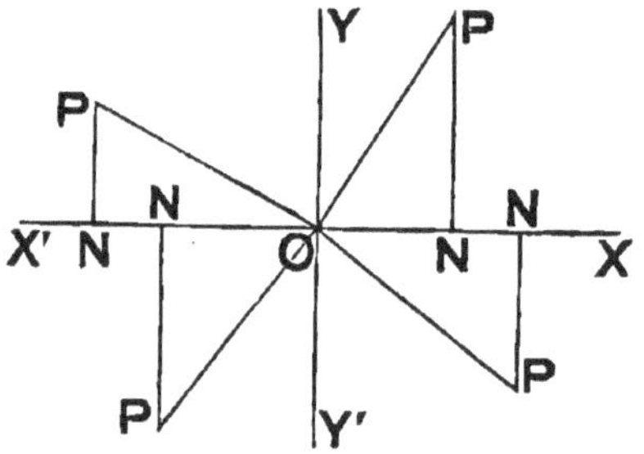

Let XOX′, YOY′ be rect'r axes; where OX, OY are pos′ direc′s. Let OP, initially along OX, revolve anti-clock-wise thro′ an $\hat{\alpha}$. Drop PN ⊥ on XOX′.

$$\text{Then} \quad \sin \alpha = \frac{NP}{OP}$$

1°, when $\alpha = 0$, $\sin \alpha = 0$.

As OP rotates thro′ the 1st quadrant XOY, NP increases continuously; until, when $\alpha = 90°$, NP coalesces with OP along OY: also NP remains pos′ thro′ the quadrant.

∴, as α changes from 0° to 90°, $\sin \alpha$ changes continuously from 0 to 1, rem′g pos′ all the way.

2°, when $\alpha = 90°$, $\sin \alpha = 1$.

As OP rotates thro′ the 2nd quadrant YOX′, NP diminishes continuously; until, when $\alpha = 180°$, NP vanishes: also NP remains pos′ thro′ the quadrant.

∴, as α changes from 90° to 180°, $\sin \alpha$ changes continuously from 1 to 0, rem′g pos′ all the way.

3°, when $\alpha = 180°$, $\sin \alpha = 0$.

As OP rotates thro′ the 3rd quadrant X′OY′, NP increases continuously; until, when $\alpha = 270°$, NP coalesces with OP along OY′: also NP, having changed sign from pos′ to neg′ as it passed thro′ the value 0, remains neg′ thro′ the quadrant.

∴, as α changes from 180° to 270°, $\sin \alpha$ changes continuously from 0 to -1, rem′g neg′ all the way.

4°, when $\alpha = 270°$, $\sin \alpha = -1$.

As OP rotates thro′ the 4th quadrant Y′OX, NP diminishes continuously; until, when $\alpha = 360°$, NP vanishes: also NP remains neg′ thro′ the quadrant.

∴, as α changes from 270° to 360°, $\sin \alpha$ changes continuously from -1 to 0, rem′g neg′ all the way.

To trace the changes of magnitude and sign in $\cos \alpha$, *as* α *changes continuously from* 0° *to* 360°.

Construct as for the sine.

$$\text{Then} \quad \cos \alpha = \frac{ON}{OP}$$

1°, when $\alpha = 0$, $\cos \alpha = 1$.

As OP rotates thro' the 1st quadrant XOY, ON diminishes continuously; until, when $\alpha = 90^\circ$, ON vanishes: also ON remains pos' thro' the quadrant.

∴, as α changes from 0° to 90°, $\cos\alpha$ changes continuously from 1 to 0, rem'g pos' all the way.

2°, when $\alpha = 90^\circ$, $\cos\alpha = 0$.

As OP rotates thro' the 2nd quadrant YOX', ON increases continuously; until, when $\alpha = 180^\circ$, ON coalesces with OP along OX': also ON, having changed sign from pos' to neg' in passing thro' the value 0, remains neg' thro' the quadrant.

∴, as α changes from 90° to 180°, $\cos\alpha$ changes continuously from 0 to -1, rem'g neg' all the way.

3°, when $\alpha = 180^\circ$, $\cos\alpha = -1$.

As OP rotates thro' the 3rd quadrant X'OY', ON diminishes continuously; until, when $\alpha = 270^\circ$, ON vanishes: also ON remains neg' thro' the quadrant.

∴, as α changes from 180° to 270°, $\cos\alpha$ changes continuously from -1 to 0, rem'g neg' all the way.

4°, when $\alpha = 270^\circ$, $\cos\alpha = 0$.

As OP rotates thro' the 4th quadrant Y'OX, ON increases continuously; until, when $\alpha = 360^\circ$, ON coalesces with OP along OX: also ON, having changed sign from neg' to pos' in passing thro' the value 0, remains pos' thro' the quadrant.

∴, as α changes from 270° to 360°, $\cos\alpha$ increases continuously from 0 to 1, rem'g pos' all the way.

To trace the changes of magnitude and sign in $\tan\alpha$, *as* α *changes continuously from* 0° *to* 360°.

Construct as for the sine.

$$\text{Then}\quad \tan\alpha = \frac{NP}{ON}$$

1°, when $\alpha = 0$, $\tan\alpha = 0$.

As OP rotates thro' the 1st quadrant XOY, NP increases and ON diminishes (both continuously) until, as α approaches 90°,

ON diminishes indefinitely; so that, ultimately, when $\alpha = 90°$, the ratio NP/ON becomes indefinitely large: also both NP and ON remain pos′ thro′ the quadrant.

∴, as α changes from $0°$ to $90°$, tan α changes continuously from 0 to ∞, rem′g pos′ all the way.

2°, when $\alpha = 90°$, $\tan\alpha = \infty$.

As OP rotates thro′ the 2nd quadrant YOX′, NP diminishes and ON increases (both continuously) until, when $\alpha = 180°$, NP vanishes; so that, ultimately, when $\alpha = 180°$, the ratio NP/ON vanishes: also NP remains pos′ thro′ the quadrant; but ON, which changed sign from pos′ to neg′ in passing thro′ the value 0, remains neg′ thro′ the quadrant.

∴, as α changes from $90°$ to $180°$, tan α changes continuously from ∞ to 0, rem′g neg′ all the way.

3°, when $\alpha = 180°$, $\tan\alpha = 0$.

As OP rotates thro′ the 3rd quadrant X′OY′, NP increases and ON diminishes (both continuously) until, as α approaches $270°$, ON diminishes indefinitely; so that, ultimately, when $\alpha = 270°$, the ratio NP/ON becomes indefinitely large: also ON remains neg′ thro′ the quadrant; and NP which changed sign from pos′ to neg′ in passing thro′ the value 0, remains neg′ thro′ the quadrant.

∴, as α changes from $180°$ to $270°$, tan α changes continuously from 0 to ∞, rem′g pos′ all the way.

4°, when $\alpha = 270°$, $\tan\alpha = \infty$.

As OP rotates thro′ the 4th quadrant Y′OX, NP diminishes and ON increases (both continuously) until, when $\alpha = 360°$, NP vanishes; so that, ultimately, when $\alpha = 360°$, the ratio NP/ON vanishes: also NP remains neg′ thro′ the quadrant; but ON, which changed sign from neg′ to pos′ in passing thro′ the value 0, remains pos′ thro′ the quadrant.

∴, as α changes from $270°$ to $360°$, tan α changes continuously from ∞ to 0, rem′g neg′ all the way.

Note—The Student should now find no difficulty in tracing the changes of the other T. F.s

§ 11. The *graphs* of the different T. F.s afford excellent aid towards recollecting their successive changes. These we now proceed to construct roughly: to draw them with any great accuracy requires a knowledge (which we shall get in Chapter V) of the numerical values of the T. F.s of several angles intermediate between those which are exact multiples of a right angle.

To draw the **Graph** *of* $\sin \alpha$*—sometimes called the* **Sinusoid**; *and sometimes the* **Curve of Sines.**

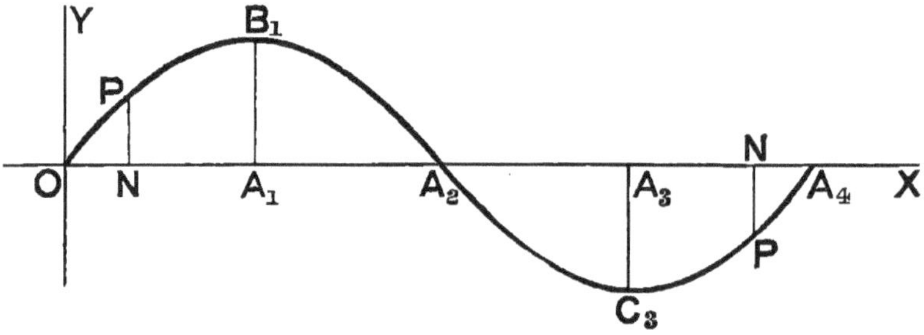

Take rect'r axes; and let OX, OY be their pos' direc's.

Let A_1, A_2, A_3, A_4 be p'ts in OX, such that

$$OA_1 = A_1A_2 = A_2A_3 = A_3A_4;$$

and let each of these equal lines represent a r't $\angle$.

At A_1 draw A_1B_1, $\perp$ to OX, and in pos' direc'; and at A_3 draw A_3C_3, $\perp$ to OX, and in neg' direc'; so that each of them is a unit of length.

Then the *graph* of $\sin \alpha$ will be a curve drawn from O thro' B_1, A_2, C_3, A_4; and if P is any p't on the graph, and PN is $\perp$ to OX, then PN represents $\sin \alpha$, for that value of α represented by ON.

Note—The filling in of the curve between the above 4 p'ts will be more easily seen when the numerical values of a few intermediate sines have been calculated.

E. g. (in § 31) $\sin 30°$ is found to be $\frac{1}{2}$, so that when $ON = \frac{1}{3}OA_1$, PN should be $\frac{1}{2}A_1B_1$. Also, for the easier practical drawing of graphs, the Student should use what is sometimes called *logarithmic paper*; viz. paper ruled in two ⊥ direc's, with the rulings rather close together—say /10th inch apart. The absolute lengths of OA_1 and $A_1 B_1$ are arbitrary, and need not *necessarily* bear any particular relation to each other; but there is a certain convenience in taking them so that $OA_1 : A_1 B_1 = 3 : 2$; for then, inasmuch as $3 : 2 = 90^\circ : 60^\circ$, and a radian is not far from 60°, it will happen that the line which represents the sine, for a particular ∧, will also nearly represent that *angle in radian measure*. This would enable us to draw the graph of a function of θ *and* $\sin\alpha$, where ${}^r\theta = \alpha^\circ$. Some Exercises in graph drawing will be given after Chapter V.

The remarks in this Note are to be taken also in connection with each of the following graphs.

To draw the **Graph** *of* $\cos\alpha$*—sometimes called the* **Cosinusoid.**

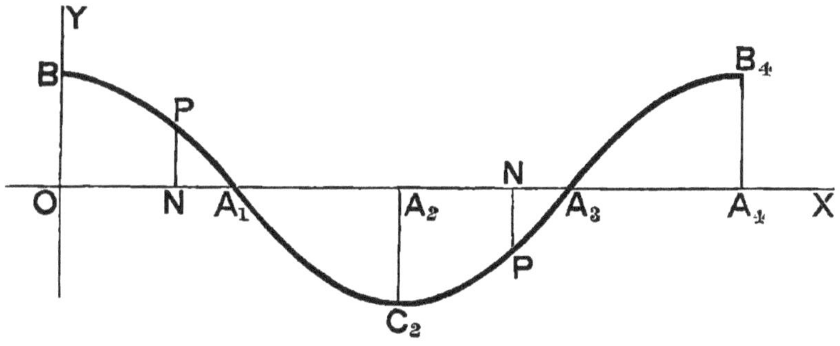

Take rect'r axes; and let OX, OY be their pos' direc's.

Let A_1, A_2, A_3, A_4 be p'ts in OX, such that

$$OA_1 = A_1 A_2 = A_2 A_3 = A_3 A_4;$$

and let each of these equal lines represent a r't ∧.

In OY take B, so that OB is a unit of length; at A_2 draw $A_2 C_2$ in the neg' direc'; and at A_4 draw $A_4 B_4$ in the pos' direc', so that $A_2 C_2 = A_4 B_4 = OB$.

Then the *graph* of $\cos\alpha$ will be a curve drawn from B thro' A_1, C_2, A_3, B_4; and, if P is any p't on the graph, and PN is ⊥ to OX, then PN represents $\cos\alpha$, for that value of α represented by ON.

To draw the **Graph** *of* tan α.

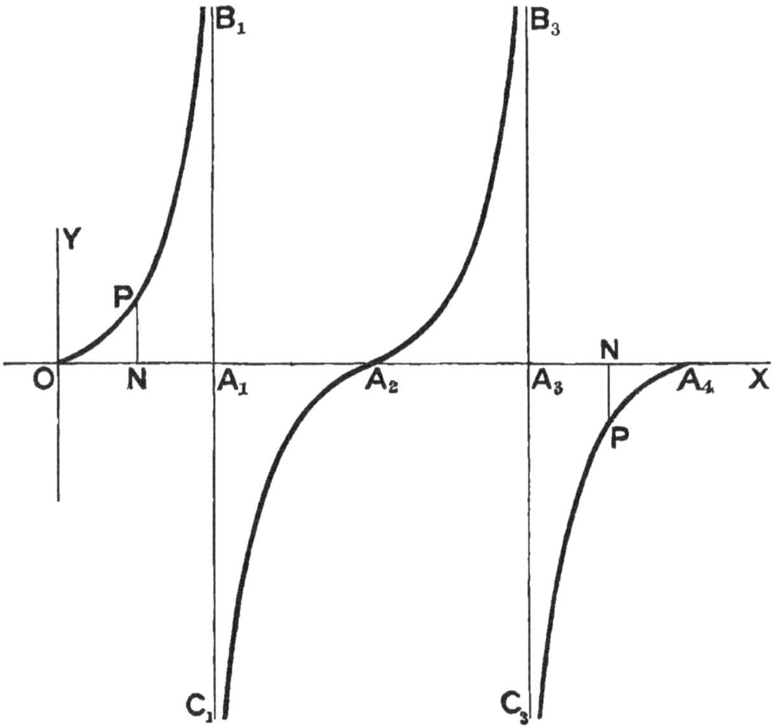

Take rect′r axes; and let OX, OY be their pos′ direc′s.

Let A_1, A_2, A_3, A_4 be p′ts in OX, such that

$$OA_1 = A_1A_2 = A_2A_3 = A_3A_4;$$

and let each of these equal lines represent a r′t ∧.

Thro′ A_1 and A_3 draw indefinitely extended lines $B_1A_1C_1$, $B_3A_3C_3$, ⊥ to OX; where A_1B_1, A_3B_3 are in pos′ direc′, but A_1C_1, A_3C_3 are in neg′ direc′.

Then the *graph* of tan α is a series of 3 curves drawn thus—

1°, from O in the pos′ direc′ so as to have A_1B_1 an asymptote;

2°, thro' A_2 in the neg' direc' so as to have $A_1 C_1$ an asymptote, and in the pos' direc' so as to have $A_3 B_3$ an asymptote;

3°, from A_4 in the neg' direc' so as to have $A_3 C_3$ an asymptote;

and if P is any p't on the graph, and PN ⊥ to OX, then PN represents tan α for that value of α represented by ON.

Sim'ly the graphs of the other 3 T. F.s can be drawn: we give the diagrams, leaving the Student to follow out the details as above. Notice that all 3 have asymptotes.

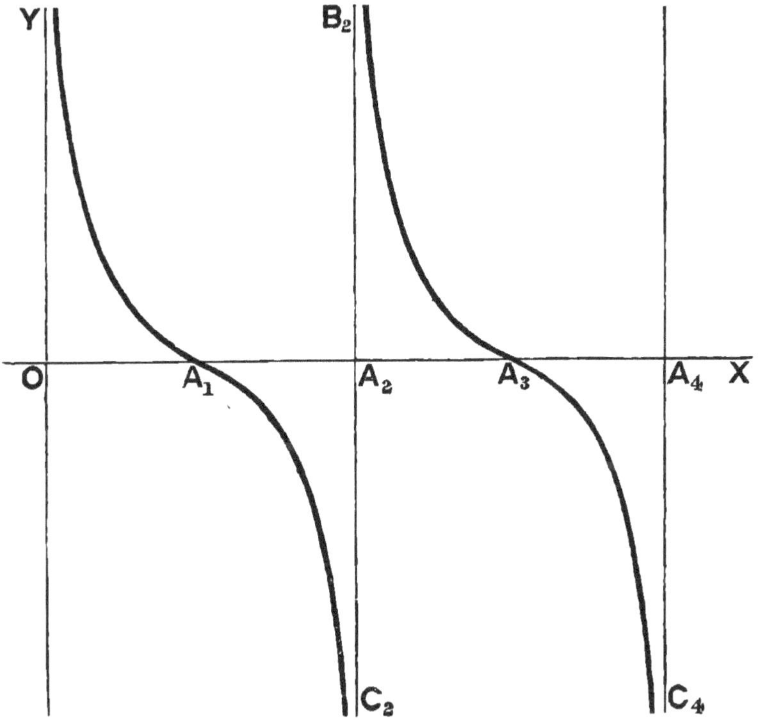

Graph *of* cot α

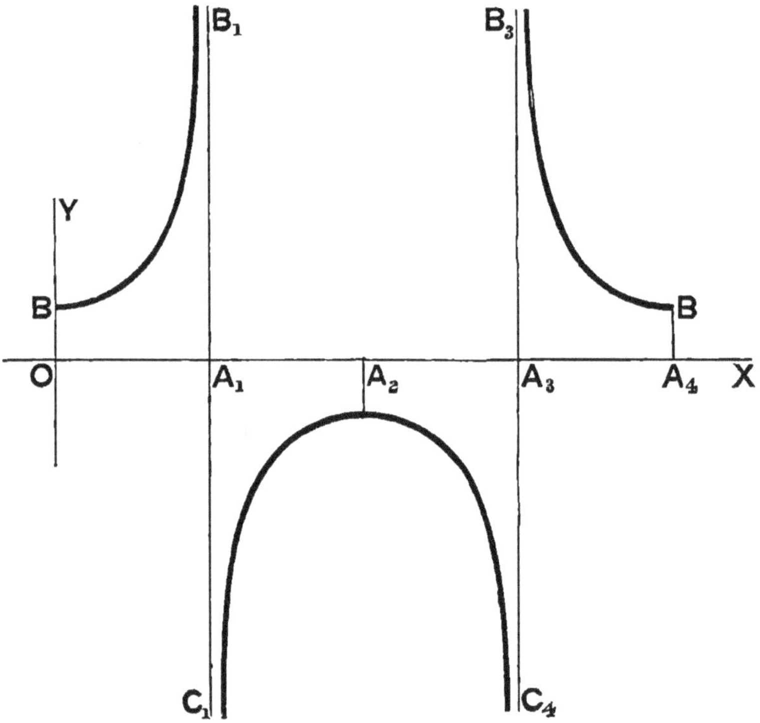

Graph *of* sec α

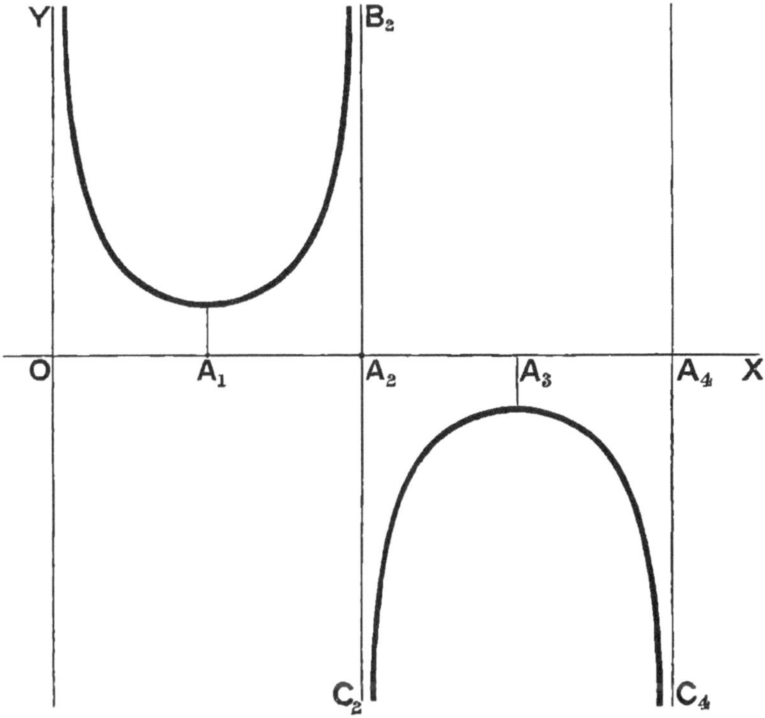

Graph *of* cosec α

CHAPTER III

On certain relations between the Trigonometrical Functions of Angles whose sum or difference is a multiple of a Right Angle

§ **12.** **Theorem**—*If any angle is* $\alpha°$, *then*

$$\sin\alpha = \cos(90° - \alpha)$$

and $$\cos\alpha = \sin(90° - \alpha)$$

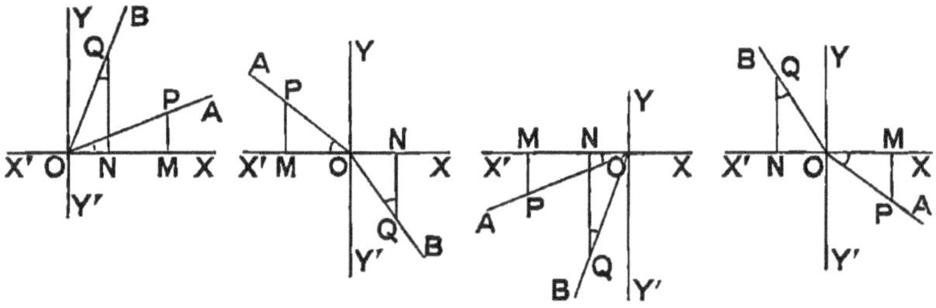

Take XOX′, YOY′ ⊥ lines, where OX, OY are pos′ direc′s. Let OA, initially along OX, revolve in pos′ direc′ (i. e. anti-clockwise) thro′ $\alpha°$.

To construct $90° - \alpha$; let OB, initially along OX, revolve in pos′ direc′ thro′ 90° (this will bring it to OY) and then back, in neg′ direc′, thro′ $\alpha°$.

Then, from the mode of construction, we see that

When OA is in the 1st quadrant, OB is in the 1st
„ „ „ 2nd „ , „ „ 4th
„ „ „ 3rd „ , „ „ 3rd
„ „ „ 4th „ , „ „ 2nd

In OA take any p′t P, and drop PM ⊥ on XOX′.
„ OB „ „ Q, „ „ QN „ „

Then, since $\hat{MOP} = \hat{YOB} = \hat{NQO}$,

and $\hat{M} = \hat{N}$;

$\therefore$ Δ^s MOP, NQO are sim'r.

Also, by examining the fig's, we see that MP and ON are always of the same sign.

$\therefore$ the ratios $\frac{MP}{OP}$ and $\frac{ON}{OQ}$ have always the same sign and magnitude.

$$\therefore \quad \sin \alpha = \cos (90^\circ - \alpha)$$

So also the ratios $\frac{OM}{OP}$ and $\frac{NQ}{OQ}$ have always the same sign and mag'.

$$\therefore \quad \cos \alpha = \sin (90^\circ - \alpha)$$

Hence, if any two $\wedge^s$ are complementary—

the **sine** of either $\equiv$ the **cosine** of the other

Sim'ly, or as a deduction from the foregoing, we should get that, if any two $\wedge^s$ are complementary—

the **tangent** of either $\equiv$ the **cotangent** of the other
and ,, **secant** ,, ,, ,, **cosecant** ,, ,,

Note—It is from these properties that are derived the words
Co-sine, equivalent to sine of *Complement*,
Co-tangent, ,, ,, tangent ,, ,, ,
and Co-secant, ,, ,, secant ,, ,,

Examples

If $A + B + C = 180^\circ$, then $\frac{A}{2}$ and $\frac{B+C}{2}$ are complementary;

so that then $\sin \frac{A}{2} = \cos \frac{B+C}{2}$

and $\cos \frac{A}{2} = \sin \frac{B+C}{2}$

§ 13. Theorem—*If any angle is* α°, *then*

$$\sin\alpha = \sin(180^\circ - \alpha)$$

and $$\cos\alpha = -\cos(180^\circ - \alpha)$$

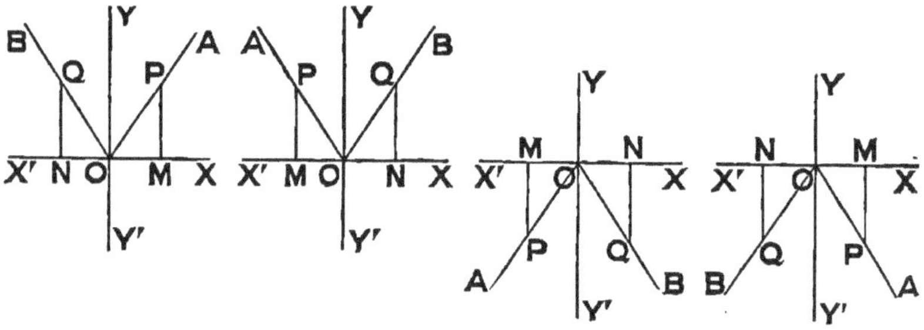

Take XOX′, YOY′ ⊥ lines, where OX, OY are pos′ direc′s. Let OA, initially along OX, revolve in the pos′ direc′ (i.e. anti-clock-wise) thro′ α°.

To construct $180^\circ - \alpha$; let OB, initially along OX, revolve in pos′ direc′ thro′ 180° (this will bring it to OX′) and then back, in the neg′ direc′, thro′ α°.

Then, from the mode of construction, we see that

when	OA	is in the	1st	quadrant,	OB	is in the	2nd
,,	,,	,,	2nd	,,	,	,,	,,	1st
,,	,,	,,	3rd	,,	,	,,	,,	4th
,,	,,	,,	4th	,,	,	,,	,,	3rd

In OA take any p′t P, and drop PM ⊥ on XOX′.
,, OB ,, ,, Q, ,, ,, QN ,, ,,

Then, since $P\hat{O}M = Q\hat{O}N$,

and $\hat{M} = \hat{N}$;

∴ Δ^s POM, QON are sim′r.

Also, by examining the fig′s, we see that MP and NQ are always of the same sign.

$\therefore$ the ratios $\frac{MP}{OP}$ and $\frac{NQ}{OQ}$ have always the same sign and magnitude.

$$\therefore \quad \sin\alpha = \sin(180^\circ - \alpha)$$

Again OM and ON have always opposite signs, so that always

$$\frac{OM}{OP} = -\frac{ON}{OQ}$$

$$\therefore \quad \cos\alpha = -\cos(180^\circ - \alpha)$$

Sim'ly, or as a deduction from the foregoing, we should get that

$$\tan\alpha = -\tan(180^\circ - \alpha)$$
$$\cot\alpha = -\cot(180^\circ - \alpha)$$
$$\sec\alpha = -\sec(180^\circ - \alpha)$$
$$\operatorname{cosec}\alpha = \operatorname{cosec}(180^\circ - \alpha)$$

So that the T. F. of any $\wedge$ is expressible in terms of the same T. F. of the supplementary $\wedge$.

Examples

$$\sin 120^\circ = \sin(180^\circ - 120^\circ) = \sin 60^\circ$$
$$\cos 150^\circ = -\cos(180^\circ - 150^\circ) = -\cos 30^\circ$$
$$\operatorname{vers}(180^\circ - \alpha) = 1 - \cos(180^\circ - \alpha) = 1 + \cos\alpha$$

Exercises

Simplify each of the following—

1. $\sin 90^\circ + \tan^2(180^\circ - \alpha) - \operatorname{cosec}^2(90^\circ - \alpha)$

2. $\cos(90^\circ - \alpha)\sin(180^\circ - \alpha) - \sin(90^\circ - \alpha)\cos(180^\circ - \alpha)$

3. $\operatorname{vers}(90^\circ - \alpha)\operatorname{vers}(180^\circ - \alpha)$

4. $\cos(B + C) - \cos\frac{B + C}{2} - \cos(180^\circ - A) + \sin\frac{A}{2}$

where A, B, C are angles of a triangle.

§ 14. **Theorem**—*If any angle is* $\alpha°$, *then*

$$\sin\alpha = -\sin(180° + \alpha)$$

and $$\cos\alpha = -\cos(180° + \alpha)$$

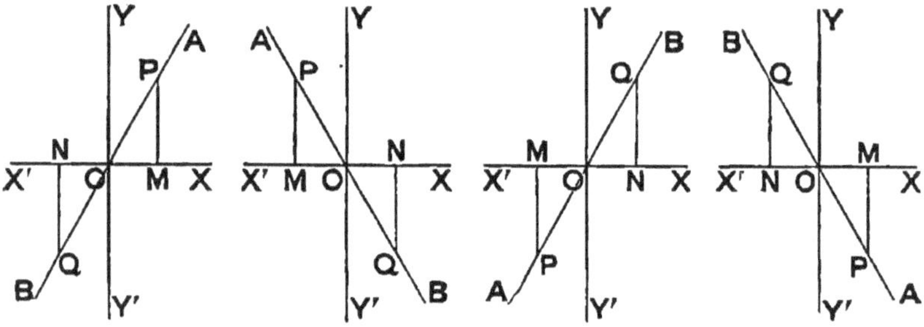

Take XOX′, YOY′, ⊥ lines, where OX, OY are pos′ direc′s. Let OA, initially along OX, revolve in the pos′ direc′ (i. e. anti-clock-wise) thro′ $\alpha°$.

To construct $180° + \alpha$; let OB, initially along OX, revolve in the pos′ direc′ thro′ $\alpha°$ (this will bring it to OA) and then on in the pos′ direc′ thro′ 180°.

Then, from the mode of construction, we see that

when OA	is in	the	1st	quadrant,	OB	is in	the	3rd
,,	,,	,,	2nd	,, ,	,,	,,		4th
,,	,,	,,	3rd	,, ,	,,	,,		1st
,,	,,	,,	4th	,, ,	,,	,,		2nd

In OA take any p′t P, and drop PM ⊥ on XOX′.
,, OB ,, ,, Q, ,, ,, QN ,, ,,

Since AOB is a st′ line, △s POM, QON are clearly sim′r.
Also MP and NQ are always of opposite sign.

∴ the ratios $\frac{MP}{OP}$ and $\frac{NQ}{OQ}$ are always of the same magnitude, but of opposite sign.

$$\therefore \quad \sin\alpha = -\sin(180° + \alpha)$$

So also the ratios $\dfrac{OM}{OP}$ and $\dfrac{ON}{OP}$ are always of the same mag′, but of opposite sign.

$$\therefore \quad \cos\alpha = -\cos(180^\circ + \alpha)$$

Sim′ly, or as a deduction from the foregoing, we should get that

$$\tan\alpha = \tan(180^\circ + \alpha)$$
$$\cot\alpha = \cot(180^\circ + \alpha)$$
$$\sec\alpha = -\sec(180^\circ + \alpha)$$
$$\operatorname{cosec}\alpha = -\operatorname{cosec}(180^\circ + \alpha)$$

Note—We saw (on p. 25) that all the T. F.s are periodic, and that 360° (or 2π) is the extent of period for each of them. But we can now find shorter periods for the tangent and cotangent.

For, since $\tan\alpha = \tan(180^\circ + \alpha)$ it follows that 180° may be added any number of times to α, without altering the value of $\tan\alpha$.

Also, anticipating the result of § 16, we have

$$\tan(-180^\circ + \alpha), \text{ which is the same as } \tan\{-(180^\circ - \alpha)\}$$
$$= -\tan(180^\circ - \alpha), \text{ by § 16}$$
$$= \tan\alpha, \text{ by § 13}$$

So that -180° may be added any number of times to α, without altering the value of $\tan\alpha$.

$$\therefore \quad \tan\alpha = \tan(n\,.\,180^\circ + \alpha)$$

where n is any integer, pos′, neg′, or zero.

$$\text{Sim'ly} \quad \cot\alpha = \cot(n\,.\,180^\circ + \alpha)$$

Expressing these important results in radian measure, we have

$$\tan\theta = \tan(n\pi + \theta)$$
$$\cot\theta = \cot(n\pi + \theta)$$

Hence the periods of the tangent and cotangent are 180° (or π).

Examples

$$\tan 915^\circ = \tan(\overline{5 \times 180^\circ} + 15^\circ) = \tan 15^\circ$$
$$\cot(-885^\circ) = \cot\{\overline{5 \times (-180^\circ)} + 15^\circ\} = \cot 15^\circ$$

§ 15. **Theorem**—*If any angle is* $\alpha°$, *then*

$$\sin\alpha = -\cos(90° + \alpha)$$
and $$\cos\alpha = \sin(90° + \alpha)$$

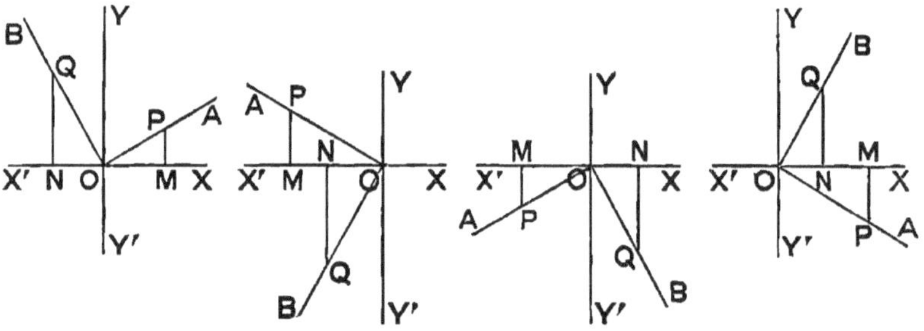

Take XOX′, YOY′ ⊥ lines, where OX, OY are pos′ direc′s. Let OA, initially along OX, revolve in the pos′ direc′ (i. e. anti-clock-wise) thro′ $\alpha°$.

To construct $90° + \alpha$; let OB, initially along OX, revolve in the pos′ direc′ thro′ 90° (this will bring it to OY) and then on, in the pos′ direc′, thro′ $\alpha°$.

Then, from the mode of construction, we see that

when OA	is in	the	1st	quadrant,	OB	is in	the	2nd
,,	,,	,,	2nd	,,	,,	,,	,,	3rd
,,	,,	,,	3rd	,,	,,	,,	,,	4th
,,	,,	,,	4th	,,	,,	,,	,,	1st

In OA take any p′t P, and drop PM ⊥ on XOX′.
,, OB ,, ,, Q, ,, ,, QN ,, ,,

Then, since $P\hat{O}M = Y\hat{O}Q = O\hat{Q}N$,

and $\hat{M} = \hat{N}$;

$\therefore$ Δ^s POM, OQN are sim′r.

Also, by examining the fig′s, we see that MP and ON are always opposite in sign.

$\therefore$ the ratios $\frac{MP}{OP}$ and $\frac{ON}{OQ}$ have the same magnitude but opposite signs.

$$\therefore \quad \sin\alpha = -\cos(90^\circ + \alpha)$$

So also the ratios $\frac{OM}{OP}$ and $\frac{NQ}{OQ}$ have always the same sign and mag′.

$$\therefore \quad \cos\alpha = \sin(90^\circ + \alpha)$$

Sim′ly, or as a deduction from the foregoing, we should get that

$$\tan\alpha = -\cot(90^\circ + \alpha)$$
$$\cot\alpha = -\tan(90^\circ + \alpha)$$
$$\sec\alpha = \operatorname{cosec}(90^\circ + \alpha)$$
$$\operatorname{cosec}\alpha = -\sec(90^\circ + \alpha)$$

Note—We know that the removal of any multiple of 4 right $\wedge^s$ from an $\wedge$ will leave any T. F. of it unaltered; and, by the foregoing results, we see that 2 right $\wedge^s$ can be removed, or the supplementary $\wedge$ substituted, if, in certain cases, a change of sign is made.

Hence the T. F.s of large $\wedge^s$ can be reduced to the same T. F.s of smaller $\wedge^s$.

Again, one right $\wedge$ can be removed, or the complementary $\wedge$ substituted, if we change the T. F. into its complementary T. F.s, and, in certain cases, make a change of sign.

Examples

Since $487^\circ = 360^\circ + 180^\circ - 53^\circ$

$$\therefore \quad \sin 487^\circ = \sin(180^\circ - 53^\circ) = \sin 53^\circ$$

Again, since $1365^\circ = \overline{3 \times 360^\circ} + 180^\circ + 90^\circ + 15^\circ$

$$\therefore \quad \cos 1365^\circ = \cos(180^\circ + 90^\circ + 15^\circ) = -\cos(90^\circ + 15^\circ) = \sin 15^\circ$$

Exercises

Simplify each of the following—

1. $\operatorname{vers}(90^\circ + \alpha)\operatorname{vers}(90^\circ - \alpha) + \operatorname{vers}\alpha \operatorname{vers}(180^\circ - \alpha)$

2. $\sin 903^\circ + \sin 643^\circ$

§ **16.** **Theorem**—*If any angle is* $\alpha°$, *then*

$$\sin\alpha = -\sin(-\alpha)$$
$$\text{and} \quad \cos\alpha = \cos(-\alpha)$$

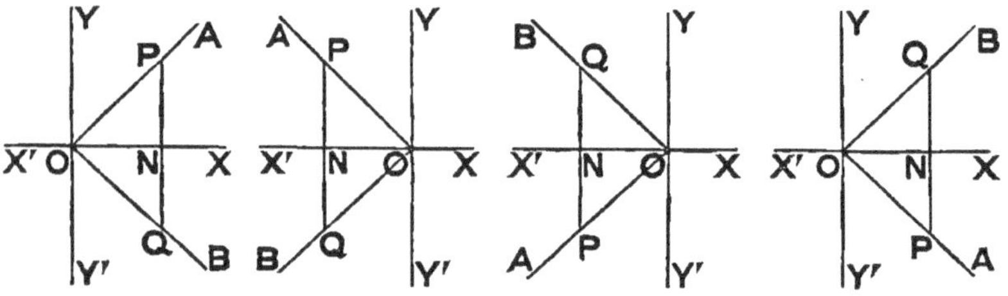

Take XOX′, YOY′ ⊥ lines, where OX, OY are pos′ direc′s. Let OA, OB, initially along OX, revolve respectively in pos′ and neg′ direc′s, thro′ $\alpha°$.

Then, from the mode of construction, we see that

when OA	is	in the	1st	quadrant,	OB	is	in the	4th,
,,	,,	,,	2nd	,,	,	,,	,,	3rd,
,,	,,	,,	3rd	,,	,	,,	,,	2nd,
,,	,,	,,	4th	,,	,	,,	,,	1st.

In OA take any p′t P, and drop PN ⊥ on XOX′, producing it to meet OB in Q.

Then obviously—

1°, $\triangle ONP \equiv \triangle ONQ$; and,

2°, NP, NQ are always of opposite sign.

∴ the ratios $\frac{NP}{OP}$ and $\frac{NQ}{OP}$ are always equal in magnitude and opposite in sign.

$$\therefore \quad \sin\alpha = -\sin(-\alpha)$$

So also the ratios $\frac{ON}{OP}$ and $\frac{ON}{OQ}$ have always the same sign and mag′.

$$\therefore \quad \cos\alpha = \cos(-\alpha)$$

Sim'ly, or as a deduction from the foregoing, we should get that

$$\tan\alpha = -\tan(-\alpha)$$
$$\cot\alpha = -\cot(-\alpha)$$
$$\sec\alpha = \sec(-\alpha)$$
$$\text{cosec}\,\alpha = -\text{cosec}(-\alpha)$$

§ **17.** Summing up all the preceding results, we have

$$\sin\alpha = \cos(90^\circ - \alpha) = \sin(180^\circ - \alpha)$$
$$= -\sin(-\alpha) = -\cos(90^\circ + \alpha) = -\sin(180^\circ + \alpha)$$

$$\cos\alpha = \sin(90^\circ - \alpha) = -\cos(180^\circ - \alpha)$$
$$= \cos(-\alpha) = \sin(90^\circ + \alpha) = -\cos(180^\circ + \alpha)$$

If $\alpha^\circ = {}^r\theta$, then these formulæ, in radian measure, are

$$\sin\theta = \cos(\pi/2 - \theta) = \sin(\pi - \theta)$$
$$= -\sin(-\theta) = -\cos(\pi/2 + \theta) = -\sin(\pi + \theta)$$

$$\cos\theta = \sin(\pi/2 - \theta) = -\cos(\pi - \theta)$$
$$= \cos(-\theta) = \sin(\pi/2 + \theta) = -\cos(\pi + \theta)$$

From these the analogous formulæ for the other T. F.s are at once deducible.

Note—The Learner should make himself familiar with the above results: they are of the utmost importance; for, by means of them the T. F. of *any* $\wedge$ can be expressed in terms of the same T.F. of a pos' $\wedge$ less than 90°. In case of forgetting any one of them, it can be easily recovered by considering its particular fig' in the simple case when $\alpha < 90^\circ$. All such simple cases are here included in the adjoining diagram, which should be carefully studied.

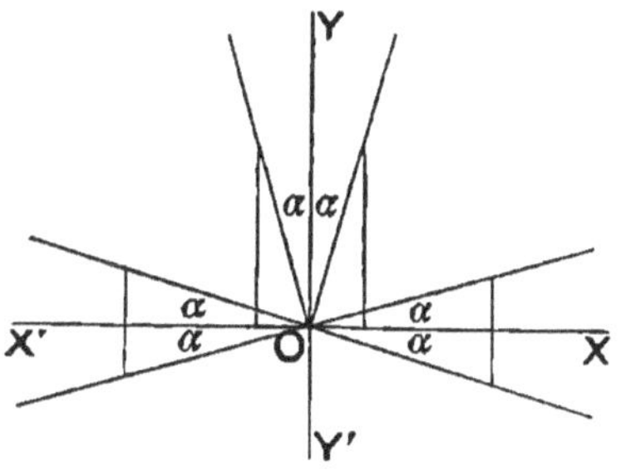

CHAPTER IV

Fundamental relations between Trigonometrical Functions of Angles and Trigonometrical Functions of parts of those Angles

§ 18. **Theorem**—*If* α *and* β *denote any angles, then*

$$\sin(\alpha+\beta) = \sin\alpha\cos\beta + \cos\alpha\sin\beta$$

and $$\cos(\alpha+\beta) = \cos\alpha\cos\beta - \sin\alpha\sin\beta$$

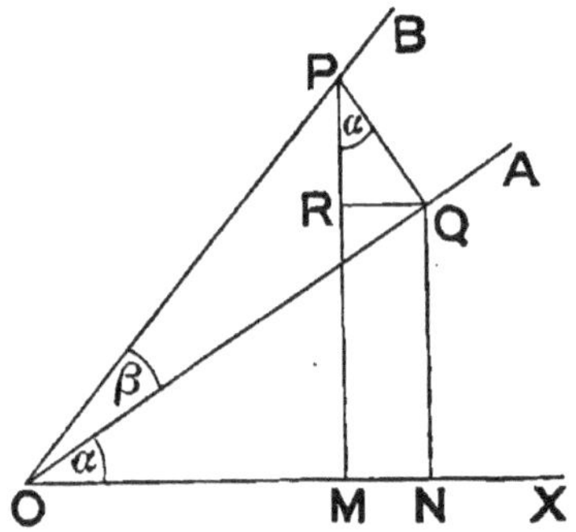

Take the case when α, β are pos′ $\wedge$s, and each of the three α, β, $\alpha+\beta$, $< 90°$.

Let a line, initially along OX, revolve in the pos′ direc′ (i.e. anti-clock-wise) thro′ α into position OA, and then on thro′ β into position OB.

Take P any p′t in OB: drop $\perp$s PM on OX, PQ on OA, QN on OX, QR on PM.

Then $\widehat{QPR} = 90° - \widehat{PQR} = \widehat{RQO} = \alpha$.

Now $$\frac{MP}{OP} = \frac{NQ + PR}{OP} = \frac{NQ}{OQ}\cdot\frac{OQ}{OP} + \frac{PR}{PQ}\cdot\frac{PQ}{OP}$$

$$\therefore\ \sin(\alpha+\beta) = \sin\alpha\cos\beta + \cos\alpha\sin\beta$$

Again $$\frac{OM}{OP} = \frac{ON - RQ}{OP} = \frac{ON}{OQ}\cdot\frac{OQ}{OP} - \frac{RQ}{PQ}\cdot\frac{QP}{OP}$$

$$\therefore\ \cos(\alpha+\beta) = \cos\alpha\cos\beta - \sin\alpha\sin\beta$$

Now $\sin(\overline{90^\circ + \alpha} + \beta)$

$= \cos(\alpha + \beta)$

$= \cos\alpha \cos\beta - \sin\alpha \sin\beta$

$= \sin(90^\circ + \alpha)\cos\beta + \cos(90^\circ + \alpha)\sin\beta$

And $\cos(\overline{90^\circ + \alpha} + \beta)$

$= -\sin(\alpha + \beta)$

$= -\sin\alpha\cos\beta - \cos\alpha\sin\beta$

$= \cos(90^\circ + \alpha)\cos\beta - \sin(90^\circ + \alpha)\sin\beta$

$\therefore$ the restriction that $\alpha < 90^\circ$ may be removed.

Sim'ly " " " β " " "

$\therefore$ the theorem is true for all pos' $\wedge$s.

Lastly, if $\alpha = -x$, and $\beta = -y$, then

$\sin(\alpha + \beta) = \sin\{-(x + y)\} = -\sin(x + y)$

$= -\sin x \cos y - \cos x \sin y$

$= \sin(-x)\cos(-y) + \cos(-x)\sin(-y)$

$= \sin\alpha\cos\beta + \cos\alpha\sin\beta$

and

$\cos(\alpha + \beta) = \cos\{-(x + y)\} = \cos(x + y)$

$= \cos x \cos y - \sin x \sin y$

$= \cos(-x)\cos(-y) - \sin(-x)\sin(-y)$

$= \cos\alpha\cos\beta - \sin\alpha\sin\beta$

$\therefore$ the theorem is true for all neg' $\wedge$s;

and, $\therefore$, is true for *all* values of α and β.

Note—By $(\alpha + \beta)$ is denoted a single angle, where α, β are *any* angles such that β added to α gives its magnitude.

For example :

$$\sin 13\alpha = \sin(8\alpha + 5\alpha) = \sin 8\alpha \cos 5\alpha + \cos 8\alpha \sin 5\alpha$$

§ **19. Theorem**—*If* α *and* β *are any angles, then*

$$\sin(\alpha - \beta) = \sin\alpha\cos\beta - \cos\alpha\sin\beta$$

and $$\cos(\alpha - \beta) = \cos\alpha\cos\beta + \sin\alpha\sin\beta$$

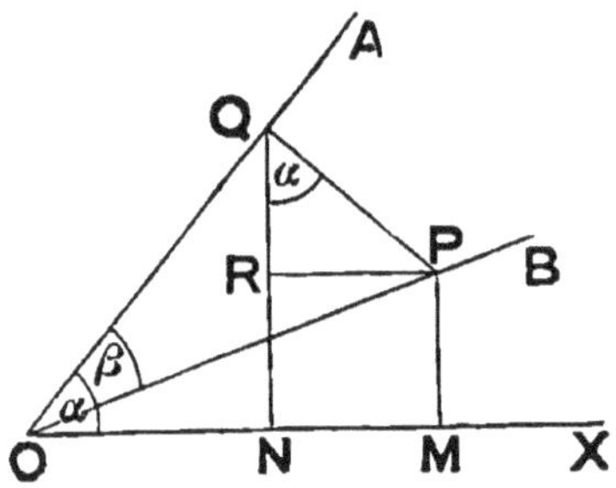

First take the case when α, β are pos′ ∧s, such that each $< 90°$, and that $\alpha > \beta$.

Let a line, initially along OX, revolve in the pos′ direc′ (i. e. anti-clock-wise) thro′ α into position OA, and then back, in the neg′ direc′, thro′ β into position OB.

Take P any p′t in OB: drop ⊥s PM on OX, PQ on OA, QN on OX, PR on QN.

$$\text{Then} \quad P\hat{Q}R = 90° - O\hat{Q}N = \alpha$$

$$\text{Now} \quad \frac{MP}{OP} = \frac{NQ - QR}{OP} = \frac{NQ}{OQ}\cdot\frac{OQ}{OP} - \frac{QR}{QP}\cdot\frac{QP}{OP}$$

$$\therefore \quad \sin(\alpha - \beta) = \sin\alpha\cos\beta - \cos\alpha\sin\beta$$

$$\text{Again} \quad \frac{OM}{OP} = \frac{ON + RP}{OP} = \frac{RP}{QP}\cdot\frac{QP}{OP} + \frac{ON}{OQ}\cdot\frac{OQ}{OP}$$

$$\therefore \quad \cos(\alpha - \beta) = \cos\alpha\cos\beta + \sin\alpha\sin\beta$$

Next suppose that $\beta > \alpha$.

$$\begin{aligned} \text{Then} \quad \sin(\alpha - \beta) &= -\sin(\beta - \alpha) \\ &= -\sin\beta\cos\alpha + \cos\beta\sin\alpha \\ &= \sin\alpha\cos\beta - \cos\alpha\sin\beta \end{aligned}$$

$$\begin{aligned} \text{Also} \quad \cos(\alpha - \beta) &= \cos(\beta - \alpha) \\ &= \cos\beta\cos\alpha + \sin\beta\sin\alpha \end{aligned}$$

∴ the theorem is true for any values of α, β less than 90°.

Again $\sin(\overline{90^\circ+\alpha}-\beta)$

$= \cos(\alpha-\beta)$

$= \cos\alpha\cos\beta + \sin\alpha\sin\beta$

$= \sin(90^\circ+\alpha)\cos\beta - \cos(90^\circ+\alpha)\sin\beta$

And $\cos(\overline{90^\circ+\alpha}-\beta)$

$= -\sin(\alpha-\beta)$

$= -\sin\alpha\cos\beta + \cos\alpha\sin\beta$

$= \cos(90^\circ+\alpha)\cos\beta + \sin(90^\circ+\alpha)\sin\beta$

∴ the restriction that $\alpha < 90^\circ$ may be removed.

Sim'ly " " " β " " " "

∴ the theorem is true for all pos' ∧s.

Lastly, if $\alpha = -x$, and $\beta = -y$, then

$\sin(\alpha-\beta) = \sin\{-(x-y)\} = -\sin(x-y)$

$= -\sin x\cos y + \cos x\sin y$

$= \sin(-x)\cos(-y) - \cos(-x)\sin(-y)$

$= \sin\alpha\cos\beta - \cos\alpha\sin\beta$

and

$\cos(\alpha-\beta) = \cos\{-(x-y)\} = \cos(x-y)$

$= \cos x\cos y + \sin x\sin y$

$= \cos(-x)\cos(-y) + \sin(-x)\sin(-y)$

$= \cos\alpha\cos\beta + \sin\alpha\sin\beta$

∴ the theorem is true for all neg' ∧s;

and, ∴, is true for *all* values of α and β.

Note—By $(\alpha-\beta)$ is denoted a single angle, where α, β are *any* angles such that β subtracted from α gives its magnitude.

For example:

$\cos 13\alpha = \cos(20\alpha - 7\alpha) = \cos 20\alpha\cos 7\alpha + \sin 20\alpha\sin 7\alpha$

§ **20.** If any one of the four results in §§ 18, 19 is assumed to be true for *all* values of the $\angle^s$, then any other one of the four can be immediately deduced from it.

Examples

1. *To deduce* $\cos(\alpha+\beta)$ *from* $\sin(\alpha-\beta)$

$$\cos(\alpha+\beta) = \sin\{90^\circ - (\alpha+\beta)\}$$
$$= \sin(\overline{90^\circ - \alpha} - \beta)$$
$$= \sin(90^\circ - \alpha)\cos\beta - \cos(90^\circ - \alpha)\sin\beta$$
$$= \cos\alpha\cos\beta - \sin\alpha\sin\beta$$

2. *To deduce* $\sin(\alpha-\beta)$ *from* $\sin(\alpha+\beta)$

$$\sin(\alpha-\beta) = \sin\{\alpha + (-\beta)\}$$
$$= \sin\alpha\cos(-\beta) + \cos\alpha\sin(-\beta)$$
$$= \sin\alpha\cos\beta - \cos\alpha\sin\beta$$

Exercises

1. Deduce $\cos(\alpha+\beta)$ from $\sin(\alpha+\beta)$

2. Deduce $\sin(\alpha+\beta)$ from $\cos(\alpha-\beta)$

Any particular case, where the magnitudes of the $\angle^s$ are approximately defined, may be proved by a geometrical construction specially adapted to it.

Example

To prove that $\cos(\alpha-\beta) = \cos\alpha\cos\beta + \sin\alpha\sin\beta$ *when* α *is about* 330°, *and* β *about* 120°.

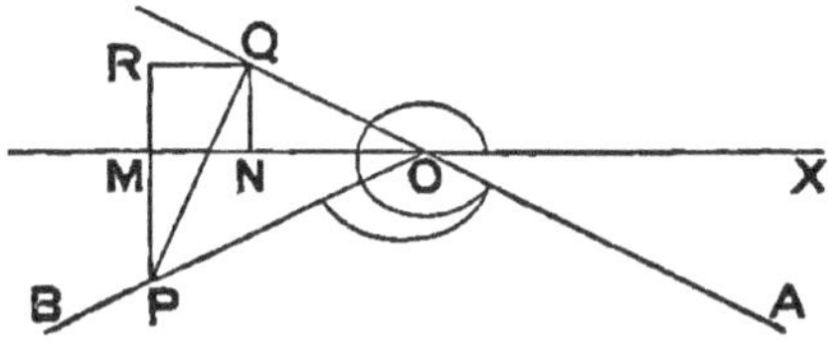

Let a line, initially along OX, revolve in the pos′ direc′ thro′ α (about 330°) into position OA; and then back, in neg′ direc′, thro′ β (about 120°) into position OB.

Take any p't P in OB: drop ⊥s PM on XO produced, PQ on AO produced, QN on XO produced, and QR on PM produced. Then

$$\frac{OM}{OP} = \frac{ON + NM}{OP} = \frac{ON - MN}{OP} = \frac{ON - RQ}{OP} = \frac{ON}{OQ} \cdot \frac{OQ}{OP} - \frac{RQ}{QP} \cdot \frac{QP}{OP}$$

Now $\frac{OM}{OP} = \cos(\alpha - \beta)$

$$\frac{ON}{OQ} = \cos XOQ = \cos(\alpha - 180°) = -\cos\alpha$$

$$\frac{OQ}{OP} = \cos QOP = \cos(180° - \beta) = -\cos\beta$$

$$\frac{RQ}{QP} = \sin RPQ = \sin QON = \sin XOA = \sin(360° - \alpha) = -\sin\alpha$$

$$\frac{QP}{OP} = \sin QOP = \sin(180° - \beta) = \sin\beta$$

$$\therefore \quad \cos(\alpha - \beta) = \cos\alpha\cos\beta + \sin\alpha\sin\beta$$

Note—In the 3rd line of the above, notice particularly the change from NM to − MN. In a case like this, great care is necessary in the substitutions for the ratios.

Exercises

Prove the truth of the theorems in § 18 and § 19, by special geometrical construction, in each of the following cases—

1. α about three-fourths of a right angle, and β about half a right angle:

2. α and β each between one and two right angles, but $\alpha + \beta$ less than three right angles:

3. α greater than two right angles, and $\alpha + \beta$ less than three right angles.

4. α greater than a right angle, and $\alpha + \beta$ less than two right angles.

5. α a little less than a right angle, and β a little greater than a right angle.

6. α between two and two and a half right angles, and β between one and one and a half right angles.

§ **21**. A number of useful developments follow immediately from the fundamental and most important theorems of §§ 18, 19.

Thus, by putting α for β in the theorem of § 18, we get

$$\sin 2\alpha = 2 \sin \alpha \cos \alpha$$

and $$\cos 2\alpha = \cos^2 \alpha - \sin^2 \alpha$$

The latter may be put into either of the forms

$$\cos 2\alpha = 2 \cos^2 \alpha - 1$$

or $$\cos 2\alpha = 1 - 2 \sin^2 \alpha$$

Which again may be written

$$\cos \alpha = \sqrt{(1 + \cos 2\alpha)/2}$$

and $$\sin \alpha = \sqrt{(1 - \cos 2\alpha)/2}$$

Note—The result $\sin 2\alpha = 2 \sin \alpha \cos \alpha$, is equivalent to this verbal formula—

The sine *of an angle is equal to twice the* sine *of half that angle, multiplied by the* cosine *of the same half angle.*

Hence we see that it covers all such equivalent forms as

$$\sin \alpha = 2 \sin \frac{\alpha}{2} \cos \frac{\alpha}{2}$$

or $$\sin \frac{\alpha}{8} = 2 \sin \frac{\alpha}{16} \cos \frac{\alpha}{16}$$

or $$\sin 16\alpha = 2 \sin 8\alpha \cos 8\alpha$$

The last result, by repeated use of the same formula, gives

$$\sin 16\alpha = 16 \sin \alpha \cos \alpha \cos 2\alpha \cos 4\alpha \cos 8\alpha$$

It should now be obvious that, by extending this to n repetitions, we should get

$$\sin 2^n \alpha = 2^n \sin \alpha \cos \alpha \cos 2\alpha \cos 4\alpha \text{ \&c. up to } \cos 2^{n-1} \alpha$$

Or, in another form,

$$\cos \frac{\alpha}{2} \cos \frac{\alpha}{4} \cos \frac{\alpha}{8} \text{ \&c. up to } \cos \frac{\alpha}{2^n} = \frac{\sin \alpha}{2^n \sin \frac{\alpha}{2^n}}$$

Again, by addition and subtraction, we get

$$\sin(\alpha+\beta)+\sin(\alpha-\beta)=2\sin\alpha\cos\beta$$

$$\sin(\alpha+\beta)-\sin(\alpha-\beta)=2\cos\alpha\sin\beta$$

$$\cos(\alpha+\beta)+\cos(\alpha-\beta)=2\cos\alpha\cos\beta$$

$$\cos(\alpha+\beta)-\cos(\alpha-\beta)=-2\sin\alpha\sin\beta$$

Also since $\alpha \equiv \frac{\alpha+\beta}{2}+\frac{\alpha-\beta}{2}$

and $\beta \equiv \frac{\alpha+\beta}{2}-\frac{\alpha-\beta}{2}$

we have

$$\sin\alpha+\sin\beta=2\sin\frac{\alpha+\beta}{2}\cos\frac{\alpha-\beta}{2}$$

$$\sin\alpha-\sin\beta=2\cos\frac{\alpha+\beta}{2}\sin\frac{\alpha-\beta}{2}$$

$$\cos\alpha+\cos\beta=2\cos\frac{\alpha+\beta}{2}\cos\frac{\alpha-\beta}{2}$$

$$\cos\alpha-\cos\beta=-2\sin\frac{\alpha+\beta}{2}\sin\frac{\alpha-\beta}{2}$$

Next, by division, we get

$$\frac{\sin(\alpha+\beta)}{\cos(\alpha+\beta)}=\frac{\sin\alpha\cos\beta+\cos\alpha\sin\beta}{\cos\alpha\cos\beta-\sin\alpha\sin\beta}$$

$$=\frac{\frac{\sin\alpha}{\cos\alpha}+\frac{\sin\beta}{\cos\beta}}{1-\frac{\sin\alpha}{\cos\alpha}\cdot\frac{\sin\beta}{\cos\beta}}$$

$$\therefore\quad \tan(\alpha+\beta)=\frac{\tan\alpha+\tan\beta}{1-\tan\alpha\tan\beta}$$

Sim'ly $\tan(\alpha-\beta)=\frac{\tan\alpha-\tan\beta}{1+\tan\alpha\tan\beta}$

So also $\cot(\alpha + \beta) = \dfrac{\cot\alpha \cot\beta - 1}{\cot\alpha + \cot\beta}$

and $\cot(\alpha - \beta) = -\dfrac{\cot\alpha \cot\beta + 1}{\cot\alpha - \cot\beta}$

Whence (or from the formulæ at the beginning of this §)

$$\tan 2\alpha = \frac{2\tan\alpha}{1 - \tan^2\alpha}$$

$$\cot 2\alpha = \frac{\cot^2\alpha - 1}{2\cot\alpha}$$

Again, by division, we get

$$\frac{\sin\alpha + \sin\beta}{\sin\alpha - \sin\beta} = \frac{\tan\dfrac{\alpha+\beta}{2}}{\tan\dfrac{\alpha-\beta}{2}}$$

Others, sim'r to the last, but of less importance, come by dividing in other ways.

Two useful formulæ are got thus—

Since $\dfrac{\sin\alpha}{\cos\alpha} + \dfrac{\sin\beta}{\cos\beta} = \dfrac{\sin\alpha\cos\beta + \cos\alpha\sin\beta}{\cos\alpha\cos\beta}$

$\therefore \quad \tan\alpha + \tan\beta = \dfrac{\sin(\alpha + \beta)}{\cos\alpha\cos\beta}$

Sim'ly $\tan\alpha - \tan\beta = \dfrac{\sin(\alpha - \beta)}{\cos\alpha\cos\beta}$

As another development of the theorem of § 18—

$$\begin{aligned}\sin(\alpha + \beta + \gamma) &= \sin(\alpha + \beta)\cos\gamma + \cos(\alpha + \beta)\sin\gamma \\ &= \sin\alpha\cos\beta\cos\gamma + \cos\alpha\sin\beta\cos\gamma \\ &\quad + \cos\alpha\cos\beta\sin\gamma - \sin\alpha\sin\beta\sin\gamma\end{aligned}$$

Sim'ly we should find that

$$\cos(\alpha+\beta+\gamma) = \cos\alpha\cos\beta\cos\gamma - \sin\alpha\sin\beta\cos\gamma$$
$$- \sin\alpha\cos\beta\sin\gamma - \cos\alpha\sin\beta\sin\gamma$$

Whence $\tan(\alpha+\beta+\gamma)$

$$= \frac{\tan\alpha + \tan\beta + \tan\gamma - \tan\alpha\tan\beta\tan\gamma}{1 - \tan\alpha\tan\beta - \tan\beta\tan\gamma - \tan\gamma\tan\alpha}$$

and $\cot(\alpha+\beta+\gamma)$

$$= \frac{\cot\alpha\cot\beta\cot\gamma - \cot\alpha - \cot\beta - \cot\gamma}{\cot\beta\cot\gamma + \cot\gamma\cot\alpha + \cot\alpha\cot\beta - 1}$$

From the last four we have

$$\sin 3\alpha = \sin\alpha\cos^2\alpha + \cos^2\alpha\sin\alpha$$
$$+ \cos^2\alpha\sin\alpha - \sin^3\alpha$$
$$= 3\sin\alpha(1 - \sin^2\alpha) - \sin^3\alpha$$
$$= 3\sin\alpha - 4\sin^3\alpha$$

$$\cos 3\alpha = \cos^3\alpha - \sin^2\alpha\cos\alpha - \sin^2\alpha\cos\alpha$$
$$- \sin^2\alpha\cos\alpha$$
$$= \cos^3\alpha - 3\cos\alpha(1 - \cos^2\alpha)$$
$$= 4\cos^3\alpha - 3\cos\alpha$$

$$\tan 3\alpha = \frac{3\tan\alpha - \tan^3\alpha}{1 - 3\tan^2\alpha}$$

$$\cot 3\alpha = \frac{\cot^3\alpha - 3\cot\alpha}{3\cot^2\alpha - 1} = \frac{3\cot\alpha - \cot^3\alpha}{1 - 3\cot^2\alpha}$$

Note—These formulæ can of course be got by considering 3α as $2\alpha + \alpha$, expanding the function, and using the already found values for the T. F.s of 2α.

From the formulæ at the beginning of this §, we have

$$2\sin^2\alpha = 1 - \cos 2\alpha$$

and $$2\cos^2\alpha = 1 + \cos 2\alpha$$

$$\therefore \quad \tan^2 \alpha = \frac{1 - \cos 2\alpha}{1 + \cos 2\alpha}$$

$$\therefore \quad \cos 2\alpha = \frac{1 - \tan^2 \alpha}{1 + \tan^2 \alpha}$$

Again, since $\dfrac{2 \sin \alpha \cos \alpha}{\cos^2 \alpha + \sin^2 \alpha} = \dfrac{\dfrac{2 \sin \alpha}{\cos \alpha}}{1 + \dfrac{\sin^2 \alpha}{\cos^2 \alpha}}$

$$\therefore \quad \sin 2\alpha = \frac{2 \tan \alpha}{1 + \tan^2 \alpha}$$

Since $\dfrac{\sin 2\alpha}{1 + \cos 2\alpha} = \dfrac{2 \sin \alpha \cos \alpha}{2 \cos^2 \alpha} = \dfrac{\sin \alpha}{\cos \alpha}$

and $\dfrac{1 - \cos 2\alpha}{\sin 2\alpha} = \dfrac{2 \sin^2 \alpha}{2 \sin \alpha \cos \alpha} = \dfrac{\sin \alpha}{\cos \alpha}$

$$\therefore \quad \frac{\sin 2\alpha}{1 + \cos 2\alpha} = \tan \alpha = \frac{1 - \cos 2\alpha}{\sin 2\alpha}$$

and $\dfrac{\sin 2\alpha}{1 - \cos 2\alpha} = \cot \alpha = \dfrac{1 + \cos 2\alpha}{\sin 2\alpha}$

From the theorems in § 18 and § 19 we have

$$\begin{aligned}
&\sin(\alpha + \beta) \sin(\alpha - \beta) \\
&\quad = \sin^2 \alpha \cos^2 \beta - \cos^2 \alpha \sin^2 \beta \\
&\quad = \sin^2 \alpha - \sin^2 \alpha \sin^2 \beta - \sin^2 \beta + \sin^2 \alpha \sin^2 \beta \\
&\quad = \sin^2 \alpha - \sin^2 \beta, \text{ or } = \cos^2 \beta - \cos^2 \alpha
\end{aligned}$$

The same results may also be obtained thus—

$$\begin{aligned}
\sin^2 \alpha - \sin^2 \beta &= \tfrac{1}{2}(1 - \cos 2\alpha) - \tfrac{1}{2}(1 - \cos 2\beta) \\
&= -\tfrac{1}{2}(\cos 2\alpha - \cos 2\beta) \\
&= \sin(\alpha + \beta) \sin(\alpha - \beta)
\end{aligned}$$

So also

$$\cos^2\beta - \cos^2\alpha = \tfrac{1}{2}(\cos 2\beta + 1) - \tfrac{1}{2}(\cos 2\alpha + 1)$$
$$= -\tfrac{1}{2}(\cos 2\alpha - \cos 2\beta)$$
$$= \sin(\alpha + \beta)\sin(\alpha - \beta)$$

Exercises

1. Similarly prove, in each of the foregoing ways, that

$$\cos(\alpha + \beta)\cos(\alpha - \beta) = \cos^2\alpha - \sin^2\beta = \cos^2\beta - \sin^2\alpha$$
$$= \cos^2\alpha + \cos^2\beta - 1$$

2. Make the following useful modifications of the preceding formulæ—

(1) $\sin\theta\sin\phi = \sin^2\frac{\theta + \phi}{2} - \sin^2\frac{\theta - \phi}{2}$

(2) $\cos\theta\cos\phi = \cos^2\frac{\theta + \phi}{2} + \cos^2\frac{\theta - \phi}{2} - 1$

3. Show that

(1) $\tan^2\alpha - \tan^2\beta = \dfrac{\sin(\alpha + \beta)\sin(\alpha - \beta)}{\cos^2\alpha\cos^2\beta}$

(2) $1 + \cos 3\alpha\cos 5\alpha = \cos^2 4\alpha + \cos^2\alpha$

Again $(\cos\alpha + \cos\beta)^2 + (\sin\alpha + \sin\beta)^2$

$$= \cos^2\alpha + \cos^2\beta + 2\cos\alpha\cos\beta$$
$$+ \sin^2\alpha + \sin^2\beta + 2\sin\alpha\sin\beta$$
$$= 2 + 2\cos(\alpha - \beta)$$
$$= 2\{1 + \cos(\alpha - \beta)\}$$
$$= 4\cos^2\frac{\alpha - \beta}{2}$$

Exercises

Similarly prove that

(1) $(\cos\alpha - \cos\beta)^2 + (\sin\alpha - \sin\beta)^2 = 4\sin^2\frac{\alpha - \beta}{2}$

(2) $(\cos\alpha + \cos\beta)^2 - (\sin\alpha + \sin\beta)^2 = 4\cos^2\frac{\alpha - \beta}{2}\cos(\alpha + \beta)$

§ **22**. The formula for the expansion of $\tan(\alpha + \beta + \gamma)$ may be inductively generalized.

For let there be n angles $\alpha_1, \alpha_2, \alpha_3$, &c, α_n.

Put k for $\alpha_1 + \alpha_2 +$ &c $+ \alpha_n$, and assume that

$$\tan k = \frac{t_1 - t_3 + t_5 - \&c}{1 - t_2 + t_4 - \&c}$$

where
$$t_1 = \Sigma(\tan\alpha_1)$$
$$t_2 = \Sigma(\tan\alpha_1 \tan\alpha_2)$$
$$t_3 = \Sigma(\tan\alpha_1 \tan\alpha_2 \tan\alpha_3)$$
$$\&c \qquad \&c$$
$$t_n = \tan\alpha_1 \tan\alpha_2 \;\&c\; \tan\alpha_n$$

Then, introducing a new ∧ λ, we get

$$\tan(k + \lambda) = \frac{\dfrac{t_1 - t_3 + t_5 - \&c}{1 - t_2 + t_4 - \&c} + \tan\lambda}{1 - \tan\lambda\left(\dfrac{t_1 - t_3 + t_5 - \&c}{1 - t_2 + t_4 - \&c}\right)}$$

$$= \frac{(t_1 + \tan\lambda) - (t_3 + t_2\tan\lambda) + (t_5 + t_4\tan\lambda) - \&c}{1 - (t_2 + t_1\tan\lambda) + (t_4 + t_3\tan\lambda) - \&c}$$

$$= \frac{t_1' - t_3' + t_5' - \&c}{1 - t_2' + t_4' - \&c}$$

Where t_1', t_2', &c are the values of t_1, t_2, &c for n+1 ∧s.
∴ if the assumption is true for n ∧s, it is true for n + 1 ∧s.
But it is true for 3 ∧s; and ∴, by induction, it is true generally.

Cor'—If $\alpha_1 = \alpha_2 =$ &c $= \alpha_n$, we get

$$\tan n\alpha = \frac{C_1 \tan\alpha - C_3\tan^3\alpha + C_5\tan^5\alpha - \&c}{1 - C_2\tan^2\alpha + C_4\tan^4\alpha - \&c}$$

where C_r = N° of comb'ns of n things r together.

Sim'ly

$$\cot n\alpha = \frac{\cot^n \alpha - C_2 \cot^{n-2} \alpha + C_4 \cot^{n-4} \alpha - \&c}{C_1 \cot^{n-1} \alpha - C_3 \cot^{n-3} \alpha + C_5 \cot^{n-5} \alpha - \&c}$$

§ **23.** Multiplying by $2 \cos \alpha$ each side of the formula

$$2 \cos^2 \alpha = \cos 2\alpha + 1$$

we get
$$\begin{aligned} 4 \cos^3 \alpha &= 2 \cos 2\alpha \cos \alpha + 2 \cos \alpha \\ &= \cos 3\alpha + \cos \alpha + 2 \cos \alpha \\ &= \cos 3\alpha + 3 \cos \alpha \end{aligned}$$

So also
$$\begin{aligned} 8 \cos^4 \alpha &= 2 \cos 3\alpha \cos \alpha + 6 \cos^2 \alpha \\ &= \cos 4\alpha + \cos 2\alpha + 3 (\cos 2\alpha + 1) \\ &= \cos 4\alpha + 4 \cos 2\alpha + 3 \end{aligned}$$

This process can be carried on to any extent.

Exercises

Prove that

1. $16 \cos^5 \alpha = \cos 5\alpha + 5 \cos 3\alpha + 10 \cos \alpha$

2. $32 \cos^6 \alpha = \cos 6\alpha + 6 \cos 4\alpha + 15 \cos 2\alpha + 10$

3. $64 \cos^7 \alpha = \cos 7\alpha + 7 \cos 5\alpha + 21 \cos 3\alpha + 35 \cos \alpha$

4. $128 \cos^8 \alpha = \cos 8\alpha + 8 \cos 6\alpha + 28 \cos 4\alpha + 56 \cos 2\alpha + 35$

Again from

$$-2 \sin^2 \alpha = \cos 2\alpha - 1$$

we get
$$\begin{aligned} -4 \sin^3 \alpha &= 2 \cos 2\alpha \sin \alpha - 2 \sin \alpha \\ &= \sin 3\alpha - \sin \alpha - 2 \sin \alpha \\ &= \sin 3\alpha - 3 \sin \alpha \end{aligned}$$

whence
$$\begin{aligned} 8 \sin^4 \alpha &= -2 \sin 3\alpha \sin \alpha + 6 \sin^2 \alpha \\ &= \cos 4\alpha - \cos 2\alpha + 3 (1 - \cos 2\alpha) \\ &= \cos 4\alpha - 4 \cos 2\alpha + 3 \end{aligned}$$

And the process can be continued.

Exercises

Prove that

1. $16 \sin^5 \alpha = \sin 5\alpha - 5 \sin 3\alpha + 10 \sin \alpha$

2. $-32 \sin^6 \alpha = \cos 6\alpha - 6 \cos 4\alpha + 15 \cos 2\alpha - 10$

3. $-64 \sin^7 \alpha = \sin 7\alpha - 7 \sin 5\alpha + 21 \sin 3\alpha - 35 \sin \alpha$

4. $128 \sin^8 \alpha = \cos 8\alpha - 8 \cos 6\alpha + 28 \cos 4\alpha - 56 \cos 2\alpha + 35$

§ **24.** Since
$$\begin{cases} \cos 2\alpha = 2\cos^2\alpha - 1 \\ \sin 2\alpha = 2 \sin\alpha \cos\alpha \end{cases}$$
and
$$\begin{cases} \cos 3\alpha = 4\cos^3\alpha - 3\cos\alpha \\ \sin 3\alpha = 3 \sin\alpha - 4\sin^3\alpha \end{cases}$$
we have
$$\begin{aligned} \cos 5\alpha &= \cos(3\alpha + 2\alpha) \\ &= (4\cos^3\alpha - 3\cos\alpha)(2\cos^2\alpha - 1) \\ &- (3\sin\alpha - 4\sin^3\alpha)\, 2\sin\alpha\cos\alpha \\ &= 8\cos^5\alpha - 6\cos^3\alpha - 4\cos^3\alpha + 3\cos\alpha \\ &= 6\cos\alpha + 6\cos^3\alpha + 8\cos\alpha(1 - \cos^2\alpha)^2 \\ &= 16\cos^5\alpha - 20\cos^3\alpha + 5\cos\alpha \end{aligned}$$

Exercises

1. Similarly prove that
$$\sin 5\alpha = 16\sin^5\alpha - 20\sin^3\alpha + 5\sin\alpha$$

2. Assuming the expansions for $\cos 5\alpha$ and $\sin 5\alpha$, prove that

(1) $\sin 5\alpha + \cos 5\alpha = (\sin\alpha + \cos\alpha)(2\cos 4\alpha + 2\sin 2\alpha - 1)$

(2) $\sin 7\alpha = 7\sin\alpha - 56\sin^3\alpha + 112\sin^5\alpha - 64\sin^7\alpha$

(3) $\cos 7\alpha = 64\cos^7\alpha - 112\cos^5\alpha + 56\cos^3\alpha - 7\cos\alpha$

(4) $\cos 9\alpha + \cos 7\alpha - 4(\cos 5\alpha + \cos 3\alpha) + 6\cos\alpha$
$$= 256\sin^4\alpha\cos^5\alpha$$

§ **25.** Since
$$\left.\begin{aligned} \sin^2\alpha + \cos^2\alpha &= 1 \\ \text{and}\quad 2\sin\alpha\cos\alpha &= \sin 2\alpha \end{aligned}\right\}$$

$$\left.\begin{aligned} \therefore\quad \sin\alpha + \cos\alpha &= \pm\sqrt{1+\sin 2\alpha} \\ \text{and}\quad \sin\alpha - \cos\alpha &= \pm\sqrt{1-\sin 2\alpha} \end{aligned}\right\}$$

$$\left.\begin{aligned} \therefore\quad 2\sin\alpha &= \pm\sqrt{1+\sin 2\alpha} \pm \sqrt{1-\sin 2\alpha} \\ \text{and}\quad 2\cos\alpha &= \pm\sqrt{1+\sin 2\alpha} \mp \sqrt{1-\sin 2\alpha} \end{aligned}\right\}$$

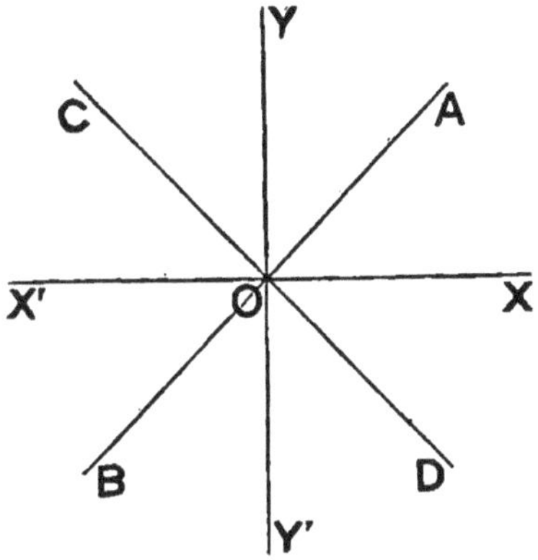

Now, if XOX′, YOY′ are rect′r axes, and AOB, COD are the respective bisectors of $X\hat{O}Y$, $Y\hat{O}X'$; then, OX being taken as the initial position of the generator of α—

1°, if the position, with respect to AOB, COD, of the generator of α is given, the ambiguities of sign before the radicals can be removed; and,

2°, if the sign of each radical is defined, the limits of α can be assigned.

Examples

1. If α is between 225° and 315°,

then $\sin\alpha > \cos\alpha$,

and $\sin\alpha$ is neg′.

$$\left.\begin{aligned} \therefore\quad \sin\alpha + \cos\alpha &= -\sqrt{1+\sin 2\alpha} \\ \text{and}\quad \sin\alpha - \cos\alpha &= -\sqrt{1-\sin 2\alpha} \end{aligned}\right\}$$

$$\left.\begin{aligned} \therefore\quad 2\sin\alpha &= -\sqrt{1+\sin 2\alpha} - \sqrt{1-\sin 2\alpha} \\ \text{and}\quad 2\cos\alpha &= -\sqrt{1+\sin 2\alpha} + \sqrt{1-\sin 2\alpha} \end{aligned}\right\}$$

2. If $2\sin\alpha = -\sqrt{1+\sin 2\alpha} + \sqrt{1-\sin 2\alpha}$

$$\left.\begin{aligned}\text{then}\quad \sin\alpha + \cos\alpha &= -\sqrt{1+\sin 2\alpha}\\ \text{and}\quad \sin\alpha - \cos\alpha &= +\sqrt{1-\sin 2\alpha}\end{aligned}\right\}$$

$$\therefore\quad \cos\alpha > \sin\alpha,$$

$$\text{and}\quad \cos\alpha \text{ is neg}'.$$

$$\therefore\quad \alpha \text{ lies between } \frac{3\pi}{4} \text{ and } \frac{5\pi}{4}.$$

Or, in general terms, between

$$2n\pi + \frac{3\pi}{4} \text{ and } 2n\pi + \frac{5\pi}{4},$$

where n is any integer, pos′, neg′, or zero.

Exercises

1. In the formula

$$2\sin\alpha = \pm\sqrt{1+\sin 2\alpha} \pm \sqrt{1-\sin 2\alpha}$$

remove the ambiguities of sign in the case when α is between 45° and 135°.

2. In the formula

$$2\cos\alpha = \pm\sqrt{1+\sin 2\alpha} \pm \sqrt{1-\sin 2\alpha}$$

remove the ambiguities of sign in the case when α is between 135° and 225°.

3. Determine the limits between which α must lie, when

$$2\sin\alpha = +\sqrt{1+\sin 2\alpha} - \sqrt{1-\sin 2\alpha}$$

4. Determine the limits between which α must lie, when

$$2\cos\alpha = -\sqrt{1+\sin 2\alpha} - \sqrt{1-\sin 2\alpha}$$

5. Show that the ambiguity of sign in the formula

$$\cos\frac{\alpha}{2} + \sin\frac{\alpha}{2} = \pm\sqrt{1+\sin\alpha}$$

may be replaced by $(-1)^m$, where m is the greatest integer in $\frac{\alpha^\circ + 90^\circ}{360^\circ}$.

6. Show that the ambiguity of sign in the formula

$$\cos\frac{\alpha}{2} - \sin\frac{\alpha}{2} = \pm\sqrt{1-\sin\alpha}$$

may be replaced by $(-1)^m$, where m is the greatest integer in $\frac{\alpha^\circ + 270^\circ}{360^\circ}$.

Wolstenholme: 408.

§ **26.** If $A + B + C = 180°$, which is the case when A, B, C are the $\wedge$s of a Δ, then (see Examples p. 49)

$$\sin\frac{A}{2} = \cos\frac{B+C}{2}$$

and $$\sin\frac{B+C}{2} = \cos\frac{A}{2}$$

$$\therefore \quad \sin A + \sin B + \sin C$$

$$= 2\sin\frac{A}{2}\cos\frac{A}{2} + 2\sin\frac{B+C}{2}\cos\frac{B-C}{2}$$

$$= 2\cos\frac{A}{2}\left\{\cos\frac{B+C}{2} + \cos\frac{B-C}{2}\right\}$$

$$= 4\cos\frac{A}{2}\cos\frac{B}{2}\cos\frac{C}{2}$$

So also $$\cos A + \cos B + \cos C$$

$$= 1 - 2\sin^2\frac{A}{2} + 2\cos\frac{B+C}{2}\cos\frac{B-C}{2}$$

$$= 1 + 2\sin\frac{A}{2}\left\{\cos\frac{B-C}{2} - \cos\frac{B+C}{2}\right\}$$

$$= 1 + 4\sin\frac{A}{2}\sin\frac{B}{2}\sin\frac{C}{2}$$

Again, on the same supposition,

$$\cos A = -\cos(B+C)$$

and $$\sin(B+C) = \sin A$$

$$\therefore \quad \sin 2A + \sin 2B + \sin 2C$$

$$= 2\sin A\cos A + 2\sin(B+C)\cos(B-C)$$

$$= 2\sin A\{\cos(B-C) - \cos(B+C)\}$$

$$= 4\sin A\sin B\sin C$$

Using π for 180°; if $A + B + C = \pi$, we have

$$\sin\frac{A}{2} + \sin\frac{B}{2} + \sin\frac{C}{2}$$

$$= \cos\left(\frac{\pi}{2} - \frac{A}{2}\right) + 2\sin\frac{B+C}{4}\cos\frac{B-C}{4}$$

$$= 1 - 2\sin^2\frac{\pi - A}{4} + 2\sin\frac{\pi - A}{4}\cos\frac{B-C}{4}$$

$$= 1 + 2\sin\frac{\pi - A}{4}\left\{\cos\frac{B-C}{4} - \cos\left(\frac{\pi}{2} - \frac{B+C}{4}\right)\right\}$$

$$= 1 + 4\sin\frac{\pi - A}{4}\sin\frac{\pi - B}{4}\sin\frac{\pi - C}{4}$$

$$= 1 + 4\sin\frac{B+C}{4}\sin\frac{C+A}{4}\sin\frac{A+B}{4}$$

Note—This last result is at once deducible from the 2nd of this §, by writing $\frac{\pi}{2} - \frac{A}{2}$ for A.

Exercises

Prove that if A, B, C are the angles of a triangle, then—

1. $\cos\frac{A}{2} + \cos\frac{B}{2} + \cos\frac{C}{2} = 4\cos\frac{\pi - A}{4}\cos\frac{\pi - B}{4}\cos\frac{\pi - C}{4}$

or $= 4\cos\frac{B+C}{4}\cos\frac{C+A}{4}\cos\frac{A+B}{4}$

2. $\cos 2A + \cos 2B + \cos 2C = -4\cos A\cos B\cos C - 1$

3. $\sin A + \sin B - \sin C = 4\sin\frac{A}{2}\sin\frac{B}{2}\cos\frac{C}{2}$

4. $\cos A + \cos B - \cos C = 4\cos\frac{A}{2}\cos\frac{B}{2}\sin\frac{C}{2} - 1$

5. $\sin 4A + \sin 4B + \sin 4C = -4\sin 2A\sin 2B\sin 2C$

6. $\cos 4A + \cos 4B + \cos 4C = 4\cos 2A\cos 2B\cos 2C - 1$

7. $\cos\frac{A}{2} = \sqrt{\frac{(\sin A + \sin B + \sin C)(\sin B + \sin C - \sin A)}{4\sin B\sin C}}$

8. $\cos\left(\frac{3A}{2} + B - 2C\right) + \cos\left(\frac{3B}{2} + C - 2A\right)$

$$+ \cos\left(\frac{3C}{2} + A - 2B\right)$$

$$= 4 \cos \frac{5A - 2B - C}{4} \cos \frac{5B - 2C - A}{4} \cos \frac{5C - 2A - B}{4}$$

Math' Tri': '73.

9. $8(\sin A + \sin B + \sin C) \sin \frac{A}{2} \sin \frac{B}{2} \sin \frac{C}{2}$

$$= \sin 2A + \sin 2B + \sin C$$

Pet' Camb': '66.

Some of the preceding formulæ may be very concisely expressed by means of the symbols Σ, Π.*

If, e. g., A, B, C are $\angle$s of a $\triangle$, we have, by what goes before,

$$\Sigma \sin A = 4 \Pi \cos \frac{A}{2}$$

$$\Sigma \cos A = 4 \Pi \sin \frac{A}{2} + 1$$

$$\Sigma \sin 2A = 4 \Pi \sin A$$

$$\Sigma \cos 2A = -4 \Pi \cos A - 1$$

A very useful additional result comes thus—

Since $\tan(A + B) = -\tan C$

$$\therefore \quad \frac{\tan A + \tan B}{1 - \tan A \tan B} = -\tan C$$

whence $\tan A + \tan B + \tan C = \tan A \tan B \tan C$

i. e. when $\Sigma A = 180°$, $\Sigma \tan A = \Pi \tan A$

* The symbol Π may perhaps be new to the Reader; for it has only lately come into use in English text-books. It is the symbolic equivalent for the words *the product of all such terms as.* Thus, with respect to a, b, c, while Σa means $a + b + c$, Πa means abc; so also

$\Pi(a - b)$ means $(a - b)(b - c)(c - a)$

Exercises

1. If $A + B + C = 180°$, prove that,

(1) $\Sigma \cot \frac{A}{2} = \Pi \cot \frac{A}{2}$

(2) $\Sigma \left(\tan \frac{A}{2} \tan \frac{B}{2}\right) = 1$

(3) $\Sigma (\cot A \cot B) = 1$

(4) $\Sigma \cot A = \Pi \cot A + \Pi \operatorname{cosec} A$

(5) $\Sigma \tan \frac{A}{2} = \Pi \tan \frac{A}{2} + \Pi \sec \frac{A}{2}$

(6) $\Sigma \tan \frac{A}{2} . \Sigma \cot \frac{A}{2} = 1 + \Pi \operatorname{cosec} \frac{A}{2}$

NOTE—*These formulæ can be deduced from the expansions of* $\cot(\alpha + \beta + \gamma)$, $\tan(\alpha + \beta + \gamma)$, $\cos(\alpha + \beta + \gamma)$, $\sin(\alpha + \beta + \gamma)$ *by considering that* $\tan 180° = 0$, $\cot 90° = 0$, $\tan 90° = \infty$, $\cot 180° = \infty$, $\cos 180° = -1$, $\sin 90° = 1$.

It is however advisable also to prove them independently, as in the Example given at the end of the preceding page.

2. In the manner indicated in the above Note; and with the notation of § 22, prove that (m being any integer)

$$1°, \text{ if } \Sigma \alpha = m\pi,$$

$$\text{then } \quad t_1 + t_5 + t_9 + \&c = t_3 + t_7 + t_{11} + \&c;$$

$$\text{and, } \quad 2°, \text{ if } \Sigma \alpha = (2m + 1)\frac{\pi}{2},$$

$$\text{then } \quad 1 + t_4 + t_8 + \&c = t_2 + t_6 + t_{10} + \&c.$$

3. If ABCDEF is a hexagon, prove that

$\Sigma(\tan A \tan B \tan C) = \Sigma \tan A + \Sigma(\tan A \tan B \tan C \tan D \tan E)$

How many terms will there be in each Σ?

E. T. xl.

4. If $\theta + \phi + \psi = \pi$, prove that

$$\Sigma \tan \left\{\left(\frac{2^n - 1}{2^n}\right)\frac{\pi}{3} + \frac{\theta}{2^n}\right\} = \Pi \tan \left\{\left(\frac{2^n - 1}{2^n}\right)\frac{\pi}{3} + \frac{\theta}{2^n}\right\}$$

Joh' Camb': '42.

§ **27.** *To find a relation between the* cosines *of the angles of a triangle* ABC ; *and also a relation between the* sines *of the same angles.*

For brevity put C_1, C_2, C_3 for cos A, cos B, cos C respectively

and S_1, S_2, S_3 „ sin A, sin B, sin C „

Then $C_1^2 + C_2^2 + C_3^2$

$$= C_1^2 + C_2^2 + \cos^2(A + B)$$

$$= C_1^2 + C_2^2 + C_1^2 C_2^2 + (1 - C_1^2)(1 - C_2^2) - 2\,C_1 C_2 S_1 S_2$$

$$= 1 + 2\,C_1^2 C_2^2 - 2\,C_1 C_2 S_1 S_2$$

$$= 1 - 2\,C_1 C_2 C_3$$

$$\therefore \quad \Sigma \cos^2 A = 1 - 2\,\Pi \cos A$$

Again

$$(S_1 + S_2 + S_3)(S_2 + S_3 - S_1)(S_3 + S_1 - S_2)(S_1 + S_2 - S_3)$$

$$= 4 \cos\frac{A}{2} \cos\frac{B}{2} \cos\frac{C}{2} \cdot 4 \cos\frac{A}{2} \sin\frac{B}{2} \sin\frac{C}{2}$$

$$\cdot\, 4 \sin\frac{A}{2} \cos\frac{B}{2} \sin\frac{C}{2} \cdot 4 \sin\frac{A}{2} \sin\frac{B}{2} \cos\frac{C}{2}$$

$$= 4\,S_1^2 S_2^2 S_3^2$$

Note—Each of these results will be obtained otherwise hereafter.

Exercises

1. If α, β, γ are the consecutive angles between three lines drawn from a point, so that $\alpha + \beta + \gamma = 360°$,

prove that $\Sigma \cos^2 \alpha = 1 + 2\,\Pi \cos \alpha$

2. If A, B, C are the angles of a triangle, prove that

(1) $\Sigma \sin^2 A = 2 + 2\,\Pi \cos A$

(2) $\Sigma \sin^2 2A = 2 - 2\,\Pi \cos 2A$

(3) $\Sigma \cos^2 2A = 1 + 2\,\Pi \cos 2A$

(4) $\Sigma \sin^2 \frac{A}{2} = 1 - 2\,\Pi \sin \frac{A}{2}$

(5) $\Sigma \cos^2 \frac{A}{2} = 2\,\Sigma \left(\sin \frac{A}{2} \cos \frac{B}{2} \cos \frac{C}{2}\right)$

(6) $\Sigma \sin^3 A = 3\,\Pi \cos \frac{A}{2} + \Pi \cos \frac{3A}{2}$

(7) $\Sigma \sin^4 A = \frac{3}{2} + 2\,\Pi \cos A + \frac{1}{2}\,\Pi \cos 2A$

Math' Tri' : '82.

(8) $\dfrac{\Sigma \sin 2A}{\Sigma \sin A} = 8\,\Pi \sin \frac{A}{2}$

(9) $\Sigma \{\sin^5 A \sin (B - C)\} + \Pi \{\sin A \sin (B - C)\} = 0$

R. U. Schol': '87.

§ **28.** The expression

$$1 + 2 \cos \alpha \cos \beta \cos \gamma - \cos^2 \alpha - \cos^2 \beta - \cos^2 \gamma$$

can be reduced to the product of four factors.

For the given expression

$$\begin{aligned}
&\equiv 1 - \cos^2 \alpha - \cos^2 \beta - (\cos \gamma - \cos \alpha \cos \beta)^2 \\
&\qquad + \cos^2 \alpha \cos^2 \beta \\
&\equiv (1 - \cos^2 \alpha)(1 - \cos^2 \beta) - (\cos \gamma - \cos \alpha \cos \beta)^2 \\
&\equiv \sin^2 \alpha \sin^2 \beta - (\cos \gamma - \cos \alpha \cos \beta)^2 \\
&\equiv (\sin \alpha \sin \beta + \cos \gamma - \cos \alpha \cos \beta) \\
&\quad \times (\sin \alpha \sin \beta - \cos \gamma + \cos \alpha \cos \beta) \\
&\equiv \{\cos \gamma - \cos (\alpha + \beta)\}\{\cos (\alpha - \beta) - \cos \gamma\} \\
&\equiv 4 \sin \frac{\alpha + \beta + \gamma}{2} \sin \frac{\alpha + \beta - \gamma}{2} \\
&\quad \times \sin \frac{\gamma + \alpha - \beta}{2} \sin \frac{\gamma + \beta - \alpha}{2}
\end{aligned}$$

If we denote $\alpha + \beta + \gamma$ by 2σ, the identity may be conveniently written

$$1 + 2\,\Pi \cos \alpha - \Sigma \cos^2 \alpha \equiv 4 \sin \sigma \sin (\sigma - \alpha) \sin (\sigma - \beta) \sin (\sigma - \gamma)$$

Exercise

Prove similarly that if α, β, γ are any angles,

$$1 - 2\Pi(\cos\alpha) - \Sigma(\cos^2\alpha)$$
$$\equiv -4\cos\sigma\cos(\sigma-\alpha)\cos(\sigma-\beta)\cos(\sigma-\gamma)$$

NOTE—*This may also be deduced from the preceding by writing* $\pi - \alpha$, $\pi - \beta$, $\pi - \gamma$ *for* α, β, γ *respectively.*

§ 29. By means of the formulæ already demonstrated, endless results can be established. The Student is urgently requested to recollect that facility in proving such can only come from *perfect familiarity* with these formulæ; and that it is as idle to attempt to work trigonometrical exercises without a thorough knowledge of the fundamental formulæ, as it would be to attempt to work arithmetical questions without a knowledge of the tables. Such knowledge will be best attained by continuous *rewriting of the foregoing proofs. Indeed (excepting perhaps in the case of a Student of transcendent abilities) there is no other way of attaining mathematical knowledge which is at once so sure and so speedy.

Examples

1. *To find the value of* $\cos 3\alpha$, *when* $\tan 2\alpha = -\frac{3}{4}$.

From the formula of Ex′ 2, p. 30 we get $\cos 2\alpha = \pm\frac{4}{5}$

$$\therefore \cos\alpha \left(\text{which} = \sqrt{\frac{1+\cos 2\alpha}{2}}\right) = \pm\frac{3}{\sqrt{10}} \text{ or } \frac{1}{\sqrt{10}}$$

$$\therefore \cos 3\alpha \text{ (which} = 4\cos^3\alpha - 3\cos\alpha)$$

$$= \pm\frac{108}{10\sqrt{10}} \mp \frac{9}{\sqrt{10}} = \pm\frac{18}{10\sqrt{10}} = \pm\frac{9}{50}\sqrt{10}$$

$$\text{or} \quad = \pm\frac{4}{10\sqrt{10}} \mp \frac{3}{\sqrt{10}} = \pm\frac{26}{10\sqrt{10}} = \pm\frac{13}{50}\sqrt{10}$$

Since $\cos\alpha \not> 1$, the other combinations of sign are inadmissible.

* "Writing makes an *exact* man." *Bacon's Essays.*

2. *To express* $\cos\alpha \cos\beta \cos\gamma$ *as the sum of four* cosines.

$$\cos\alpha \cos\beta \cos\gamma = \tfrac{1}{2}\{\cos(\alpha+\beta) + \cos(\alpha-\beta)\}\cos\gamma$$
$$= \tfrac{1}{4}\cos(\alpha+\beta+\gamma) + \tfrac{1}{4}\cos(\alpha+\beta-\gamma)$$
$$+ \tfrac{1}{4}\cos(\gamma+\alpha-\beta) + \tfrac{1}{4}(\cos\beta+\gamma-\alpha)$$

3. *To prove that if* $A + B + C = 180^\circ$, *then*

$$4\cos^2 A\cos^2 B - \cos(A-B)\{3\cos A\cos B - \sin A\sin B\} = \cos^2 C.$$

We have $4x^2 - (x+y)(3x-y) \equiv (x-y)^2$

Put $\cos A\cos B$ for x

and $\sin A\sin B$ for y

and the result follows, since then

$$x + y = \cos(A-B)$$

and $(x-y)^2 = \cos^2(A+B) = \cos^2 C$

E. T. xlvii.

NOTE—*The above is given as an Example of the transformation of an algebraical into a trigonometrical result. The result can easily be proved without using the algebraic identity.*

4. *To prove that* $\operatorname{cosec} 2\alpha + \cot 4\alpha \equiv \cot\alpha - \operatorname{cosec} 4\alpha$.

$$\operatorname{cosec} 2\alpha + \cot 4\alpha \equiv \frac{1}{\sin 2\alpha} + \frac{\cos 4\alpha}{\sin 4\alpha}$$
$$\equiv \frac{2\cos 2\alpha + 2\cos^2 2\alpha - 1}{\sin 4\alpha}$$
$$\equiv \frac{2\cos 2\alpha(1+\cos 2\alpha)}{2\sin 2\alpha\cos 2\alpha} - \frac{1}{\sin 4\alpha}$$
$$\equiv \cot\alpha - \operatorname{cosec} 4\alpha$$

5. *To prove that*

$$\frac{\cos\theta}{1-\sin\theta} + \frac{\cos\phi}{1-\sin\phi} \equiv \frac{2(\sin\theta - \sin\phi)}{\sin(\theta-\phi) + \cos\theta - \cos\phi}.$$

This will afford a good example of the mode (4° in § 6) of proving an identity by the *conditional* assumption of its truth.

For since the dexter den'r $\cos\theta(1-\sin\phi) - \cos\phi(1-\sin\theta)$

∴ the statement of identity is true *if*

$$\{\cos\theta(1-\sin\phi)\}^2 - \{\cos\phi(1-\sin\theta)\}^2$$
$$\equiv 2(\sin\theta - \sin\phi)(1-\sin\theta)(1-\sin\phi)$$

i. e. *if* $(1 - \sin^2\theta)(1 - \sin\phi)^2 - (1 - \sin^2\phi)(1 - \sin\theta)^2$

$$\equiv 2(\sin\theta - \sin\phi)(1 - \sin\theta)(1 - \sin\phi)$$

i. e. *if* $(1 + \sin\theta)(1 - \sin\phi) - (1 + \sin\phi)(1 - \sin\theta)$

$$\equiv 2(\sin\theta - \sin\phi)$$

which is the case. *E. T.* xlviii.

6. *If* $x = \frac{\pi}{7}$, *prove that* $\cos 3x - \cos 2x + \cos x = \frac{1}{2}$.

R. U. Matric': '87.

If we put X for $\cos 3x - \cos 2x + \cos x$

then $X = \cos 3x + \cos 5x + \cos x$, since $2x + 5x = \pi$

$$\therefore 2X^2 = 3 + \cos 10x + \cos 6x + \cos 2x + 2\cos 8x + 2\cos 2x + 2\cos 6x + 2\cos 4x + 2\cos 4x + 2\cos 2x$$

$$= 3 - 5\cos 3x - 5\cos x + 5\cos 2x$$

$$= 3 - 5X$$

$$\therefore (2X - 1)(X + 3) = 0$$

But $X \neq -3$

$$\therefore X = \tfrac{1}{2}$$

7. *If* A, B, C *are angles of a triangle, and if*

$$\sin^3\omega = \sin(A - \omega)\sin(B - \omega)\sin(C - \omega)$$

prove that $\cot\omega = \cot A + \cot B + \cot C$;

and deduce that $\operatorname{cosec}^2\omega = \operatorname{cosec}^2 A + \operatorname{cosec}^2 B + \operatorname{cosec}^2 C$.

We have $\Pi \operatorname{cosec} A = \Pi \left\{\dfrac{\sin(A - \omega)}{\sin A \sin\omega}\right\}$

$$= \Pi(\cot\omega - \cot A)$$

$$\therefore \cot^3\omega - \cot^2\omega\,\Sigma\cot A + \cot\omega\,\Sigma(\cot A\cot B) - \Pi\cot A - \Pi(\operatorname{cosec} A) = 0$$

But (see p. 78, Ex's 3, 4) $\Pi\cot A + \Pi\operatorname{cosec} A = \Sigma\cot A$

and $\Sigma(\cot A\cot B) = 1$

$$\therefore \cot^3\omega + \cot\omega = (1 + \cot^2\omega)\,\Sigma\cot A$$

$\therefore$ rejecting the factor $1 + \cot^2\omega$, which would give only imaginary values for $\cot\omega$, we get

$$\cot\omega = \Sigma\cot A$$

Then squaring, and recollecting that $2\,\Sigma\,(\cot A \cot B) = 2$, we get

$$1 + \cot^2\omega = \Sigma \cot^2 A + 3$$

$$\therefore\quad \operatorname{cosec}^2\omega = \Sigma \operatorname{cosec}^2 A$$

NOTE—*The Student should carefully notice these results.* ω *is called* the Brocard Angle* *of the* Δ *whose* $\wedge^s$ *are* A, B, C; *and has numerous important properties: some of these will be given in Chapter XIV.*

8. *Given that*

$$\left.\begin{aligned} \tan\frac{\theta}{2} &= \sqrt{\frac{1+e}{1-e}}\,.\tan\frac{u}{2} \\ \textit{and}\quad r &= \frac{a(1-e^2)}{1+e\cos\theta} \end{aligned}\right\}$$

to prove that

$$\left.\begin{aligned} \sqrt{r}\,.\sin\frac{\theta}{2} &= \sqrt{a(1+e)}\,.\sin\frac{u}{2} \\ \textit{and}\quad \sqrt{r}\,.\cos\frac{\theta}{2} &= \sqrt{a(1-e)}\,.\cos\frac{u}{2} \end{aligned}\right\}$$

Christ's Camb': '47.

The 2nd of the methods given in Example 2, p. 30 will be useful here. Thus assume that

$$\left.\begin{aligned} \lambda\sin\frac{\theta}{2} &= \sqrt{1+e}\,\sin\frac{u}{2} \\ \text{and}\quad \lambda\cos\frac{\theta}{2} &= \sqrt{1-e}\,\cos\frac{u}{2} \end{aligned}\right\} \text{where } \lambda \text{ has to be found.}$$

Then, squaring and adding, we get

$$\lambda^2 = (1+e)\sin^2\frac{u}{2} + (1-e)\cos^2\frac{u}{2} = 1 - e\cos u$$

whence $\lambda = \sqrt{1 - e\cos u}$

Again since $\tan^2\dfrac{\theta}{2} = \dfrac{1+e}{1-e}\tan^2\dfrac{u}{2}$

$$\therefore\quad \frac{1-\cos\theta}{1+\cos\theta} = \frac{1+e}{1-e}\cdot\frac{1-\cos u}{1+\cos u}$$

$$\therefore\quad \cos\theta = \frac{(1-e)(1+\cos u) - (1+e)(1-\cos u)}{(1-e)(1+\cos u) + (1+e)(1-\cos u)}$$

$$= \frac{\cos u - e}{1 - e\cos u}$$

* See *Euclid Revised*, pp. 394-402.

$$\therefore \quad 1 + e \cos\theta = \frac{1 - e^2}{1 - e\cos u}$$

$$\therefore \quad 1 - e\cos u = \frac{1 - e^2}{1 + e\cos\theta} = \frac{r}{a}$$

$$\therefore \quad \lambda = \sqrt{r/a}$$

Whence the req'd results are seen to be true.

9. ABCD *is a line trisected by the points* B, C : P *is any point on the circle whose diameter is* BC : *if the angles* APB, CPD *are respectively denoted by* θ, ϕ, *it is required to show that* $\tan\theta\tan\phi = \frac{1}{4}$

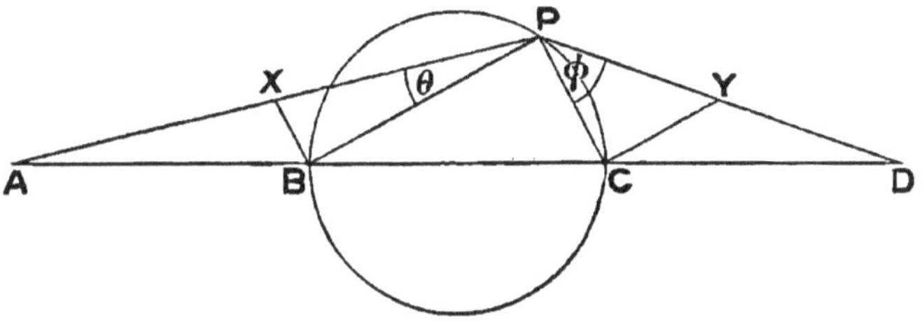

Draw BX, CY respectively ‖ to PC, PB ; and meeting PAP, D in X, Y. Then PC = 2 BX, and PB = 2 CY*.

$$\therefore \quad \tan\theta\tan\phi = \frac{BX}{BP}\cdot\frac{CY}{CP} = \frac{PC}{2\,PB}\cdot\frac{PB}{2\,PC} = \frac{1}{4}$$

Exercises

1. Given that—

(1) $\sin\theta = \frac{1}{3}$, find $\tan\theta$, and $\sin\frac{\theta}{2}$

(2) $\cos\theta = \frac{5}{11}$, find $\tan\frac{\theta}{2}$, $\tan\theta$, and $\tan 2\theta$.

(3) $\tan\theta = \frac{7}{24}$, find $\cos 2\theta$, $\tan 3\theta$, and $\tan\frac{\theta}{2}$.

(4) $\sin\theta = \frac{120}{169}$, find $\tan\frac{\theta}{2}$, and $\cos\frac{3\theta}{2}$.

Q. C. B. '74, '76, '78, '79.

* See *Euclid Revised*, p. 66

(5) $\sin\theta = \frac{5}{7}$, find $\sin 2\theta$ and $\sin\frac{\theta}{2}$.

(6) $\tan\theta + \cot\theta = 2\left(\frac{m^2+n^2}{m^2-n^2}\right)$, find $\cos 2\theta$.

2. Prove that if—

(1) $\tan\alpha = \frac{1}{7}$, and $\tan\beta = \frac{1}{2}$, then $\tan(\beta - 2\alpha) = \frac{2}{11}$.

(2) $\tan\alpha = \frac{5}{12}$, and $\cos 2\beta = \frac{527}{625}$, then $\operatorname{cosec}\frac{\alpha-\beta}{2} = 5\sqrt{13}$.

3. Prove that—

(1) $\tan\alpha + \cot\alpha \equiv 2\operatorname{cosec} 2\alpha$

(2) $\tan\alpha - \cot\alpha \equiv -2\cot 2\alpha$

(3) $\frac{\sin\alpha + \sin\beta}{\sin\alpha - \sin\beta} \equiv \tan\frac{\alpha+\beta}{2}\cot\frac{\alpha-\beta}{2}$

(4) $\frac{\sin 19\alpha + \sin 17\alpha}{\sin 10\alpha + \sin 8\alpha} \equiv 2\cos 9\alpha$

(5) $\tan 4\alpha \equiv \frac{4\tan\alpha(1-\tan^2\alpha)}{1 - 6\tan^2\alpha + \tan^4\alpha}$

(6) $\cos 4\alpha \equiv 8\cos^4\alpha - 8\cos^2\alpha + 1$

(7) $\cot\frac{\alpha}{2} - \cot\alpha \equiv \operatorname{cosec}\alpha$

(8) $\frac{\cot\alpha - \tan\alpha}{\cot\alpha + \tan\alpha} \equiv \cos 2\alpha$

(9) $\sec\alpha \equiv 1 + \tan\alpha\tan\frac{\alpha}{2}$

(10) $\sin 2\alpha + \sin 4\alpha + \sin 6\alpha + \sin 8\alpha$

$\equiv 4\sin 5\alpha\cos 2\alpha\cos\alpha$

(11) $\frac{\cos 3\alpha + \sin 3\alpha}{\cos\alpha - \sin\alpha} \equiv 1 + 2\sin 2\alpha$

(12) $\frac{\sin 3\alpha\sin 2\beta - \sin 3\beta\sin 2\alpha}{\sin 2\alpha\sin\beta - \sin 2\beta\sin\alpha} \equiv 1 + 4\cos\alpha\cos\beta$

(13) $\frac{\sin\alpha + \sin 3\alpha + \sin 5\alpha + \sin 7\alpha}{\cos\alpha + \cos 3\alpha + \cos 5\alpha + \cos 7\alpha} \equiv \tan 4\alpha$

(14) $\sin(\beta-\alpha)\sin(\delta-\gamma) + \sin(\gamma-\beta)\sin(\delta-\alpha)$

$\equiv \sin(\alpha-\gamma)\sin(\beta-\delta)$

(15) $\sqrt{1 + \sin\alpha} \equiv 1 + 2\sin\frac{\alpha}{4}\sqrt{1 - \sin\frac{\alpha}{2}}$

(16) $\sin^2(\alpha + \beta) - \sin^2(\alpha - \beta) \equiv \sin 2\alpha \sin 2\beta$

(17) $\cos^2(\alpha + \beta) - \sin^2(\alpha - \beta) \equiv \cos 2\alpha \cos 2\beta$

(18) $\operatorname{cosec}\alpha \operatorname{cosec} 2\alpha + \operatorname{cosec} 2\alpha \operatorname{cosec} 3\alpha$
$$\equiv 2\cot\alpha \operatorname{cosec} 3\alpha$$

(19) $128\sin^8\alpha \equiv \sin 8\alpha \,.\, \tan 4\alpha \,.\, \tan^2 2\alpha \,.\, \tan^4\alpha$

(20) $\cos 10\alpha + \cos 8\alpha + 3\cos 4\alpha + 3\cos 2\alpha$
$$\equiv 8\cos\alpha \cos^3 3\alpha$$

(21) $\sin\theta - 3\sin(\theta + \alpha) + 3\sin(\theta + 2\alpha) - \sin(\theta + 3\alpha)$
$$\equiv 8\sin^3\frac{\alpha}{2}\cos\left(\theta + \frac{3\alpha}{2}\right)$$

(22) $\sin 3\alpha \sin^3\alpha + \cos 3\alpha \cos^3\alpha \equiv \cos^3 2\alpha$

(23) $\cos^3\alpha \sin 3\alpha + \sin^3\alpha \cos 3\alpha \equiv \frac{3}{4}\sin 4\alpha$

(24) $2\cos\alpha \equiv \sqrt{2 + \sqrt{2 + \sqrt{2 + \&c + \sqrt{2 + 2\cos 2^n\alpha}}}}$

where n is any positive integer, and the $\sqrt{\ }$ is repeated n times.

(25) $\dfrac{1 + (\operatorname{cosec}\alpha \tan x)^2}{1 + (\operatorname{cosec}\beta \tan x)^2} \equiv \dfrac{1 + (\cot\alpha \sin x)^2}{1 + (\cot\beta \sin x)^2}$

(26) $\operatorname{vers}(180^\circ - \alpha) \equiv 2\operatorname{vers}\dfrac{180^\circ + \alpha}{2}\operatorname{vers}\dfrac{180^\circ - \alpha}{2}$

(27) $\sqrt{\operatorname{vers}\alpha \operatorname{vers}\beta} \equiv \operatorname{vers}\dfrac{\alpha + \beta}{2} - \operatorname{vers}\dfrac{\alpha - \beta}{2}$

(28) $8(\cos^8\alpha - \sin^8\alpha) \equiv \cos 6\alpha + 7\cos 2\alpha$

(29) $64(\cos^8\alpha + \sin^8\alpha) \equiv \cos 8\alpha + 28\cos 4\alpha + 35$

(30) $\sin\alpha \sin\beta + \sin\gamma \sin(\alpha + \beta + \gamma)$
$$\equiv \sin(\alpha + \gamma)\sin(\beta + \gamma)$$

(31) $\sin(\alpha - \beta) + \sin(\beta - \gamma) + \sin(\gamma - \alpha)$
$$\equiv -4\sin\frac{\alpha - \beta}{2}\sin\frac{\beta - \gamma}{2}\sin\frac{\gamma - \alpha}{2}$$

(32) $\sec^2\dfrac{\alpha}{2}\sec\alpha\left(\cot^2\dfrac{\alpha}{2} - \cot^2\dfrac{3\alpha}{2}\right) \equiv 8\left(1 + \cot^2\dfrac{3\alpha}{2}\right)$

(33) $$\frac{(\sec\alpha\sec\beta+\tan\alpha\tan\beta)^2-(\tan\alpha\sec\beta+\sec\alpha\tan\beta)^2}{2(1+\tan^2\alpha\tan^2\beta)-\sec^2\alpha\sec^2\beta} \equiv \frac{\sec 2\alpha\sec 2\beta}{\sec^2\alpha\sec^2\beta}$$

(34) $$\{\sec\alpha+\operatorname{cosec}\alpha(1+\sec\alpha)\}\left(1-\tan^2\frac{\alpha}{2}\right)\left(1-\tan^2\frac{\alpha}{4}\right) \equiv \left(\sec\frac{\alpha}{2}+\operatorname{cosec}\frac{\alpha}{2}\right)\sec^2\frac{\alpha}{4}$$

(35) $$\left(\tan^2\frac{\alpha}{2}\operatorname{cosec}^2\frac{\alpha}{6}-\sec^2\frac{\alpha}{2}\right)\cot\frac{2\alpha}{3} \equiv \left(\operatorname{cosec}^2\frac{\alpha}{6}-\sec^2\frac{\alpha}{2}\right)\tan\frac{\alpha}{3}$$

(36) $$\Sigma\sin\alpha\sin\beta\sin(\alpha-\beta) \equiv -\sin(\alpha-\beta)\sin(\beta-\gamma)\sin(\gamma-\alpha)$$

(37) $$\Sigma\{\sin\alpha\sin\beta\sin(\alpha-\beta)\} \equiv \tfrac{1}{2}\Sigma\{\sin 2\alpha\sin(\gamma+\beta)\sin(\gamma-\beta)\}$$

Pet' Camb': '66.

(38) $$1+\sin\alpha\sin\beta+\sin\alpha\sin\gamma+\sin\beta\sin\gamma-\cos\alpha\cos\beta\cos\gamma \equiv 2\left\{\sin\frac{\alpha+\beta}{2}\cos\frac{\gamma}{2}+\cos\frac{\alpha-\beta}{2}\sin\frac{\gamma}{2}\right\}^2$$

(39) $$\Sigma\sin\alpha-\sin(\alpha+\beta+\gamma)\equiv 4\,\Pi\sin\frac{\alpha+\beta}{2}$$

(40) $$\Sigma\cos\alpha+\cos(\alpha+\beta+\gamma)\equiv 4\,\Pi\cos\frac{\alpha+\beta}{2}$$

(41) $$\Sigma\tan\alpha-\frac{\sin(\alpha+\beta+\gamma)}{\cos\alpha\cos\beta\cos\gamma}\equiv\Pi\tan\alpha$$

(42) $$\Sigma\cot\alpha+\frac{\cos(\alpha+\beta+\gamma)}{\sin\alpha\sin\beta\sin\gamma}\equiv\Pi\cot\alpha$$

NOTE—*The Student should compare these last four identities with the particular conditional cases of them given on* pp. 77, 78.

(43) $$\Sigma\sin 2\alpha\sin(\beta-\gamma)\equiv\Sigma\sin(\alpha+\beta)\times\Sigma\sin(\beta-\alpha)$$

(44) $$\Sigma\cos 2\alpha\sin(\beta-\gamma)\equiv\Sigma\cos(\alpha+\beta)\times\Sigma\sin(\beta-\alpha)$$

Wolstenholme: 467.

4. If $\sin\beta = m\sin(2\alpha+\beta)$, prove that

$$\tan(\alpha+\beta) = \frac{1+m}{1-m}\tan\alpha$$

5. If $\tan\frac{z}{2} = \tan\frac{x}{2}\tan\frac{y}{2}$, prove that

$$\tan z = \frac{\sin x \sin y}{\cos x + \cos y}$$

6. If

$$\left.\begin{aligned} \sin x \sin y &= \sin(\alpha+\beta)\sin\gamma \\ \cos x \cos y &= \cos(\alpha+\beta)\cos\gamma \\ \text{and}\quad \cos^2 x + \cos^2 y &= 1 + \cos^2(\alpha+\beta+\gamma) \end{aligned}\right\}$$

prove that $\sin^2(\alpha+\beta) + \sin^2\gamma = \sin^2(\alpha+\beta+\gamma)$

7. If $\tan\theta = m$, and $\tan\phi = n$, prove that

$$\sin 2(\theta+\phi) = \frac{2(m+n)(1-mn)}{(1+m^2)(1+n^2)}$$

8. If $\cos\theta = \dfrac{\cos\alpha - \cos\beta}{1-\cos\alpha\cos\beta}$, prove that

$$\tan\frac{\theta}{2} = \pm\frac{\tan\frac{\alpha}{2}}{\tan\frac{\beta}{2}}$$

9. If $\sin^2\alpha + \sin^2\beta - 2\sin\alpha\sin\beta\cos(\alpha-\beta) = n^2(\sin\alpha - \sin\beta)^2$

prove that $\tan\frac{\alpha}{2} = \frac{1\pm n}{1\mp n}\cot\frac{\beta}{2}$

10. If $\operatorname{vers}\alpha = x$, $\operatorname{vers}\beta = mx$, $\operatorname{vers}\gamma = 1-m$, and $\alpha+\beta=\gamma$,

show that $x = 1 \pm \sqrt{\left(\frac{2m}{1+m}\right)}$

11. If

$$\left.\begin{aligned} \sin x \cos y &= \tan\alpha\cot\gamma \\ \sin y \cos x &= \tan\beta\cot\gamma \\ \text{and}\quad \cos^2 y - \cos^2 x &= \cos^2\gamma \end{aligned}\right\}$$

prove that $\sec^2\alpha - \sec^2\beta = \sin^2\gamma$

12. If $\tan\frac{\alpha}{2} = \tan^3\frac{\beta}{2}$, and $\tan\beta = 2\tan\phi$, show that $\frac{\alpha+\beta}{2}$ is *one* value of ϕ.

13. If $\sigma = \alpha + \beta + \gamma + \delta$, and

$$\cos(\sigma - 2\alpha) + \cos(\sigma - 2\beta) = \cos(\sigma - 2\gamma) + \cos(\sigma - 2\delta)$$

prove that $\tan\alpha \tan\beta = \tan\gamma \tan\delta$

14. If $2\sigma = \alpha + \beta + \gamma$, prove that

$$\cos^2\sigma + \cos^2(\sigma - \alpha) + \cos^2(\sigma - \beta) + \cos^2(\sigma - \gamma)$$
$$= 2 + 2\cos\alpha\cos\beta\cos\gamma$$

15. If

$$\left.\begin{aligned} a &= b\sin x + c\sin y \\ o &= b\cos x - c\cos y \\ \text{and}\quad b\sin(x + \theta) &= c\sin(y - \theta) \end{aligned}\right\}$$

prove that $2\tan\theta = \tan y - \tan x$;

and find $\sin x$, $\sin y$, $\sin\theta$ in terms of a, b, c.

16. If $\sin\theta = \frac{a}{b}\sin\phi$, and $\cos\theta = \frac{a'}{b'}\cos\phi$,

prove that $\sin(\theta \pm \phi) = \sqrt{(b^2 - a^2)(a'^2 - b'^2)} \Big/ \overline{ab' \mp a'b}$

and that $\cos(\theta \pm \phi) = \overline{aa' \mp bb'} \Big/ \overline{ab' \pm a'b}$

17. If $\tan\theta = \frac{x\sin\alpha}{y - x\cos\alpha}$, and $\tan\phi = \frac{y\sin\alpha}{x - y\cos\alpha}$,

show that $\tan(\theta + \phi) = -\tan\alpha$

18. If $\tan^2\theta = \tan(\theta - \alpha)\tan(\theta - \beta)$

prove that $\tan 2\theta = \frac{2\sin\alpha\sin\beta}{\sin(\alpha + \beta)}$

R. U. Schol': '87.

19. If

$$\left.\begin{aligned} a\cos\theta + b\sin\theta &= c \\ \text{and}\quad a\cos\phi + b\sin\phi &= c \end{aligned}\right\}$$

prove that $\tan(\theta + \phi) = 2ab \Big/ (a^2 - b^2)$

Joh' Camb': '87.

20. If $\tan\phi = \frac{\sin\alpha\sin\theta}{\cos\theta - \cos\alpha}$

prove that $\tan\theta = \frac{\sin\alpha\sin\phi}{\cos\phi \pm \cos\alpha}$

Math' Tri': '68.

21. If
$$\left.\begin{aligned}\tan\theta &= \frac{\sin\alpha\cos\gamma - \sin\beta\sin\gamma}{\cos\alpha\cos\gamma - \cos\beta\sin\gamma}\\ \text{and}\quad \tan\phi &= \frac{\sin\alpha\sin\gamma - \sin\beta\cos\gamma}{\cos\alpha\sin\gamma - \cos\beta\cos\gamma}\end{aligned}\right\}$$
find $\tan(\theta + \phi)$ in its simplest form

22. If $\sin\alpha = \dfrac{\cos\alpha}{\sqrt{1 - m^2\sin^2\alpha}}$, find $\tan\alpha$

23. If $\tan 466° \, 15' \, 38'' = -\frac{24}{7}$, find the sine and cosine of $233° \, 7' \, 49''$ *Joh' Camb'*: '50.

24. If $\sin A$, $\sin B$, $\sin C$ are in arithmetical progression; and A, B, C are angles of a triangle; prove that
$$\tan\frac{A}{2}\tan\frac{C}{2} = \frac{1}{3}$$

25. If $\tan A$, $\tan B$, $\tan C$ are in arithmetical progression; and A, B, C are angles of a triangle; prove that
$$\cos(B + C - A) = \frac{4 + 5\cos 2C}{5 + 4\cos 2C}$$

26. If $\alpha + \beta + \gamma = 90°$, prove that
$$\frac{\cos\alpha + \sin\gamma - \sin\beta}{\cos\beta + \sin\gamma - \sin\alpha} = \frac{1 + \tan\frac{\alpha}{2}}{1 + \tan\frac{\beta}{2}}$$

27. If $\tan 2A = 2\tan B$, and $\tan C = \tan^3 A$ (where A, B, C are angles of a triangle), show that B is a right angle.

If also $m^2 = 3\cos^2 2A + 1$, prove that
$$(\sin C)^{\frac{2}{3}} + (\cos C)^{\frac{2}{3}} = \left(\frac{2}{m}\right)^{\frac{2}{3}}$$

28. If $\sin\theta + \sin\phi = a$, and $\cos\theta + \cos\phi = b$, find each of the following in terms of a and b.

(1) $\sin\theta\sin\phi$ (3) $\tan\theta + \tan\phi$ (5) $\cos 2\theta + \cos 2\phi$

(2) $\cos\theta\cos\phi$ (4) $\tan\frac{\theta}{2} + \tan\frac{\phi}{2}$ (6) $\cos 3\theta + \cos 3\phi$

Wolstenholme: 462.

29. If $A + B + C = 360^\circ$, prove that

$$\Sigma \sin A (1 + 2\cos B) = -4 \Pi \sin \frac{A - B}{2}$$

Math' Tri': '64.

30. If

$$\left.\begin{aligned} \sqrt{2}\cos A &= \cos B + \cos^3 B \\ \text{and} \quad \sqrt{2}\sin A &= \sin B - \sin^3 B \end{aligned}\right\}$$

prove that $\pm \sin (B - A) = \cos 2B = \frac{1}{3}$

Math' Tri': '71.

31. If

$$\left.\begin{aligned} x \sin \omega &= X \sin (\omega - \alpha) + Y \sin (\omega - \beta) \\ \text{and} \quad y \sin \omega &= X \sin \alpha + Y \sin \beta \end{aligned}\right\}$$

show that $x^2 + y^2 + 2xy\cos\omega = X^2 + Y^2 + 2XY\cos(\alpha - \beta)$

NOTE—*The given eq'ns are those connecting the coord's* (x, y) *of a p't, referred to axes* Ox, Oy, *inclined at* ω, *with its coord's* (X, Y) *referred to axes* OX, OY, *making* $\wedge^s$ α, β *with* Ox. Cf. *Salmon's Conics*, pp. 7, 9. *The result shows that the expression* $x^2 + y^2 + 2xy \cos xOy$ *is invariable, no matter how the axes are turned about* O.

32. If $\alpha, \beta, \gamma, \delta$ are the angles of a quadrilateral; prove that

(1) $$\Sigma \sin \alpha = 4 \sin \frac{\alpha + \beta}{2} \sin \frac{\beta + \gamma}{2} \sin \frac{\gamma + \alpha}{2}$$

(2) $$\Sigma \cos \alpha = 4 \cos \frac{\alpha + \beta}{2} \cos \frac{\beta + \gamma}{2} \cos \frac{\gamma + \alpha}{2}$$

33. If $\alpha, \beta, \gamma, \delta$ are the angles of a quadrilateral; and if θ, ϕ are the angles formed by producing each pair of opposite sides to meet; prove that

$$\cos \frac{\alpha}{2} \cos \frac{\gamma}{2} + \cos \frac{\beta}{2} \cos \frac{\delta}{2} = \cos \frac{\theta}{2} \cos \frac{\phi}{2}$$

Q. C. B. '66.

34. In a convex quadrilateral ABCD, prove that

$$\cos \frac{A}{2} \cos \frac{B}{2} + \cos \frac{C}{2} \cos \frac{D}{2} = \sin \frac{A}{2} \sin \frac{B}{2} + \sin \frac{C}{2} \sin \frac{D}{2}$$

E. T. lii.

35. If $\alpha + \beta + \gamma = (2n + 1)\pi$, prove that

$$\Sigma \left(1 - \sin \frac{\beta}{2}\right)\left(1 - \sin \frac{\gamma}{2}\right) \cos \frac{\alpha}{2} = \Pi \left(\cos \frac{\alpha}{2}\right)$$

Christ's Camb': '49.

36. If $\alpha + \beta + \gamma = \frac{\pi}{2}$, prove that

$$\Sigma \{\cot\alpha\,(\tan\beta + \tan\gamma)\} + 2 = \Pi \operatorname{cosec}\alpha$$

37. If $\Sigma \cos\theta + \Pi \cos\theta = 0$, prove that

$$\Sigma \operatorname{cosec}^2\theta \pm 2\,\Pi \operatorname{cosec}\theta = 1$$

where the symbols refer to three angles θ, ϕ, ψ

Math' Tri': '62.

38. If $\Pi (1 + \sin\theta) = \Pi \cos\theta$, prove that

$$\Sigma \sec^2\theta = 1 + 2\,\Pi \sec\theta$$

where the symbols refer to three angles θ, ϕ, ψ

Wolstenholme: 447.

NOTE—*By squaring the given relation, we get* $\Sigma \sin\theta + \Pi \sin\theta = 0$: *then square again.*

39. If $\tan x_1 \tan x_2 \tan x_3$ &c $\tan x_n = \tan\alpha$; and if S_r denote the sum of the products of $\cos 2x_1$, $\cos 2x_2$, &c taken r together; prove that

$$\cos 2\alpha = \overline{S_1 + S_3 + S_5 + \&c} \Big/ \overline{1 + S_2 + S_4 + \&c}$$

E. T. xviii.

NOTE—*Square given relation; and substitute by formula*

$$\tan^2\alpha = \overline{1 - \cos 2\alpha} \Big/ \overline{1 + \cos 2\alpha}:$$

then componendo et dividendo gives result.

40. In a triangle ABC, the angle C is obtuse, and $B = 2A$: if $\sin A = \sqrt{\lambda}/2$; and BC is produced to X; prove that

$$\cot ACX = \frac{1 - \lambda}{3 - \lambda} \sqrt{\frac{4}{\lambda} - 1}$$

Which sign of the radical must be taken?

41. ACB, ADB are fixed right-angled triangles on the same side of a common hypotenuse AB: if X is a variable point in AB, prove that

$$\tan ACX \,.\, \tan BDX \text{ is constant}$$

42. If $\theta + \phi + \alpha = \pi$, show that

$$\sin^2\theta + \sin^2\phi - 2\sin\theta\sin\phi\cos\alpha = \sin^2\alpha$$

Hence prove that if two fixed circles, radii R, r, cut at an angle α; and, if the segments intercepted on a variable line through one point of section are X in circle radius R, and x in circle radius r, then

$$\left(\frac{X}{R}\right)^2 + \left(\frac{x}{r}\right)^2 - \frac{2Xx}{Rr}\cos\alpha = 4\sin^2\alpha$$

43. A circle, radius r, touches the chord and arc of a segment of a circle, radius R; and the chord is divided at the point of contact into segments h, k: if α is the angle in the segment, prove that

$$r = hk\cot\frac{\alpha}{2} \Big/ \overline{h+k}$$

Genese: *E. T.* xli.

NOTE— $(R - r)^2 = (r - R\cos\alpha)^2 + (h - R\sin\alpha)^2$

and $h + k = 2R\sin\alpha$: *substitute for* R *from 2nd of these in 1st.*

44. G is the centroid of a triangle ABC, and M the mid point of AB: if A, B, C denote the angles at those respective corners; and if α, β, θ denote the angles GAB, GBA, GMA respectively; prove that

$$\tan\theta = \frac{\tan A \tan B(\tan\alpha + \tan\beta) + \tan\alpha\tan\beta(\tan A + \tan B)}{\tan A\tan\alpha - \tan B\tan\beta}$$

$$\text{or} \quad = \frac{\tan A\tan B(\tan\alpha + \tan\beta) - \tan\alpha\tan\beta(\tan A + \tan B)}{\tan A\tan\beta - \tan B\tan\alpha}$$

NOTE—*Drop* GL, CN $\perp^s$ *on* AB; *and recollect that* $CN = 3GL$: *the results come readily by putting down the def's of the* tan's *in the dexter fractions* *Walker*: *E. T.* xxi.

45. On BA (or BA produced) of a triangle ABC, E is taken so that BE equals BC; and on BC (or BC produced) D is taken so that BD equals BA: on ED any triangle EFD is constructed; AP is drawn to meet BF in P, and so that the angle PAB equals the angle EFB: PS is drawn *perpendicular to the plane of* ABC; and SA, SB, SC are joined: if α, β, γ are angles such that

$$EB : EF = \tan\alpha : \tan\beta,$$

$$\text{and} \quad DB : DF = \tan\gamma : \tan\beta;$$

and if of the set of angles α, β, γ, and the set SAP, SBP, SCP, any one of one set equals the one corresponding in order of the other set, prove that the other two are equal each to each. *Queens' Camb'*: '51.

NOTE—*Since* AP *is *anti-*$\parallel$ *to* FE,

$$\therefore \quad BP \cdot BF = BA \cdot BE = BC \cdot BD$$

$\therefore$ Δ^s BPC, BDF *are *inversely sim'r. The results easily follow.*

---

* See *Euclid Revised*, p. 378.

CHAPTER V

Numerical values of the Trigonometrical Functions of some special Angles

We have found (in Chapter II) numerical values for the T. F.s of all $\wedge$s which are exact multiples of a right $\wedge$: we now proceed to find numerical values for the T. F.s of certain other $\wedge$s.

§ **30**. *To find the* T. F.s *of* 45°; *and thence of* 135°, 225° *and* 315°.

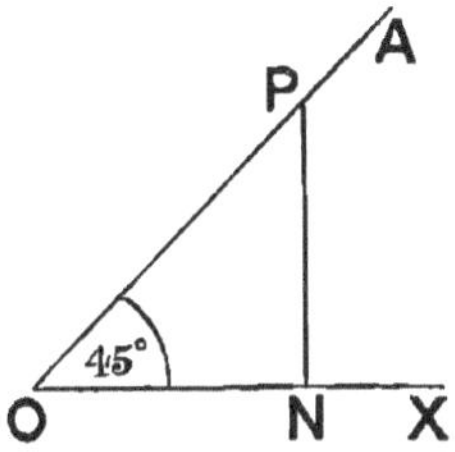

If $X\hat{O}A = 45°$; and PN is dropped, from any p't P in OA, ⊥ to OX; then, in the auxil' △ PON, we have

$$ON = OP$$

$\therefore$ OP (which $= \sqrt{ON^2 + NP^2}$) $= NP\sqrt{2}$ or $ON\sqrt{2}$

$$\sin 45° \left(\text{which} = \frac{NP}{OP}\right) = \frac{1}{+\sqrt{2}}$$

$$\text{and} \quad \cos 45° \left(\text{which} = \frac{ON}{OP}\right) = \frac{1}{+\sqrt{2}}$$

$$\text{Whence also} \quad \tan 45° = \quad 1 = \cot 45°$$

$$\text{and} \quad \sec 45° = +\sqrt{2} = \operatorname{cosec} 45°$$

$$\text{Now} \quad 135° = 180° - 45°$$

$$\sin 135^\circ = \sin 45^\circ = \frac{1}{+\sqrt{2}}$$

$$\cos 135^\circ = -\cos 45^\circ = \frac{1}{-\sqrt{2}}$$

$$\tan 135^\circ = -\tan 45^\circ = -1$$

So also $\cot 135^\circ = -1$, $\sec 45^\circ = -\sqrt{2}$, $\operatorname{cosec} 45^\circ = +\sqrt{2}$

Again $225^\circ = 180^\circ + 45^\circ$

$$\therefore \sin 225^\circ = -\sin 45^\circ = \frac{1}{-\sqrt{2}}$$

$$\cos 225^\circ = -\cos 45^\circ = \frac{1}{-\sqrt{2}}$$

$$\tan 225^\circ = \tan 45^\circ = 1$$

So also $\cot 225^\circ = 1$, $\sec 225^\circ = -\sqrt{2}$, $\operatorname{cosec} 225^\circ = -\sqrt{2}$

Lastly $315^\circ = 360^\circ - 45^\circ$

$$\therefore \sin 315^\circ = \sin(-45^\circ) = -\sin 45^\circ = \frac{1}{-\sqrt{2}}$$

$$\cos 315^\circ = \cos 45^\circ = \frac{1}{+\sqrt{2}}$$

$$\tan 315^\circ = \tan(-45^\circ) = -\tan 45^\circ = -1$$

So also $\cot 315^\circ = -1$, $\sec 315^\circ = +\sqrt{2}$, $\operatorname{cosec} 315^\circ = -\sqrt{2}$

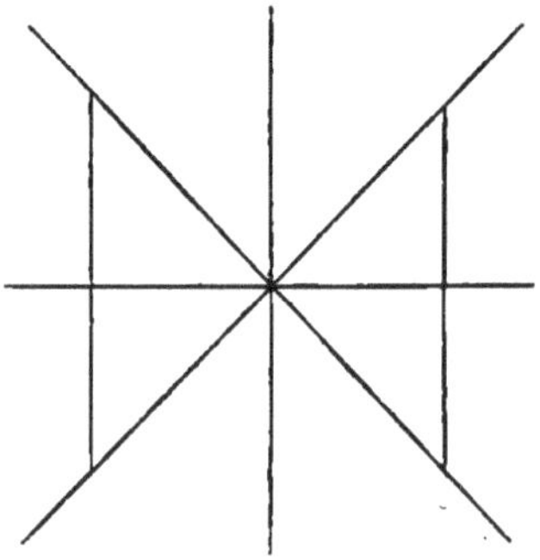

Of course all the results which have been deduced from $\sin 45^\circ$ and $\cos 45^\circ$ may be obtained directly by such a fig' as the accompanying.

§ **31.** *To find the* T. F.s *of* 30°; *and thence of* 60°, 120°, 150°, 210°, 300° *and* 330°.

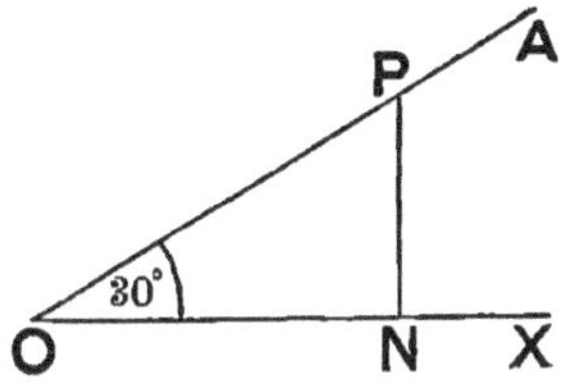

If $X\hat{O}A = 30°$; and PN is dropped, from any p't P in OA, ⊥ to OX; then in the auxil' △ PON, we have that

$$2\,P\hat{O}N = O\hat{P}N$$

∴, by a well known geometrical* result, $NP = \frac{1}{2}OP$

$$\therefore\quad ON \text{ (which} = \sqrt{OP^2 - NP^2}) = \frac{\sqrt{3}}{2}OP$$

So that $\sin 30° \left(\text{which} = \frac{NP}{OP}\right) = \frac{1}{2}$

and $\cos 30° \left(\text{which} = \frac{ON}{OP}\right) = \frac{+\sqrt{3}}{2}$

Whence also $\tan 30° = \frac{1}{+\sqrt{3}}$

$$\cot 30° = +\sqrt{3}$$

$$\sec 30° = \frac{2}{+\sqrt{3}}$$

and $\operatorname{cosec} 30° = 2$

Now $60° = 90° - 30°$

$$\therefore\quad \sin 60° = \cos 30° = \frac{+\sqrt{3}}{2}$$

$$\cos 60° = \sin 30° = \frac{1}{2}$$

$$\tan 60° = \cot 30° = +\sqrt{3}$$

* See *Euclid Revised*, p. 68.

So also $\cot 60^\circ = \frac{1}{+\sqrt{3}}$, $\sec 60^\circ = 2$, $\operatorname{cosec} 60^\circ = \frac{2}{+\sqrt{3}}$

$$\text{Again} \quad 120^\circ = 180^\circ - 60^\circ, \quad \text{or} \quad = 90^\circ + 30^\circ$$

Whence the T. F.s of 120° can be at once written down.

$$\text{Next} \quad 150^\circ = 180^\circ - 30^\circ$$

From which the T. F.s of 150° are known.

So also $210^\circ = 180^\circ + 30^\circ$ gives the T. F.s of 210°
$240^\circ = 180^\circ + 60^\circ$ „ „ „ 240°
$300^\circ = 360^\circ - 60^\circ$ „ „ „ 300°
$330^\circ = 360^\circ - 30^\circ$ „ „ „ 330°

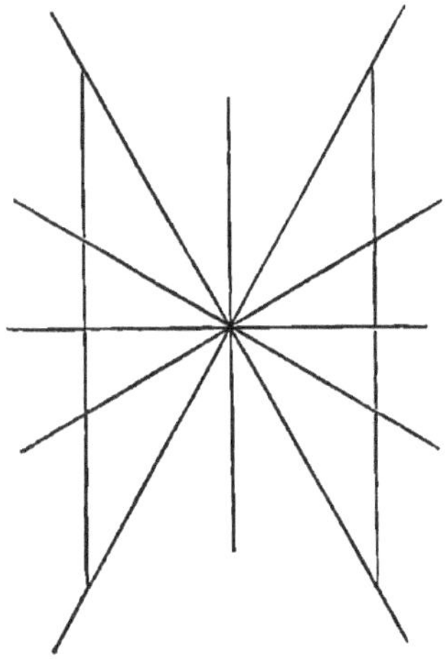

Of course all the results which have been deduced from sin 30° and cos 30° may be obtained directly from such a fig′ as the accompanying.

Here is subjoined a complete table of the T. F.s of the ∧s enumerated: this table the Student should verify and commit to memory. He should also accustom himself to think of the ∧s in radians as well as in degrees: thus the idea of half a right ∧ should suggest $\pi/4$ as readily as 45°.

Note—If it be recollected that $\sin 30^\circ = 1/2$, and $\sin 45^\circ = 1/\sqrt{2}$, all the rest of the table is at once deducible mentally by means of the formulæ in § 17.

α	30° $\pi/6$	45° $\pi/4$	60° $\pi/3$	120° $2\pi/3$	135° $3\pi/4$	150° $5\pi/6$	210° $7\pi/6$	225° $5\pi/4$	240° $4\pi/3$	300° $5\pi/3$	315° $7\pi/4$	330° $11\pi/6$
sin α	$\frac{1}{2}$	$\frac{1}{+\sqrt{2}}$	$\frac{+\sqrt{3}}{2}$	$\frac{+\sqrt{3}}{2}$	$\frac{1}{+\sqrt{2}}$	$\frac{1}{2}$	$-\frac{1}{2}$	$\frac{1}{-\sqrt{2}}$	$\frac{-\sqrt{3}}{2}$	$\frac{-\sqrt{3}}{2}$	$\frac{1}{-\sqrt{2}}$	$-\frac{1}{2}$
cos α	$\frac{+\sqrt{3}}{2}$	$\frac{1}{+\sqrt{2}}$	$\frac{1}{2}$	$-\frac{1}{2}$	$\frac{1}{-\sqrt{2}}$	$\frac{-\sqrt{3}}{2}$	$\frac{-\sqrt{3}}{2}$	$\frac{1}{-\sqrt{2}}$	$-\frac{1}{2}$	$\frac{1}{2}$	$\frac{1}{+\sqrt{2}}$	$\frac{+\sqrt{3}}{2}$
tan α	$\frac{1}{+\sqrt{3}}$	1	$+\sqrt{3}$	$-\sqrt{3}$	-1	$\frac{1}{-\sqrt{3}}$	$\frac{1}{+\sqrt{3}}$	1	$+\sqrt{3}$	$-\sqrt{3}$	-1	$\frac{1}{-\sqrt{3}}$
cot α	$+\sqrt{3}$	1	$\frac{1}{+\sqrt{3}}$	$\frac{1}{-\sqrt{3}}$	-1	$-\sqrt{3}$	$+\sqrt{3}$	1	$\frac{1}{+\sqrt{3}}$	$\frac{1}{-\sqrt{3}}$	-1	$-\sqrt{3}$
sec α	$\frac{2}{+\sqrt{3}}$	$+\sqrt{2}$	2	-2	$-\sqrt{2}$	$\frac{2}{-\sqrt{3}}$	$\frac{2}{-\sqrt{3}}$	$-\sqrt{2}$	-2	2	$+\sqrt{2}$	$\frac{2}{+\sqrt{3}}$
cosec α	2	$+\sqrt{2}$	$\frac{2}{+\sqrt{3}}$	$\frac{2}{+\sqrt{3}}$	$+\sqrt{2}$	2	-2	$-\sqrt{2}$	$\frac{2}{-\sqrt{3}}$	$\frac{2}{-\sqrt{3}}$	$-\sqrt{2}$	-2

Exercises

1. Find *all* the values of $\sin\{n\,.\,90° + (-1)^n\, 30°\}$ where n is any integer, including zero.

NOTE—*Consider separately the cases when* n *has one of the forms* $4m$, $4m+1$, $4m+2$, $4m+3$.

2. Find *all* the values of $\sec m\pi/3$, where m is any integer.

3. Find *all* the values of $\text{vers}\ m\,.\,45°$, where m is any integer.

From the foregoing values those of the T. F.s of many other ∧s can be deduced by the aid of the formulæ in Chapter IV.

§ 32. *To find the* T. F.s *of* $15°$ *and* $75°$; *and thence those of* $7°\tfrac{1}{2}$, $22°\tfrac{1}{2}$, *and many similar angles.*

$$\sin 15° = \sin(45° - 30°)$$
$$= \sin 45° \cos 30° - \cos 45° \sin 30°$$
$$= \frac{\sqrt{3}}{2\sqrt{2}} - \frac{1}{2\sqrt{2}}$$
$$= \tfrac{1}{4}(\sqrt{6} - \sqrt{2})$$

Sim'ly $\cos 15° = \tfrac{1}{4}(\sqrt{6} + \sqrt{2})$

Then $\tan 15° = \dfrac{\sqrt{6} - \sqrt{2}}{\sqrt{6} + \sqrt{2}} = \dfrac{\sqrt{3} - 1}{\sqrt{3} + 1} = 2 - \sqrt{3}$

Or the last three might be deduced from $\sin 30°$ only.

For $\sin 15° = \tfrac{1}{2}\{\sqrt{1 + \sin 30°} - \sqrt{1 - \sin 30°}\}$

$$= \tfrac{1}{2}\left\{\sqrt{\tfrac{3}{2}} - \sqrt{\tfrac{1}{2}}\right\}$$
$$= \tfrac{1}{4}(\sqrt{6} - \sqrt{2})$$

Sim'ly for $\cos 15°$.

Also $\tan 15° = \dfrac{\sin 30°}{1 + \cos 30°} = \dfrac{1}{2 + \sqrt{3}} = 2 - \sqrt{3}$

In either of the preceding ways we should find

$$\sin 75^\circ = \tfrac{1}{4}(\sqrt{6} + \sqrt{2})$$

$$\cos 75^\circ = \tfrac{1}{4}(\sqrt{6} - \sqrt{2})$$

$$\tan 75^\circ = 2 + \sqrt{3}$$

Of course, as $75^\circ = 90^\circ - 15^\circ$, these could be deduced from the T. F.s of 15°.

Then, as $105^\circ = 180^\circ - 75^\circ$, the T. F.s of 105° are known.

Next $\tan 22^\circ\tfrac{1}{2} = \dfrac{\sin 45^\circ}{1 + \cos 45^\circ} = \dfrac{1}{1 + \sqrt{2}} = \sqrt{2} - 1$

Whence $\sin 22^\circ\tfrac{1}{2}$ and $\cos 22^\circ\tfrac{1}{2}$ can be written down.

or $\sin 22^\circ\tfrac{1}{2} = \tfrac{1}{2}\{\sqrt{1 + \sin 45^\circ} - \sqrt{1 - \sin 45^\circ}\}$

or again $\sin 22^\circ\tfrac{1}{2} = +\sqrt{(1 - \cos 45^\circ)/2}$

For the T. F.s of $7^\circ\tfrac{1}{2}$ we have

$$\sin 7^\circ\tfrac{1}{2} = \tfrac{1}{2}\{\sqrt{1 + \sin 15^\circ} - \sqrt{1 - \sin 15^\circ}\}$$

$$\text{or} \quad = +\sqrt{(1 - \cos 15^\circ)/2}$$

And sim'ly for $\cos 7^\circ\tfrac{1}{2}$.

Also $\tan 7^\circ\tfrac{1}{2} = \dfrac{\sin 15^\circ}{1 + \cos 15^\circ}$

But the reduction of the surds in the last expression is rather tedious; and the following method will give $\tan 7^\circ\tfrac{1}{2}$ more easily.

Put x for $\tan 7^\circ\tfrac{1}{2}$: then

$$\frac{2x}{1 - x^2} = \tan 15^\circ = \frac{1}{2 + \sqrt{3}}$$

$$\therefore \quad x^2 + 2(2 + \sqrt{3})x - 1 = 0$$

$$\therefore \quad x + 2 + \sqrt{3} = \pm\sqrt{8 + 4\sqrt{3}}$$

$$= \pm(\sqrt{6} + \sqrt{2})$$

$$\therefore \quad \tan 7^\circ\tfrac{1}{2} = \sqrt{6} + \sqrt{2} - \sqrt{3} - 2$$
$$= (\sqrt{3} - \sqrt{2})(\sqrt{2} - 1)$$

The neg′ sign of $\sqrt{8 + 4\sqrt{3}}$ is inadmissible, $\because$ $\tan 7^\circ\frac{1}{2}$ is pos′.

As above the T. F.s of $52^\circ\frac{1}{2}$ come from those of 105°.

Since $\tan x \tan y = 1$, when $x + y = 90^\circ$, we can *in that case* deduce either of $\tan x$, $\tan y$ from the other.

Thus, since $82^\circ\frac{1}{2} + 7^\circ\frac{1}{2} = 90^\circ$, we have

$$\tan 82^\circ\tfrac{1}{2} = \frac{1}{\tan 7^\circ\frac{1}{2}} = \frac{1}{(\sqrt{3} - \sqrt{2})(\sqrt{2} - 1)}$$
$$= (\sqrt{3} + \sqrt{2})(\sqrt{2} + 1)$$

Exercises

1. If $\tan \alpha = \sqrt{3}$, and $\tan \beta = 2 - \sqrt{3}$, find $\tan(4\beta - \alpha)$

2. Show that $\cot 11^\circ\frac{1}{4} = 1 + \sqrt{2} + \sqrt{2}\sqrt{2 + \sqrt{2}}$

3. Find $\tan 11^\circ\frac{1}{4}$

4. Prove that—

(1) $\sin 22^\circ\frac{1}{2} = \frac{1}{2}\sqrt{2 - \sqrt{2}}$

(2) $\sin 67^\circ\frac{1}{2} = \frac{1}{2}\sqrt{2 + \sqrt{2}}$

(3) $\cos 33^\circ 45' = \frac{1}{2}\sqrt{2 + \sqrt{2 - \sqrt{2}}}$

5. Find $\tan 37^\circ\frac{1}{2}$; and deduce $\tan 52^\circ\frac{1}{2}$

6. Deduce $\sin 7^\circ\frac{1}{2}$ both from $\tan 7^\circ\frac{1}{2}$, and from $\sin 15^\circ$; and show that the two surd results are equal.

7. Prove that

$$(\tan 7^\circ\tfrac{1}{2} + \tan 37^\circ\tfrac{1}{2} + \tan 67^\circ\tfrac{1}{2})$$
$$\times (\tan 22^\circ\tfrac{1}{2} + \tan 52^\circ\tfrac{1}{2} + \tan 82^\circ\tfrac{1}{2}) = 17 + 8\sqrt{3}$$

E. T. xlix.

8. Prove that

$$\frac{(\tan 67^\circ\tfrac{1}{2} - \tan 7^\circ\tfrac{1}{2})(\tan 127^\circ\tfrac{1}{2} + \tan 22^\circ\tfrac{1}{2})}{(\tan 22^\circ\tfrac{1}{2} + \tan 7^\circ\tfrac{1}{2})(\tan 127^\circ\tfrac{1}{2} - \tan 67^\circ\tfrac{1}{2})} = 1$$

9. Prove that

$$\frac{\tan 52^\circ\tfrac{1}{2} + \tan 7^\circ\tfrac{1}{2}}{\tan 82^\circ\tfrac{1}{2} + \tan 37^\circ\tfrac{1}{2}} = \frac{\tan 52^\circ\tfrac{1}{2} - \tan 7^\circ\tfrac{1}{2}}{\tan 82^\circ\tfrac{1}{2} - \tan 37^\circ\tfrac{1}{2}} = \frac{\tan 22^\circ\tfrac{1}{2}}{\tan 67^\circ\tfrac{1}{2}} = (\sqrt{2} - 1)^2$$

E. T. xlviii.

10. The $\widehat{ACB}$ is right: a transversal XPY is drawn across it so that X is in CA, Y in CB; $\widehat{PXC} = 7^\circ\tfrac{1}{2}$, and $\widehat{PCX} = 52^\circ\tfrac{1}{2}$: find two angles into which the last two may be changed without altering the ratio of XP to PY.

11. By using the formulæ

$$\cot\theta + \tan\theta = 2\operatorname{cosec} 2\theta$$

$$\cot\theta - \tan\theta = 2\cot 2\theta$$

find the values of $\tan\frac{\pi}{24}$ and $\cot\frac{\pi}{24}$

§ **33.** *To find the* T. F.s *of* 18°; *and thence those of* 36°, 54°, 3°, 9°, 81°, *and many similar angles.*

Put α for 18°, so that $2\alpha = 36^\circ$, $3\alpha = 54^\circ$

$$\therefore \quad \sin 2\alpha\ [\text{which} = \cos(90^\circ - 2\alpha)] = \cos 3\alpha$$

$$\therefore \quad 2\sin\alpha\cos\alpha = 4\cos^3\alpha - 3\cos\alpha$$

Divide by $\cos\alpha$, which by hypothesis $\neq 0$, and put x for $\sin\alpha$: then

$$4x^2 + 2x - 1 = 0$$

$$\therefore \quad 4x + 1 = \pm\sqrt{5}$$

$$\therefore \quad \sin 18^\circ = \frac{+\sqrt{5} - 1}{4}$$

The neg' sign of the radical is inadmissible, $\because \sin 18^\circ \not> 1$.

From this $\cos 18^\circ = \sqrt{1 - \sin^2 18^\circ} = \frac{1}{4}\sqrt{10 + 2\sqrt{5}}$

$$\text{Then}\quad \tan 18^\circ = \frac{\sqrt{5}-1}{\sqrt{10+2\sqrt{5}}} = \sqrt{\frac{6-2\sqrt{5}}{2\sqrt{5}(\sqrt{5}+1)}}$$

$$= \sqrt{\left\{\frac{(3-\sqrt{5})(\sqrt{5}-1)\sqrt{5}}{20}\right\}}$$

$$= \sqrt{1-\tfrac{2}{5}\sqrt{5}}$$

Since $72^\circ = 90^\circ - 18^\circ$, the T. F.s of 72° are at once deducible from those of 18°.

$$\text{Next}\quad \cos 36^\circ = 1 - 2\sin^2 18^\circ = 1 - \frac{6-2\sqrt{5}}{8} = \frac{+\sqrt{5}+1}{4}$$

$$\therefore\quad \sin 36^\circ = \sqrt{1-\cos^2 36^\circ} = \tfrac{1}{4}\sqrt{10-2\sqrt{5}}$$

$$\text{then}\quad \tan 36^\circ = \frac{\sqrt{10-2\sqrt{5}}}{\sqrt{5}+1} = \sqrt{\frac{2\sqrt{5}(\sqrt{5}-1)(3-\sqrt{5})}{2(3+\sqrt{5})(3-\sqrt{5})}}$$

$$= \sqrt{\frac{20-8\sqrt{5}}{4}} = \sqrt{5-2\sqrt{5}}$$

Since $54^\circ = 90^\circ - 36^\circ$, the T. F.s of 54° are deducible.
And, since $3^\circ = 18^\circ - 15^\circ$, the T. F.s of 3° can be found.

$$\text{Again}\quad \sin 9^\circ = \tfrac{1}{2}\{\sqrt{1+\sin 18^\circ} - \sqrt{1-\sin 18^\circ}\}$$

$$= \tfrac{1}{4}\{\sqrt{3+\sqrt{5}} - \sqrt{5-\sqrt{5}}\}$$

$$\text{Sim'ly}\quad \cos 9^\circ = \tfrac{1}{4}\{\sqrt{3+\sqrt{5}} + \sqrt{5-\sqrt{5}}\}$$

It now becomes clear that the T. F.s of numbers of other $\wedge$s, such as 27°, 42°, 63°, 81°, 87°, &c can be deduced from the foregoing.

Exercises

1. Prove that—

(1) $\sin^2 24^\circ - \sin^2 6^\circ = \dfrac{\sqrt{5}-1}{8}$

(2) $\tan 54^\circ = \sqrt{1+\tfrac{2}{5}\sqrt{5}}$

(3) $\sec 54^\circ = \sqrt{\dfrac{2(\sqrt{5}+1)}{\sqrt{5}}}$

(4) $\tan 9^\circ = \sqrt{5} + 1 - \sqrt{5 + 2\sqrt{5}}$

(5) $\sin 3^\circ = \frac{1}{8}\{(\sqrt{5}-1)\sqrt{2+\sqrt{3}} - \sqrt{(10 + 2\sqrt{5})(2 - \sqrt{3})}\}$

(6) $\cos 12^\circ = \frac{1}{8}\{\sqrt{5} - 1 + \sqrt{30 + 6\sqrt{5}}\}$

(7) $\cos 27^\circ = \frac{1}{4}\{\sqrt{5+\sqrt{5}} + \sqrt{3 - \sqrt{5}}\}$

(8) $\tan 27^\circ = \sqrt{5} - 1 - \sqrt{5 - 2\sqrt{5}}$

(9) $\sin 63^\circ = \dfrac{\sqrt{10 + 2\sqrt{5}} + \sqrt{5} - 1}{4\sqrt{2}}$

(10) $\sin 87^\circ = \frac{1}{8}\{(\sqrt{5}-1)\sqrt{2-\sqrt{3}} + \sqrt{(10 + 2\sqrt{5})(2 + \sqrt{3})}\}$

(11) $\tan \dfrac{\pi}{10} \tan \dfrac{3\pi}{10} = \dfrac{1}{\sqrt{5}}$

2. Find $\cos 42^\circ$ to five decimal places.

3. If $\tan\theta = \dfrac{1}{+\sqrt{3}}$, and $\tan\phi = \dfrac{1}{+\sqrt{15}}$, prove that

$$\sin(\theta + \phi) = \sin 60^\circ \cos 36^\circ$$

4. If $\left.\begin{aligned} &\cos 60^\circ = \sin 36^\circ \cos\alpha \\ \text{and}\quad &\cos 36^\circ = \sin 60^\circ \cos\beta \end{aligned}\right\}$

prove that $\tan\alpha + \tan\beta = 1$.

And, if also $\cos\gamma = \cos\alpha\cos\beta$, show that *one* value of $\alpha + \beta + \gamma$ is 90°.

5. Show that $\dfrac{\cos 27^\circ - \sin 27^\circ}{\cos 27^\circ + \sin 27^\circ} = \dfrac{\sqrt{25 - 10\sqrt{5}}}{5}$

NOTE—*Reduce the sinister to* $\tan 18^\circ$

6. If $\cos 3x = -\dfrac{3\sqrt{3}}{4\sqrt{2}}$, show that the three values of $\cos x$ are

$$\sqrt{\frac{3}{2}}\,.\sin\frac{\pi}{10},\quad \sqrt{\frac{3}{2}}\,.\sin\frac{\pi}{6},\ \text{and}\ -\sqrt{\frac{3}{2}}\,.\sin\frac{3\pi}{10}$$

Wolstenholme : 405.

§ **34.** In the last paragraph we found, by starting with the eq'n $\sin 2\alpha = \cos 3\alpha$, that $\sin 18°$ is a root of the eq'n

$$4x^2 + 2x - 1 = 0$$

If we had started with the equation $\sin 3\alpha = \cos 2\alpha$, we should have found sim'ly that $\sin 18°$ is a root of the eq'n

$$4x^3 - 2x^2 - 3x + 1 = 0$$

The process employed may be generalized.

For, when n is even, the expressions

$$\sin n\alpha/\cos\alpha \quad \text{and} \quad \cos n\alpha;$$

and when n is odd, the expressions

$$\sin n\alpha \quad \text{and} \quad \cos n\alpha/\cos\alpha$$

are each expressible in terms of powers of $\sin\alpha$ only. This has been seen to be the case for small values of n; and will hereafter (in Chapter XVIII) be proved to be true in all cases.

Now if $m\alpha + n\alpha = \frac{\pi}{2}$, where $\overline{m+n}$ is an odd integer, one of the two m and n must be odd.

∴, since we have the two eq'ns

$$\sin m\alpha = \cos n\alpha$$

$$\text{and} \quad \cos m\alpha = \sin n\alpha$$

each of these is reducible to terms involving powers of $\sin\alpha$ only.

∴ each of them will give an eq'n in x of which $\sin\alpha$ is a root.

So also if $m\alpha + n\alpha = \pi$

$$\text{then} \quad \sin m\alpha = \sin n\alpha$$

$$\text{and} \quad \cos m\alpha = -\cos n\alpha$$

Now, when m, n are both odd, $\sin m\alpha$, $\sin n\alpha$ are each expressible in terms of $\sin\alpha$.

And when m, n are both even, $\cos m\alpha$, $\cos n\alpha$ are each expressible in terms of $\sin\alpha$.

$\therefore$ in either of these cases we should get an eq'n in x of which $\sin\alpha$ is a root.

Sim'ly if $m\alpha + n\alpha = 2\pi$, &c.

Examples

If $\alpha = 10^\circ$, then $5\alpha = 50^\circ$ and $4\alpha = 40^\circ$

$$\therefore \quad \sin 4\alpha = \cos 5\alpha$$

$$\therefore \quad 4\sin\alpha\cos\alpha\cos 2\alpha = \cos^5\alpha - 10\cos^3\alpha\sin^2\alpha + 5\cos\alpha\sin^4\alpha$$

Divide by $\cos\alpha$ (which by hypothesis $\neq 0$) and put x for $\sin\alpha$: then

$$4x(1 - 2x^2) = (1 - x^2)^2 - 10x^2(1 - x^2) + 5x^4$$

$\therefore$ $\sin 10^\circ$ is a root of the eq'n

$$16x^4 + 8x^3 - 12x^2 - 4x + 1 = 0$$

Or, starting with $\sin 5\alpha = \cos 4\alpha$, we should get that

$$5\sin\alpha - 20\sin^3\alpha + 16\sin^5\alpha = 8\sin^4\alpha - 8\sin^2\alpha + 1$$

$\therefore$ $\sin 10^\circ$ is also a root of the eq'n

$$16x^5 - 8x^4 - 20x^3 + 8x^2 + 5x - 1 = 0$$

NOTE—*It appears above, and the same will readily be seen to be the case always, that when we start with the eq'n*

$$\sin(\textit{even multiple of } \alpha) = \cos(\textit{odd mult' } \alpha)$$

we get an eq'n one degree lower, than when we start with

$$\sin(\textit{odd mult' } \alpha) = \cos(\textit{even mult' } \alpha)$$

Exercises

1. Show that $\sin \pi/14$ is a root of the equation

$$8x^4 + 4x^3 - 8x^2 - 3x + 1 = 0;$$

Q. C. B. '85.

and also of the equation

$$8x^3 - 4x^2 - 4x + 1 = 0.$$

R. U. Schol': '88.

2. Find an equation of the fifth degree of which $\sin \pi/22$ is a root.

Q. C. B. '86.

3. Find an equation of the third degree of which $\cos 2\pi/7$ is a root.

4. Find an equation of the seventh degree of which $\sin 6^\circ$ is a root.

§ **35.** The values of the **T. F.**s of certain $\wedge$s which have now been found, will enable us to trace the changes in functions of the **T. F.**s more accurately than could be done before. In doing this it is convenient to denote any such function by $f(\theta)$ and then (as was stated in the Preliminary Chapter) $f(\alpha)$ denotes the value of $f(\theta)$ for the particular case when $\theta = \alpha$.

Example

To trace the changes in sign and magnitude of $\dfrac{\sin(\pi\cos\theta)}{\cos(\pi\sin\theta)}$, *as* θ *changes from* o *to* 2π.

Denote the function by $f(\theta)$

1°, let the generator of θ pass thro' the 1st quadrant

$$\text{Initially } f(0) = \frac{\sin\pi}{\cos 0} = \frac{0}{1} = 0$$

As θ changes from o to $\pi/6$,

$f(\theta)$ „ „ o „ ∞, and is pos' :

as θ „ „ $\pi/6$ „ $\pi/2$,

$f(\theta)$ „ „ ∞ „ o, and is neg'; a change of sign taking place as $f(\theta)$ goes thro' the value ∞; and another change when $f(\theta)$ passes thro' the value o, as the generator enters the 2nd quadrant.

2°, let the generator of θ pass thro' the 2nd quadrant.

$$\text{Initially } f(\pi/2) = \frac{\sin 0}{\cos\pi} = \frac{0}{1} = 0$$

As θ changes from $\pi/2$ to $5\pi/6$,

$f(\theta)$ „ „ o „ ∞, and is pos' :

as θ „ „ $5\pi/6$ „ π,

$f(\theta)$ „ „ ∞ „ o, a change of sign to neg' taking place in passing thro' the value ∞; and another change to pos' in passing thro' the value o, as the generator of θ enters the 3rd quadrant.

The Law of change is now obvious : $f(\theta)$ changes from o to ∞, as θ goes from π to $7\pi/6$, rem'g pos', and then becomes neg', rem'g so until $f(3\pi/2) = 0$; and so on.

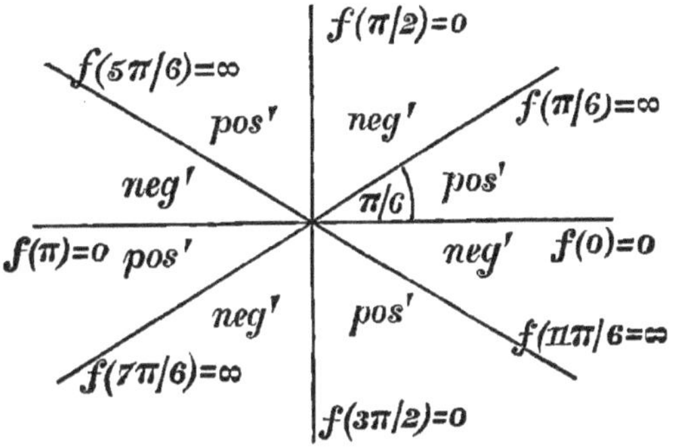

The successive changes of $f(\theta)$, as the generator of θ passes thro' the four quadrants, are shown in the accompanying fig'.

The graph of $f(\theta)$ is given below; where OX, OY are rect'r axes; and A_1, A_2, &c B_1, B_2, &c are taken so that

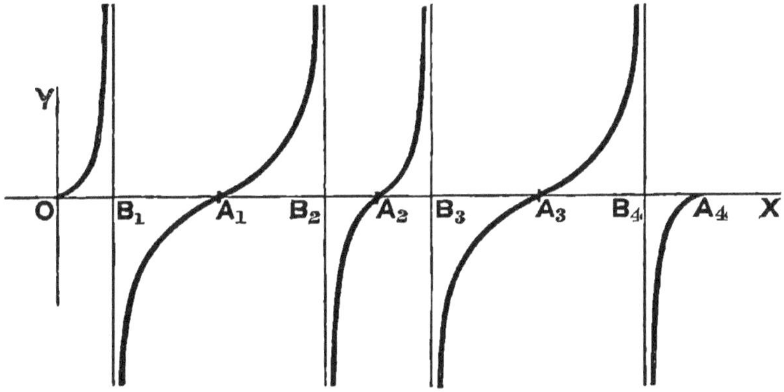

$OA_1 = A_1A_2 = A_2A_3 = A_3A_4$, and each represents a r't $\wedge$.

$OB_1 = \frac{1}{3}A_1A_2 = B_2A_2 = A_2B_3 = B_4A_4$

$B_1A_1 = \frac{2}{3}A_1A_2 = A_1B_2 = B_3A_3 = A_3B_4$

The ordinates of p'ts on the curve, corresponding to the abscissæ for $\wedge^s$ α, β, give the relative values of $f(\alpha)$, $f(\beta)$.

Exercises

Trace the changes in sign and magnitude of $f(\theta)$, as θ changes from o to 2π, and draw its Graph, in each of the following cases—

1. $f(\theta) = \sin\theta - \cos\theta$

2. $f(\theta) = \sin(\pi \cos\theta)$

3. $f(\theta) = \tan\left\{\frac{\pi}{2}(\sin\theta + \cos\theta)\right\}$

4. $f(\theta) = \sin\theta + \sin 2\theta$

5. $f(\theta) = \theta \sin\theta$

6. $f(\theta) = \dfrac{\sin 3\theta}{\cos\theta}$

7. $f(\theta) = \sin\theta \,.\, \sin 2\theta$

8. $f(\theta) = \tan\theta + \tan 2\theta$

9. $f(\theta) = \sin\theta + \cos 2\theta$

10. $f(\theta) = \dfrac{2\sin\theta - \sin 2\theta}{2\sin\theta + \sin 2\theta}$

'*Previous*,' *Camb*: '79.

11. $f(\theta) = \dfrac{\sin\theta + 2\sin\frac{\theta}{2}}{\sin\theta - 2\sin\frac{\theta}{2}}$

§ **36.** By using the numerical values of T. F.s of special ∧s, we may—

1°, modify some of the formulæ already proved; and,

2°, prove important new formulæ.

Examples

1. Since $\tan 45° = 1$, we have

$$\tan(45° \pm \alpha) = \frac{1 \pm \tan\alpha}{1 \mp \tan\alpha}$$

From this we get that

$$\tan(45° + \alpha) - \tan(45° - \alpha)$$

$$= \frac{1 + \tan\alpha}{1 - \tan\alpha} - \frac{1 - \tan\alpha}{1 + \tan\alpha}$$

$$= \frac{4\tan\alpha}{1 - \tan^2\alpha}$$

$$= 2\tan 2\alpha$$

2. Since $\sin(36° + \alpha) - \sin(36° - \alpha) = 2\cos 36° \sin\alpha$

$$= \frac{+\sqrt{5}+1}{2}\sin\alpha$$

and $\sin(72° + \alpha) - \sin(72° - \alpha) = 2\cos 72° \sin\alpha$

$$= \frac{+\sqrt{5}-1}{2}\sin\alpha$$

∴, subtracting and rearranging, we get

$$\sin\alpha + \sin(72° + \alpha) - \sin(72° - \alpha) = \sin(36° + \alpha) - \sin(36° - \alpha)$$

(*Euler's formula*)

Again $\cos(36° + \alpha) + \cos(36° - \alpha) = 2\cos 36° \cos\alpha$

$$= \frac{+\sqrt{5}+1}{2}\cos\alpha$$

and $\cos(72° + \alpha) + \cos(72° - \alpha) = 2\cos 72° \cos\alpha$

$$= \frac{+\sqrt{5}-1}{2}\cos\alpha$$

Whence

$$\cos\alpha + \cos(72° + \alpha) + \cos(72° - \alpha) = \cos(36° + \alpha) + \cos(36° - \alpha)$$

NOTE—These two results are called *formulæ of verification*; because they are specially useful as a means of verifying the accuracy of a table of numerical values of the sines and cosines of ∧s. The verification will be effected by giving any value to α, and then substituting, from the table, the numerical values of the corresponding sines and cosines.

Notice that either formula is deducible from the other by writing for α its complement.

3. *Two circles touch each other, and each touches each arm of an angle* ϕ : *if* R *is the radius of the larger, and* r *of the smaller, prove that*

$$R = r\tan^2\frac{\phi + \pi}{4}$$

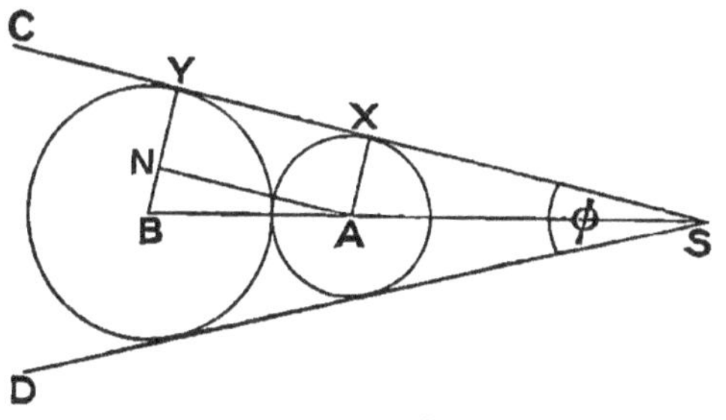

Let $\widehat{CSD}$ be ϕ; A, B the centres of the ⊙s whose respective radii are r, R; and let SC touch ⊙ A in X, and ⊙ B in Y.

Drop AN ⊥ on BY.

$$\text{Then}\quad \sin\frac{\phi}{2} = \sin BAN = \frac{BN}{BA} = \frac{R - r}{R + r}$$

$$\therefore \quad \frac{R}{r} = \frac{1 + \sin\frac{\phi}{2}}{1 - \sin\frac{\phi}{2}} = \left(\frac{\cos\frac{\phi}{4} + \sin\frac{\phi}{4}}{\cos\frac{\phi}{4} - \sin\frac{\phi}{4}}\right)^2$$

$$= \left(\frac{1 + \tan\frac{\phi}{4}}{1 - \tan\frac{\phi}{4}}\right)^2$$

$$\therefore R = r \tan^2 \frac{\phi + \pi}{4}$$

Exercises

Prove that—

1. $\tan \alpha + \sec \alpha \equiv \tan\left(45^\circ + \frac{\alpha}{2}\right)$

2. $\frac{1 \pm \sin \alpha}{\cos \alpha} \equiv \tan\left(45^\circ \pm \frac{\alpha}{2}\right)$

3. $\tan(45^\circ + \alpha) + \tan(45^\circ - \alpha) \equiv 2 \sec 2\alpha$

4. $\sin(60^\circ + \alpha) - \sin(60^\circ - \alpha) \equiv \sin \alpha$

5. $\sin(45^\circ + \alpha) - \sin(45^\circ - \alpha) \equiv \sqrt{2} \,.\, \sin \alpha$

6. $\sec^2(\alpha + 45^\circ) - \sec^2(\alpha - 45^\circ) \equiv 4 \tan 2\alpha \sec 2\alpha$

7. $\sin 2\alpha \equiv \frac{1 - \cot^2(45^\circ + \alpha)}{1 + \cot^2(45^\circ + \alpha)}$

8. $\frac{\tan \alpha + \sec \alpha}{\cot \alpha + \operatorname{cosec} \alpha} \equiv \tan\frac{\alpha}{2} \tan\left(45^\circ + \frac{\alpha}{2}\right)$

9. $\frac{\sin 3\alpha + \cos 3\alpha}{\sin 3\alpha - \cos 3\alpha} \equiv \frac{2 \sin 2\alpha + 1}{2 \sin 2\alpha - 1} \tan(45^\circ - \alpha)$

10. $\sec \theta \equiv \frac{1}{2}\left[\tan\left(\frac{\pi}{4} - \frac{\theta}{2}\right) + \cot\left(\frac{\pi}{4} - \frac{\theta}{2}\right)\right]$

11. $\tan(30^\circ + \alpha) \tan(30^\circ - \alpha) \equiv \frac{2 \cos 2\alpha - 1}{2 \cos 2\alpha + 1}$

12. $\sin 80^\circ = \sin 40^\circ + \sin 20^\circ$

13. $\sin 70^\circ = \sin 10^\circ + \sin 50^\circ$

14. $\dfrac{\sin 60^\circ - \sin 30^\circ}{\sin 60^\circ + \sin 30^\circ} = \dfrac{\tan 60^\circ - \tan 45^\circ}{\tan 60^\circ + \tan 45^\circ}$

15. $\sin^2 10^\circ + \cos^2 20^\circ - \sin 10^\circ \cos 20^\circ$

$= \frac{3}{4} = \sin^2 10^\circ + \cos^2 40^\circ + \sin 10^\circ \cos 40^\circ$

Math' Tri': '81.

16. $\tan 1^\circ = \cdot 01745$, assuming that

$\cos 29^\circ = \cdot 87462$, and $\cos 31^\circ = \cdot 85717$

Q. C. B.'73.

17. $\cos 20^\circ \cos 40^\circ \cos 80^\circ = \left(\frac{1}{2}\right)^3$

Q. C. B.'65.

18. If $x + y = 90^\circ$, and $\tan x = 4 \tan y$, then $\tan \frac{x}{2} = 2 \sin 18^\circ$

E. T. xlix.

19. $\cos^4 x + \cos^4 \left(\frac{2\pi}{3} - x\right) + \cos^4 \left(\frac{4\pi}{3} - x\right) = \frac{9}{8}$

Q. C. B.'82.

20. $\cos^4 \frac{\pi}{8} + \cos^4 \frac{3\pi}{8} + \cos^4 \frac{5\pi}{8} + \cos^4 \frac{7\pi}{8} = \frac{3}{2}$

Mag' Camb': '50.

21. $\cos^4 \frac{\pi}{9} + \cos^4 \frac{2\pi}{9} + \cos^4 \frac{3\pi}{9} + \cos^4 \frac{4\pi}{9} = \frac{19}{16}$

Math' Tri': '84.

22. $\sqrt{\text{vers}\,\alpha}\left\{\sin\left(45^\circ - \frac{\alpha}{2}\right) + \cos\left(45^\circ - \frac{\alpha}{2}\right)\right\} = \sin \alpha$

23. $\sin(90^\circ - \alpha) + \sin(18^\circ + \alpha) + \sin(18^\circ - \alpha)$

$\equiv \sin(54^\circ + \alpha) + \sin(54^\circ - \alpha)$

(*Legendre's formula*)

24. $\cos(90^\circ - \alpha) + \cos(18^\circ - \alpha) - \cos(18^\circ + \alpha)$

$\equiv \cos(54^\circ - \alpha) - \cos(54^\circ + \alpha)$

NOTE— *The last two are formulæ of verification*

25. $\sin(45^\circ - \alpha) \quad \sqrt{(1 - \sin 2\alpha)/2}$; and hence that

$$2 \cos\left(60^\circ \pm \frac{30^\circ}{2^n}\right) = \sqrt{2 - \sqrt{2 - \sqrt{2 - \&c - \sqrt{2}}}}$$

where $\sqrt{\ }$ occurs n times; and the + or − sign is taken as n is even or odd.

26. $2 \cos \frac{45^\circ}{2^n} = \sqrt{2 + \sqrt{2 + \&c \text{ to n terms}}}$

27. $\sin 2^\circ \sin 14^\circ \sin 22^\circ \sin 26^\circ \sin 34^\circ \sin 38^\circ \sin 46^\circ$
$\times \sin 58^\circ \sin 62^\circ \sin 74^\circ \sin 82^\circ \sin 86^\circ = 2^{-12}$

28. If $\tan \frac{\beta}{2} = \frac{1 - \tan^3 \alpha}{1 + \tan^3 \alpha}$, then

$$2 \cot 2\alpha = \cot^{\frac{1}{3}} \left(\frac{\pi}{4} - \frac{\beta}{2}\right) - \tan^{\frac{1}{3}} \left(\frac{\pi}{4} - \frac{\beta}{2}\right)$$

29. If $\tan \phi = \frac{\sin \alpha \cos \beta}{\cos \alpha + \sin \beta}$, then

$$\tan \frac{\phi}{2} = \tan \frac{\alpha}{2} \tan \left(\frac{\pi}{4} - \frac{\beta}{2}\right), \text{ or } = -\cot \frac{\alpha}{2} \cot \left(\frac{\pi}{4} - \frac{\beta}{2}\right)$$

NOTE—*Put* x *for* $\tan \frac{\phi}{2}$, *and solve the quadratic*

$$2x \big/ (1 - x^2) = \sin \alpha \cos \beta \big/ (\cos \alpha + \sin \beta)$$

30. $\frac{\Sigma \sin \alpha - 1}{\Sigma \cos \alpha} = \Pi \left(\frac{1 - \tan \frac{\alpha}{2}}{1 + \tan \frac{\alpha}{2}} \right)$

where the symbols apply to α, β, γ; and $\alpha + \beta + \gamma = \pi/2$

Math' Tri': '66.

31. $\pm \sqrt{\frac{1}{2} \pm \frac{1}{2} \sqrt{1 - \sin^2 \left(\frac{\pi}{2} + 2\alpha\right)}}$ equally represents

$$\sin \left(\frac{\pi}{4} + \alpha\right) \text{ and } \cos \left(\frac{\pi}{4} + \alpha\right);$$

and remove the ambiguity of signs when α is between $\pi/4$ and $\pi/2$

32. $\cos \frac{\pi}{15} \cos \frac{2\pi}{15} \cos \frac{3\pi}{15} \cos \frac{4\pi}{15} \cos \frac{5\pi}{15} \cos \frac{6\pi}{15} \cos \frac{7\pi}{15} = \left(\frac{1}{2}\right)^7$

Math' Tri': '66.

33. $\left(\sec^2 \frac{\pi}{7} + \sec^2 \frac{2\pi}{7} + \sec^2 \frac{3\pi}{7}\right)$
$\times \left(\operatorname{cosec}^2 \frac{\pi}{7} + \operatorname{cosec}^2 \frac{2\pi}{7} + \operatorname{cosec}^2 \frac{3\pi}{7}\right) = 192$

Ox' Jun' Math' Schol': '87.

34. If $\tan\left(\frac{\pi}{4}+\frac{\theta}{2}\right) = \tan^5\left(\frac{\pi}{4}+\frac{\phi}{2}\right)$, then

$$\sin\theta = 5\sin\phi\left[\frac{\left(1+\sin^2\phi\cot^2\frac{\pi}{5}\right)\left(1+\sin^2\phi\cot^2\frac{2\pi}{5}\right)}{\left(1+\sin^2\phi\tan^2\frac{\pi}{5}\right)\left(1+\sin^2\phi\tan^2\frac{2\pi}{5}\right)}\right]$$

T. C. D. Schol' '72, and Math' Tri': '85.

NOTE—*Square given equality; and on both sides use the formula*

$$\tan^2\left(\frac{\pi}{4}+\frac{\theta}{2}\right) = (1+\sin\theta)\,/\,(1-\sin\theta)$$

35. If r is the radius of the in-circle of a triangle ABC; and ρ_1, ρ_2, ρ_3 the radii of circles touching this circle and the arms of the angles A, B, C respectively; then

$$\Sigma\sqrt{\rho_1\rho_2} = r$$

NOTE—*Use the results of Example* **3** *of this* §, *and Exercise* **1.** (3) *on* p. 78.

CHAPTER VI

General Expressions for all Angles satisfying a given value of a Trigonometrical Function

WE have seen (in Chapter III) that more than one $\wedge$ may give the same value to a T. F.: e.g. θ and $\pi - \theta$ will give the same value for the sine. We proceed to enquire whether there are still more $\wedge^s$ for which this is true. It will be found that for each T. F. there is an infinite number of $\wedge^s$ which give it the same value; and that, in each case, *all* these $\wedge^s$ can be included in a simple formula.

§ **37. Problem**—*If* θ *is any known angle, it is required to find, in terms of* θ, *all angles, positive and negative, whose* sines *have the same value as* $\sin\theta$.

Since a sine is determined—

1°, as to *magnitude* by the species of the auxil′ Δ;

and 2°, as to *sign* " $\perp$ " " " ;

$\therefore$, in order that the sines of two $\wedge^s$ may be identically equal, their corresponding auxil′ Δ^s must—

1°, be sim′r, so that their $\perp^s$ are homologous;

and, 2°, have their $\perp^s$ both pos′ or both neg′.

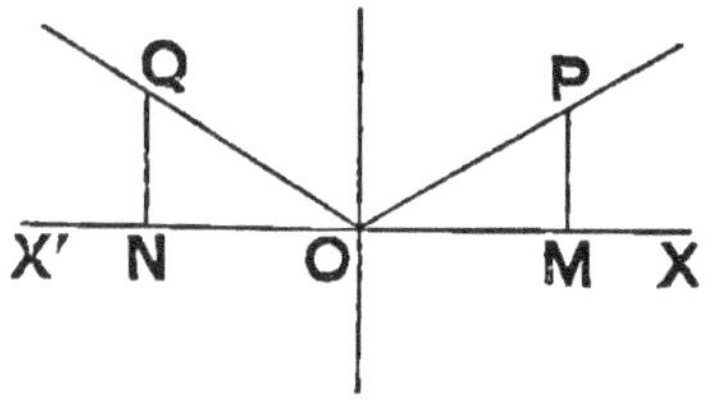

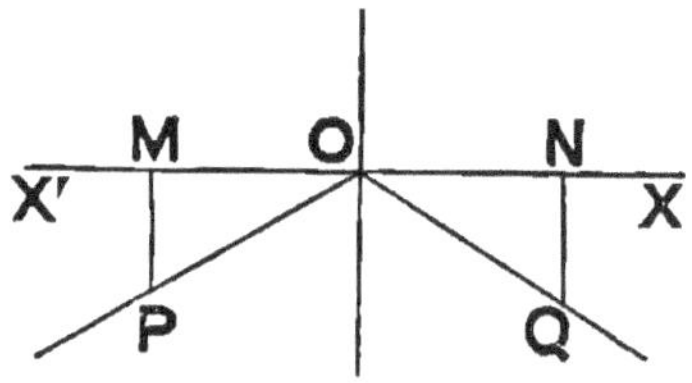

Thus; if XOX′ is the initial line; OP, OQ two lines which by revolving either way round, from coincidence with OX, generate the ∠s XOP, XOQ; PM, QN ⊥s on XOX′; then the sines of XOP, XOQ will be identically equal, when, and only when—

1°, the acute ∠s POM, QON are equal;

and, 2°, △s POM, QON are on the same side of XOX′.

Now, if (as in the fig′s) these cond′ns are satisfied, when the generator gets into either of the positions OP or OQ by revolving thro′ θ—where θ is any ∠, pos′ or neg′—it will get into position OQ or OP, respectively, by revolving thro′ $\pi - \theta$.

Also the position of the generator is not altered by supposing it to revolve thro′ $2n\pi$, where n is any pos′ or neg′ integer.

∴ *all* ∠s for which the generator, revolving as above, has the positions OP or OQ are included in

$$2n\pi + \theta \quad \text{and} \quad 2n\pi + \pi - \theta$$

i.e. in $2n\pi + \theta$ and $(2n+1)\pi - \theta$

i.e. in ± an even mult′ of $\pi + \theta$

and ± an odd mult′ of $\pi - \theta$

Hence, recollecting that an even power of (-1) is $+1$

and " *odd* " " -1

the form $n\pi + (-1)^n\theta$, *where* n *is any integer, pos′, neg′, or zero, includes all the* ∠s *whose* sines *have the same value as* $\sin\theta$.

This result may be expressed either by the formula

$$\sin\theta = \sin\{n\pi + (-1)^n\theta\}$$

where the ∠s are measured by the radian;

$$\text{or} \quad \sin\alpha = \sin\{n.180 + (-1)^n\alpha\}$$

where the ∠s are measured by the degree; and, in each case n may be any integer, pos′, neg′, or zero.

All the foregoing holds verbatim if for the word sine is substituted cosecant. Hence, or by taking the reciprocals of the formulæ for the sine, we should get

$$\text{cosec}\,\theta = \text{cosec}\,\{n\pi + (-1)^n \theta\}$$

$$\text{or} \quad \text{cosec}\,\alpha = \text{cosec}\,\{n \,.\, 180 + (-1)^n \alpha\}$$

Examples

$$\tfrac{1}{2} = \sin\,\{n\pi + (-1)^n \pi/6\}$$

$$1 = \text{cosec}\,\{n \,.\, 180^\circ + (-1)^n\, 90^\circ\}$$

§ **38.** **Problem**—*If θ is any known angle, it is required to find, in terms of θ, all angles, positive and negative, whose* cosines *have the same value as* $\cos\theta$.

Since a cosine is determined—

1°, as to *magnitude* by the species of the auxil′ △;

and, 2°, as to *sign* „ base „ „ „;

∴, in order that the cosines of two ∧s may be identically equal, the corresponding auxil′ △s must—

1°, be sim′r, so that their bases are homologous;

and, 2°, have their bases both pos′ or both neg′.

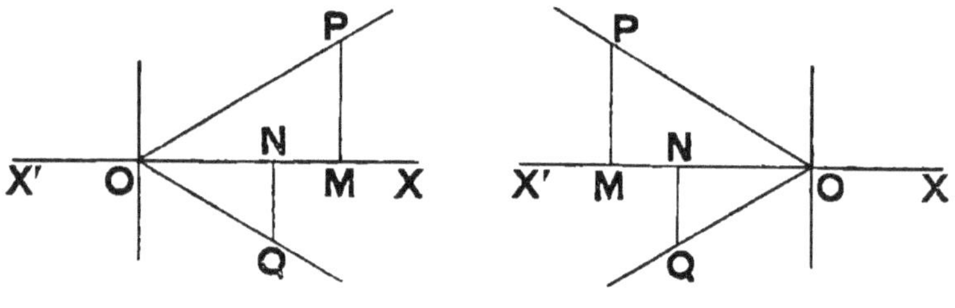

Thus; if XOX′ is the initial line; OP, OQ two lines which, by revolving either way round from coincidence with OX, generate the ∧s XOP, XOQ; PM, QN ⊥s on XOX′; then the cosines of XOP, XOQ will be identically equal, when, and only when—

1°, the acute ∧s POM, QON are equal;

and, 2°, OM, ON are both along OX, or both along OX′.

Now, if (as in the fig′s) these conditions are satisfied, when the generator gets into either of the positions OP or OQ, by revolving thro′ θ—where θ is any ∧, pos′ or neg′—it will get into position OQ or OP, respectively, by revolving thro′ $2\pi - \theta$.

Also the position of the generator is not altered by supposing it to revolve thro′ $2n\pi$, where n is any pos′ or neg′ integer.

∴ *all* ∧s for which the generator, revolving as above, has the positions OP or OQ are included in

$$2n\pi + \theta \quad \text{and} \quad 2n\pi + 2\pi - \theta$$

$$\text{i.e. in} \quad 2n\pi + \theta \quad \text{and} \quad 2(n+1)\pi - \theta$$

$$\text{i.e. in} \quad \pm \text{ an even mult}' \text{ of } \pi \pm \theta$$

Hence the form $2n\pi \pm \theta$, *where* n *is any integer, pos′, neg′, or zero, includes all the* ∧s *whose* cosines *have the same value as* $\cos\theta$.

This result may be expressed either by the formula

$$\cos\theta = \cos(2n\pi \pm \theta)$$

where the ∧s are measured by the radian;

$$\text{or} \quad \cos\alpha = \cos(2n \cdot 180 \pm \alpha)$$

where the ∧s are measured by the degree; and, in each case, n may be any integer, pos′, neg′, or zero.

All the foregoing holds verbatim if for the word cosine is substituted secant. Hence, or by taking the reciprocals of the formulæ for the cosine, we should get that

$$\sec\theta = \sec(2n\pi \pm \theta)$$

$$\text{or} \quad \sec\alpha = \sec(2.n.180 \pm \alpha)$$

§ **39. Problem**—*If θ is any known angle, it is required to find, in terms of θ, all angles, positive and negative, whose* tangents *have the same value as* tan θ.

Since a tangent is determined—

1°, as to *magnitude* by the species of the auxil′ Δ;

and, 2°, as to *sign* „ ⊥ and base „ „ „ ;

∴, in order that the tangents of two ∧s may be identically equal, their corresponding auxil′ Δs must—

1°, be sim′r so that their ⊥s are homologous;

and, 2°, have the ⊥ and base of either Δ of the same or different signs, according as the ⊥ and base of the other Δ are of the same or different signs.

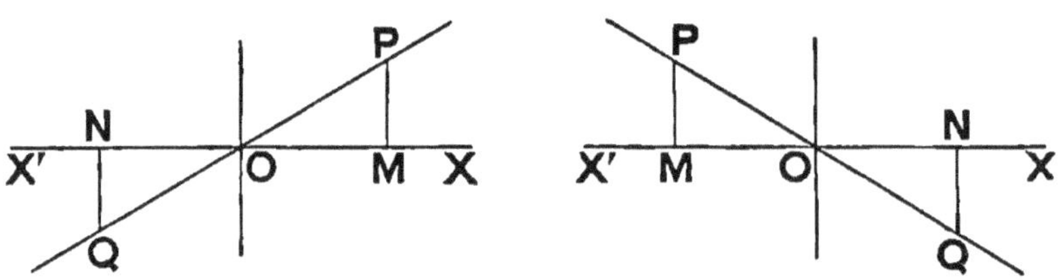

Thus; if XOX′ is the initial line; OP, OQ two lines which, by revolving either way round from coincidence with OX, generate the ∧s XOP, XOQ; PM, QN ⊥s on XOX′; then the tangents of XOP, XOQ will be identically equal when, and only when—

1°, the acute ∧s POM, QON are equal;

and, 2°, Δs POM, QON are on opposite sides of XOX′, and have OM, ON in opposite direc′s.

Now, if (as in the fig′s) these conditions are satisfied, when the generator gets into either of the positions OP or OQ, by revolving thro′ θ—where θ is any ∧, pos′ or neg′—it will get into position OQ or OP, respectively, by revolving thro′ $\pi + \theta$.

Also the position of the generator is not altered by supposing it to revolve thro′ $2\,n\,\pi$, where n is any pos′ or neg′ integer.

∴ *all* ∧s for which the generator, revolving as above, has the positions OP or OQ are included in

$$2n\pi + \theta \quad \text{and} \quad 2n\pi + \pi + \theta$$

$$\text{i.e. in} \quad 2n\pi + \theta \quad \text{and} \quad (2n+1)\pi + \theta$$

$$\text{i.e. in} \quad \pm \text{ any mult}' \text{ of } \pi + \theta$$

Hence the form $n\pi + \theta$, *where* n *is any integer pos′, neg′, or zero, includes all the* ∧s *whose* tangents *have the same value as* $\tan\theta$.

This result may be expressed either by the formula

$$\tan\theta = \tan(n\pi + \theta)$$

where the ∧s are measured by the radian;

$$\text{or} \quad \tan\alpha = \tan(n \,.\, 180 + \alpha)$$

where the ∧s are measured by the degree; and, in each case, n may be any integer, pos′, neg′, or zero.

All the foregoing holds verbatim if for the word tangent is substituted cotangent. Hence, or by taking the reciprocals of the formulæ for the tangent, we should get that

$$\cot\theta = \cot(n\pi + \theta)$$

$$\text{or} \quad \cot\alpha = \cot(n \,.\, 180 + \alpha)$$

§ **40.** The chief use of these results is to enable us when, in solving a trigonometrical equation, we arrive at a numerical value for one of the T. F.s of the ∧ which has to be found, to write down an expression containing all ∧s which satisfy the original equation.

Thus, if $f(\theta)$ is a T. F. of θ,

and if from a trig′l eq′n we deduce that

$$f(\theta) = k$$

then, if α is any one particular $\wedge$ such that

$$f(\alpha) = \mathrm{k}$$

all values of θ are included

in	$n\pi + (-1)^n\alpha$,	if $f(\ )$ stands for	sine or cosecant
or in	$2n\pi \pm \alpha$,	„ „ „	cosine „ secant
or in	$n\pi + \alpha$,	„ „ „	tangent „ cotangent

Note—In particular cases these results can be simplified. See Example 1.

Hence conditional eq'ns in trigonometry differ from those in algebra in that, whereas the latter *can* have only a limited number of roots, the former *must* have an infinite number of roots; and the formulæ just found enable us when we know a root of a conditional trig'l eq'n to write down a result containing the infinite number of corresponding roots.

Note (1)—The Student should carefully notice the difference between an *identical equation* (usually called an *identity*) which is true for all values of the $\wedge$ (or $\wedge^s$) involved, and a *conditional equation* (usually called an *equation*) which is true only for certain particular values of the $\wedge$ (or $\wedge^s$) involved.

Note (2)—The *forms* of solution of a trig'l eq'n may differ according as we express the given eq'n in terms of one or other of the T. F.s. This difference however is only apparent; and, by properly modifying n, could be shown to be not essential but formal merely. See Example 4.

Note (3)—It will probably at once occur to the Student that when the result $f(\theta) = \mathrm{k}$ is arrived at, there may be no known special value of θ satisfying that eq'n. In such cases the general solution is only expressible by means of what are called *inverse trigonometrical functions*: these will be treated of in the next Chapter. Meanwhile we note that the number of $\wedge^s$ whose T. F.s are known accurately is infinitely small compared with the number whose T. F.s are unknown, and can only be calculated approximately. This is at once obvious from the fact that the T. F.s are ratios; and that among ratios incommensurability* is the *rule* and commensurability the *exception*. It is clear

* See *Euclid Revised*, pp. 230–235.

∴ that among trig'l eq'ns those which require the inverse trigonometrical functions to express their roots will be the rule, and those which do not the exception. We ∴ here only give a few Exercises in the solution of trig'l eq'ns; and postpone a more complete set until Chapter VIII.

Examples

1. *To show that, if* $\sin\theta = 1$, *then all values of* θ *are included in* $2m\pi + \frac{\pi}{2}$; *where* m *is any integer, including zero.*

$$\text{Since} \quad \sin\theta = 1 = \sin\frac{\pi}{2}$$

$$\therefore \quad \theta = n\pi + (-1)^n \frac{\pi}{2}$$

1°, if n is even (say 2 m) then $\theta = 2m\pi + \frac{\pi}{2}$

2°, if n is odd (say 2 m + 1) then $\theta = 2m\pi + \pi - \frac{\pi}{2} = 2m\pi + \frac{\pi}{2}$

∴ $\theta = 2m\pi + \frac{\pi}{2}$, includes all values of θ satisfying the eq'n $\sin\theta = 1$

2. *To find all the values of* θ *which satisfy the equation* $\sin^2\theta = \sin^2\alpha$; *where* α *is a known angle.*

$$\text{We have} \quad \sin\theta = \pm\sin\alpha = \sin\alpha \text{ or } \sin(-\alpha)$$

$$\therefore \quad \theta = n\pi + (-1)^n\alpha$$

$$\text{or} \quad \theta = n\pi - (-1)^n\alpha$$

These alternatives become $\left.\begin{array}{l} n\pi + \alpha \\ \text{or } n\pi - \alpha \end{array}\right\}$ when n is even

but $\left.\begin{array}{l} n\pi - \alpha \\ \text{or } n\pi + \alpha \end{array}\right\}$ when n is odd

∴ *all* solutions are included in $n\pi \pm \alpha$

NOTE—$\theta = n\pi \pm \alpha$ *also solves each of the eq'ns* $\cos^2\theta = \cos^2\alpha$, and $\tan^2\theta = \tan^2\alpha$. *The geometrical interpretation in each case is that only the species of the auxiliary* Δ *is involved.*

3. *To show that, if* $\sin\theta = \sin\alpha$, *and* $\cos\theta = \cos\alpha$, *simultaneously; where* α *is a known angle; then all values of* θ *are included in*

$$2n\pi + \alpha$$

$1°$, if $\sin\theta = \sin\alpha$,

then $\theta = n\pi + (-1)^n\alpha$

$2°$, if $\cos\theta = \cos\alpha$,

then $\theta = 2n\pi \pm \alpha$

From the 2nd of these we see that the multiple of π taken must be an *even* mult′; and then, from the 1st, it follows that the sign before α is pos′.

$\therefore\ \theta = 2n\pi + \alpha$ includes all values of θ

NOTE—*The geometrical interpretation of the result is that only one position of the generator will satisfy the cond′ns.*

4. *To find all the solutions of the equation*

$$\cos\theta + \sin\theta = \frac{1}{+\sqrt{2}}$$

This will afford an example of apparently different results issuing from different modes of treatment.

For, since $\frac{1}{\sqrt{2}}\cos\theta + \frac{1}{\sqrt{2}}\sin\theta = \frac{1}{2}$,

the eq′n may be written either as

$$\sin\left(\theta + \frac{\pi}{4}\right) = \sin\frac{\pi}{6}$$

or as $$\cos\left(\theta - \frac{\pi}{4}\right) = \cos\frac{\pi}{3}$$

the solutions are, either

$$\theta + \frac{\pi}{4} = m\pi + (-1)^m\frac{\pi}{6}$$

or $$\theta - \frac{\pi}{4} = 2n\pi \pm \frac{\pi}{3}$$

Now if we give m, and n all pos′ and neg′ integral values in succession, we shall find that the solutions are really the same in each case, viz $\frac{7\pi}{12}$ and $-\frac{\pi}{12}$, together with all $\wedge^s$ got by adding $\pm$ mult′s of 2π to these.

Note—It will be instructive to notice that, by a different method of solution, we shall introduce extraneous roots, which are not solutions of

$$\cos\theta + \sin\theta = \frac{1}{+\sqrt{2}}.$$

For, by squaring, we get $\sin 2\theta = -\frac{1}{2} = \sin(\pi + \pi/6)$

$$\therefore\quad 2\theta = r\pi + (-1)^r(\pi + \pi/6)$$

Whence, by giving r different values, we get as before, the $\wedge^s$ $7\pi/12$ and $-\pi/12$, and *also* the $\wedge^s$ $11\pi/12$ and $13\pi/12$.

Now neither of these latter $\wedge^s$ is a root of

$$\cos\theta + \sin\theta = \frac{1}{+\sqrt{2}};$$

but they are roots of the cognate eq'n

$$\cos\theta + \sin\theta = \frac{1}{-\sqrt{2}}$$

The extraneous solutions were introduced when we squared the given eq'n; for this is an *irreversible step*, and, as we have seen in * Algebra, any irreversible step in the solution of an eq'n will introduce extraneous roots.

5. *To find all the values of θ which satisfy the equation*

$$4\cos\theta - 3\sec\theta = 2\tan\theta$$

Woolwich: '85.

Multiplying both sides by $\cos\theta$, we get

$$4\cos^2\theta - 3 = 2\sin\theta$$

Put x for $\sin\theta$, then $4x^2 + 2x - 1 = 0$

$$\therefore\quad 4x + 1 = \pm\sqrt{5}$$

$$\therefore\quad x = \frac{\sqrt{5}-1}{4} \text{ or } -\frac{\sqrt{5}+1}{4}$$

$$\therefore\quad \sin\theta = \sin\pi/10 \text{ or } \sin(-3\pi/10)$$

$$\therefore\quad \theta = n\pi + (-1)^n\pi/10$$

$$\text{or}\quad = n\pi - (-1)^n 3\pi/10$$

* See *Chrystal's Algebra*; 1st ed', pp. 285, 291.

6. *If* $\tan(\pi\cot\theta) = \cot(\pi\tan\theta)$, *show that*

$$\tan\theta = \tfrac{1}{4}\{2n + 1 \pm \sqrt{4n^2 + 4n - 15}\}$$

where n *is any integer not between* 2 *and* − 3.

The hypothesis gives us that

$$\tan(\pi\cot\theta) = \tan\left(\frac{\pi}{2} - \pi\tan\theta\right)$$

$$\therefore\quad \pi\cot\theta = n\pi + \left(\frac{\pi}{2} - n\tan\theta\right)$$

$$\therefore\quad 2\tan^2\theta - (2n + 1)\tan\theta + 2 = 0$$

$$\therefore\quad 4\tan\theta - (2n + 1) = \pm\sqrt{4n^2 + 4n - 15}$$

$$= \pm\sqrt{(2n + 5)(2n - 3)}$$

Unless n, if pos′, $> \frac{3}{2}$

or, if neg′, $< -\frac{5}{2}$

the expression under the radical will be neg′, and ∴ the value of $\tan\theta$ imaginary

$$\therefore\quad \tan\theta = \tfrac{1}{4}\{2n + 1 \pm \sqrt{4n^2 + 4n - 15}\}$$

where n may be any integer, *excepting* 1, 0, − 1, − 2.

Exercises

1. Show that, if $\cos\theta = 0$, then all values of θ are included in $m\pi + \pi/2$; where m is any integer, zero included.

2. Show that, if $\cos\alpha = -1$, then all values of α are included in $m \,.\, 180°$; where m is any *odd* number.

3. If $\tan\alpha = 1$, show that $\alpha = (4n + 1)\,45°$.

4. Find all values of θ which satisfy the equation $\tan^2\theta = \tan^2 \pi/4$.

5. If $\cos\theta = \cos\alpha$, and $\tan\theta = \tan\alpha$, *simultaneously* (where α is a known angle) find an expression containing all values of θ.

6. If $\sec\theta = \sec\alpha$, and $\operatorname{cosec}\theta = \operatorname{cosec}\alpha$, *simultaneously* (where α is a known angle) find an expression containing all values of θ.

7. Are there any angles which have both the same cosine and the same cosecant as a known angle α?

8. Find all the values of θ satisfying each of the following equations—

(1) $\sin^2\theta = \frac{3}{2}\cos\theta$

(2) $\sec^2\theta - \frac{5}{2}\sec\theta + 1 = 0$

(3) $6\cot^2\theta = 1 + 4\cos^2\theta$

(4) $\sin 3\theta + \cos 3\theta = \dfrac{1}{\sqrt{2}}$

(5) $\sin\theta + \sin 2\theta + \sin 3\theta = 0$

(6) $\cos n\theta + \cos(n-2)\theta = \cos\theta$

(7) $2\cot 2\theta - \tan 2\theta = 3\cot 3\theta$

(8) $8\cot\theta = \sec^2\dfrac{\theta}{2} + \operatorname{cosec}^2\dfrac{\theta}{2}$

(9) $\sec^2\theta + \operatorname{cosec}^2\theta = 3\sec^4\theta$

(10) $\operatorname{cosec} 3\theta + \operatorname{cosec} 2\theta = \sin\theta \operatorname{cosec} 2\theta \operatorname{cosec} 3\theta$

(11) $\sin(\theta + \alpha) = \cos(\theta - \alpha)$

(12) $\sin(\theta + \alpha) + \cos(\theta + \alpha) = \sin(\theta - \alpha) + \cos(\theta - \alpha)$

(13) $\sin\alpha + \sin(\theta - \alpha) + \sin(2\theta + \alpha)$
$= \sin(\theta + \alpha) + \sin(2\theta - \alpha)$

(14) $\operatorname{cosec}^2\dfrac{\theta}{2} - \sec^2\dfrac{\theta}{2} = 2\sqrt{3}\operatorname{cosec}^2\theta$

(15) $\sin\theta - \cos\theta = 4\cos^2\theta\sin\theta$

(16) $\cos 3\theta = \sin\theta$

(17) $\sin 3\theta = \cos\theta$; and find which of its solutions satisfies

$$1 + \sin^2\theta = 3\sin\theta\cos\theta$$

(18) $3\cos^2\theta + 2\sqrt{3}\cos\theta = 5\frac{1}{4}$

Woolwich: '81.

(19) $\sec 4\theta - \sec 2\theta = 2$

Math' Tri': '77.

(20) $\operatorname{cosec} 4\alpha - \operatorname{cosec} 4\theta = \cot 4\alpha - \cot 4\theta$

Math' Tri': '76.

(21) $(1 + \sin\theta)(1 - 2\sin\theta)^2 = (1 - \cos\alpha)(1 + 2\cos\alpha)^2$

Math' Tri': '62.

(22) $\dfrac{1}{\sin 3\theta} + \dfrac{1}{\sin 2\theta} = \dfrac{\sin 2\theta}{\sin\theta \sin 3\theta}$

(23) $\cot\theta - \tan\theta = \cos\theta + \sin\theta$

(24) $\tan\left(\dfrac{\pi}{2\sqrt{2}}\sin\theta\right) = \cot\left(\dfrac{\pi}{2\sqrt{2}}\cos\theta\right)$

Caius Camb': '82.

(25) $7\cos 3\theta = \sin^2\theta + \cos 2\theta$

Clare and Caius Camb': '74.

(26) $4(1 + 8\operatorname{cosec}^2 4\theta) = (\tan^2\theta + \cot^2\theta)(\tan^2\alpha + \cot^2\alpha)$

(27) $\tan 5\theta = 5\tan\theta$

9. If $\operatorname{vers}\theta = \operatorname{vers}(\theta - \alpha)$, where α is a known angle (not a multiple of π) find θ.

10. Show that all the solutions of

$$\tan\theta + \cot\theta = 4$$

are contained in $(n + \frac{1}{4} \pm \frac{1}{6})\pi$

11. Show that all the solutions of

$$\sin^2 2\theta - \sin^2\theta = \tfrac{1}{4}$$

are contained in $(n \pm \frac{1}{5} \pm \frac{1}{10})\pi$

12. If n lies between certain values, the equation $\tan 3\theta = n\tan\theta$ can only be satisfied by $\theta = m\pi$: find those values.

Solve the equation when $n = -1$.

Show that if $\theta = \pi/12$, then $n = 2 + \sqrt{3}$

13. If $\tan(\cot\theta) = \cot(\tan\theta)$ show that

$$\sin 2\theta = 4/(2n + 1)\pi$$

where n is any integer, *excepting* -1, or 0.

14. Show that the equation

$$\sin\theta + \sin\phi = \sin(\theta + \phi)$$

can only be true when one of the quantities θ, ϕ, $\theta + \phi$ is a multiple of 2π.

John's Camb': '33.

15. Five angles are in arithmetical progression; and the cosine of the mid angle is equal to the sum of all their cosines: find the difference of the angles.

16. If A is the greatest angle of a triangle and

$$2 \sec A \operatorname{cosec} A - \operatorname{cosec} A \cot A = \sec A,$$

find how many degrees there are in A.

I. I.: '88.

17. Find the six positive angles, less than four right angles, satisfying

$$8 \sin^3 \theta - 6 \sin \theta + 1 = 0.$$

John's Camb' : '83.

18. Show that all the values of θ satisfying

$$\sin 3\theta = \sin \theta \cos 2\theta$$

are contained in $n\pi/2$, where n is any integer.

19. If $\sin \theta + \sin \phi = \sqrt{3}(\cos \phi - \cos \theta)$

prove that $\sin 3\theta + \sin 3\phi = 0.$

John's Camb': '83.

20. Find x and y from the equations

$$\left.\begin{aligned} \sin x &= \sqrt{2} \,.\, \sin y \\ \tan x &= \sqrt{3} \,.\, \tan y \end{aligned}\right\}$$

21. If

$$\left.\begin{aligned} \frac{\sin \theta}{x} &= \frac{\cos \theta}{y} \\ \text{and} \quad \frac{\cos^2 \theta}{x^2} + \frac{\sin^2 \theta}{y^2} &= \frac{10}{3(x^2 + y^2)} \end{aligned}\right\} \text{find } \tan \theta.$$

22. If

$$\left.\begin{aligned} \sin \theta + \sin 2\theta + \sin 3\theta &= \sin \alpha \\ \text{and} \quad \cos \theta + \cos 2\theta + \cos 3\theta &= \cos \alpha \end{aligned}\right\}$$

prove that α is a multiple of π.

Caius Camb': 78'.

NOTE—*Square and add.*

23. If $b^2 = 4ac$, $x^2 + y^2 = (a + c)^2$

$$\left.\begin{aligned} \text{and} \quad x &= a \cos \theta + b \cos 2\theta + c \cos 3\theta \\ y &= a \sin \theta + b \sin 2\theta + c \sin 3\theta \end{aligned}\right\}$$

prove that $\theta = (2n + 1)\pi/2$; and that there are no other solutions.

Pet' Camb': '66.

24. If $\cos 2m\theta \{\tan (m + n)\theta + 1\} = \cos 2n\theta \{\tan (m + n)\theta - 1\}$ show that all values of θ are contained in the two forms $r\pi/\overline{2(m + n)}$ and $(4r + 1)\pi/4(m - n)\pi$, where r may have any integral value, including zero.

25. Solve each of the equations—

(1) $16^{\cos^2 x + 2\sin^2 x} + 4^{2\cos^2 x} = 40$

(2) $7346 \times 7^{\sec x} + 7^{1 + \sec x} - 7010 \times 7^{2\sec x}$
$- 7^{3 + 2\sec x} + 3 \times 7^{2 + 3\sec x} = 147.$

Bordage: *E. T.* xlviii.

26. If $\sin\alpha + \sin\beta + \sin\gamma = 0$
and $\cos\alpha + \cos\beta + \cos\gamma = 0$
prove that $\Sigma \sin^2\alpha = 3/2 = \Sigma \cos^2\alpha.$

Wolstenholme: 435.

NOTE—*Put* $\sin\gamma$, $\cos\gamma$ *on other sides: divide corresponding sides, giving* $\alpha + \beta = 2m\pi + 2\gamma$: *two sim'r results follow. Then square given eq'ns and subtract. Result now follows easily.*

§ **41.** The preceding general results may also be used to explain some ambiguities which have occurred in formulæ already proved, or which may occur in similar formulæ; and indeed to show *à priori* that such ambiguities are to be expected.

The general principle of the explanation is this—

If $f(\theta)$ represents a T. F. of θ; and if we are given the value of $f(\theta)$ to be a known quantity (k say) we are not given θ, but only that it is *one* of the $\wedge^s$ contained in the general expression corresponding to $f(\theta)$. Hence, in finding another T. F., say $F(m\theta)$, where m is integral or fractional, we have an infinite number of $\wedge^s$ which, though they give the same value for $f(\theta)$, do not necessarily give the same value for $F(m\theta)$. To get the values of $F(m\theta)$ corresponding to the $\wedge^s$ which satisfy $f(\theta) = k$, we must substitute from the general value of θ, satisfying $f(\theta) = k$, in $F(m\theta)$. When this is done we shall arrive at an eq'n connecting $f(\theta)$ and $F(m\theta)$, from which for a given value of the one we can find the corresponding values of the other. We can also apply to such an eq'n the algebraic connection between its roots and coefficients, and thus deduce formulæ which might otherwise be more troublesome to arrive at. The discussion of some cases will make all this clearer.

Examples

1. *To explain the ambiguity in the formula*

$$\sin\theta = \pm\sqrt{1-\cos^2\theta}$$

where it is implied that $\cos\theta$ *is given, and* $\sin\theta$ *wanted.*

If we are given the numerical value (k say) of $\cos\theta$, we are not given θ, but only that it is one of the $\wedge^s$ included in

$$2n\pi \pm \theta$$

where θ is any one definite $\wedge$ satisfying the eq'n $\cos\theta = k$.

Now, in finding $\sin\theta$ from $\cos\theta$, we have an infinite number of $\wedge^s$, which, though they give the same value for the cosine, do not necessarily give the same value for the sine.

$\therefore$, in the formula $\sqrt{1-\cos^2\theta} = \sin\theta$, we may expect *à priori* to get not only $\sin\theta$ but also the sines of all $\wedge^s$ included in $2n\pi \pm \theta$.

$$\text{But } \sin(2n\pi \pm \theta) = \pm\sin\theta$$

i. e. we may expect *à priori* to get *two* values for $\sin\theta$, equal in magnitude, but opposite in sign, corresponding to any given value of $\cos\theta$; and hence the ambiguity in $\sin\theta = \pm\sqrt{1-\cos^2\theta}$ is explained and justified.

2. *To explain the ambiguity in the formula*

$$\cos\theta = \pm\sqrt{\frac{1+\cos 2\theta}{2}}$$

where it is implied that $\cos 2\theta$ *is given, and* $\cos\theta$ *is wanted.*

If we are given the numerical value (k say) of $\cos 2\theta$, we are not given 2θ but only that it is one of the $\wedge^s$ included in

$$2n\pi \pm 2\theta$$

where 2θ is any one definite $\wedge$ satisfying the eq'n $\cos 2\theta = k$.

$\therefore$, when $\cos\theta$ is expressed in terms of k, all its values are included in

$$\cos(n\pi \pm \theta)$$

Now, if n is even this is $\cos\theta$; but, if n is odd it is $-\cos\theta$.

Hence, *à priori* we see that $\cos\theta$ when expressed in terms of $\cos 2\theta$ must have *two* values, equal in magnitude, but opposite in sign; and so the ambiguity of sign in the formula $\cos\theta = \pm\sqrt{\left(\frac{1+\cos 2\theta}{2}\right)}$ is explained and justified.

3. *To show how it is that four values arise out of the expression*

$$\tfrac{1}{2}\{\sqrt{1 + \sin 2\theta} + \sqrt{1 - \sin 2\theta}\}$$

If we are given the value (k say) of $\sin 2\theta$, we are not given 2θ, but only that it is one of the $\wedge^s$ included in

$$n\pi + (-1)^n 2\theta$$

where 2θ is any one particular $\wedge$ satisfying the eq'n $\sin 2\theta = k$

$\therefore$, when $\sin\theta$ is expressed in terms of k, all its values must be contained in

$$\sin\{n\pi/2 + (-1)^n\theta\}$$

Now, 1°, if n is even (say 2 m) then this is

$$\sin(m\pi + \theta)$$

Expanding; and recollecting that $\sin m\pi = 0$, and $\cos m\pi = \pm 1$, we get $\pm \sin\theta$.

Also, 2°, if n is odd (say $2m + 1$), then above becomes

$$\sin(m\pi + \pi/2 - \theta)$$

Expanding; and recollecting that $\sin m\pi = 0$, $\cos m\pi = \pm 1$, and $\sin(\pi/2 - \theta) = \cos\theta$; we get $\pm \cos\theta$.

Hence we see that the *four* values of the expression

$$\{\sqrt{1 + \sin 2\theta} + \sqrt{1 - \sin 2\theta}\}/2$$

which we found (in § 25) to be $\pm \sin\theta$ and $\pm \cos\theta$, may be expected *à priori*.

4. *To see what can be deduced from the formula*

$$4\cos^3\theta - 3\cos\theta = \cos 3\theta$$

If $\cos 3\theta$ has a given value (k say) and if x is put for $\cos\theta$, then the formula gives

$$4x^3 - 3x - k = 0$$

as an eq'n to determine x in terms of k.

About this eq'n we know, by theory of eq'ns, that

1°, it has three roots; and that if these are denoted by $x_1\ x_2\ x_3$, then

2°, $x_1 + x_2 + x_3 = 0$

3°, $x_1x_2 + x_2x_3 + x_3x_1 = -\frac{3}{4}$

4°, $x_1\,x_2\,x_3 = k/4$

Whence we see that $\cos\theta$ expressed in terms of $\cos 3\theta$ should have three values; and then that three eq'ns connect these values.

Now we are not given 3θ but only that it is one of the $\wedge^s$ contained in

$$2n\pi \pm 3\theta$$

where 3θ is any one particular $\wedge$ satisfying $\cos 3\theta = k$

$\therefore$ all values of x are contained in $\cos(2n\pi/3 \pm \theta)$

If $n = 3m$, this is $\cos(2m\pi \pm \theta)$, i.e. $\cos\theta$

" $3m - 1$, " $\cos(2m\pi - 2\pi/3 \pm \theta)$, i.e. $\cos(2\pi/3 \pm \theta)$

" $3m + 1$, " $\cos(2m\pi + 2\pi/3 \pm \theta)$, " $\cos(2\pi/3 \pm \theta)$

But n must be one of these forms.

$\therefore$ the values of x are

$$\cos\theta,\quad \cos(2\pi/3 + \theta),\quad \cos(2\pi/3 - \theta)$$

Whence it follows that

$$4x^3 - 3x - \cos 3\theta \equiv (x - \cos\theta)(x - \cos\overline{2\pi/3 + \theta})(x - \cos\overline{2\pi/3 - \theta})$$

Also (as may easily be proved independently)

$$\cos\theta + \cos(2\pi/3 + \theta) + \cos(2\pi/3 - \theta) = 0$$

$$\cos\theta\cos(2\pi/3 + \theta) + \cos(2\pi/3 + \theta)\cos(2\pi/3 - \theta)$$
$$+ \cos(2\pi/3 - \theta)\cos\theta = -\tfrac{3}{4}$$

$$\cos\theta\cos(2\pi/3 + \theta)\cos(2\pi/3 - \theta) = (\cos 3\theta)/4$$

Again $\quad \Sigma x_1^3 = (\Sigma x_1)^3 - 3\Sigma x_1^2(x_2 + x_3) - 6x_1x_2x_3$

$$= 3\Sigma x_1^3 - \tfrac{3}{2}\cos 3\theta$$

$$\therefore\quad \Sigma x_1^3 = \tfrac{3}{4}\cos 3\theta$$

i.e. $\cos^3\theta + \cos^3(2\pi/3 + \theta) + \cos^3(2\pi/3 - \theta) = \tfrac{3}{4}\cos 3\theta$

Again $\quad \Sigma x_1^2 = (\Sigma x_1)^2 - 2\Sigma x_1x_2 = \tfrac{3}{2}$

$$\therefore\quad \Sigma x_1^4 = (\Sigma x_1^2)^2 - 2\Sigma x_1^2x_2^2$$

$$= (\tfrac{9}{4} - 2\{(\Sigma x_1x_2)^2 - 4x_1x_2x_3\Sigma x_1\}$$

$$= \tfrac{9}{4} - \tfrac{9}{8} = \tfrac{9}{8}$$

i.e. $\cos^4\theta + \cos^4(2\pi/3 + \theta) + \cos^4(2\pi/3 - \theta) = \tfrac{9}{8}$

Note—This last result has been given (see p. 113) to be proved independently. It is clear that the formula

$$3\sin\theta - 4\sin^3\theta = \sin 3\theta$$

and all analogous formulæ may be treated sim'ly.

5. *To give another Example.*

$$\text{Since}\quad \tan n\theta = \frac{c_1 \tan\theta - c_3 \tan^3\theta + \&c}{1 - c_2 \tan^2\theta + c_4 \tan^4\theta - \&c}$$

∴, if we put x for $\tan\theta$, and suppose (e. g.) $n\theta = \pi$, then

$$c_1 x - c_3 x^3 + c_5 x^5 - \&c \pm c_n x^n = 0$$

is an eq'n whose roots are the values got by letting m have, in succession, the values

$$0,\ 1,\ 2,\ \&c \ldots \overline{n-1} \text{ in } \tan(m+1)\pi/n$$

So also we may treat the formula

$$\cot n\theta = \frac{\cot^n\theta - c_2 \cot^{n-2}\theta + c_4 \cot^{n-4}\theta - \&c}{c_1 \cot^{n-1}\theta - c_3 \cot^{n-3}\theta + c_5 \cot^{n-5}\theta - \&c}$$

Exercises

1. Show *à priori* that when $\tan\alpha$ is expressed in terms of $\sin\alpha$, two values should result.

2. Determine *à priori* the values $\tan\alpha$ has when expressed in terms of $\tan 3\alpha$.

Form the equation whose roots are those values; and deduce that

$$\tan\alpha \tan(60^\circ + \alpha) \tan(60^\circ - \alpha) = \tan 3\alpha$$

$$\text{and}\quad \tan\alpha + \tan(60^\circ + \alpha) - \tan(60^\circ - \alpha) = 3\tan 3\alpha$$

3. Show that if $\sin\alpha$ is determined from a given value of $\sin 3\alpha$ it should have three values; and that the sum of these is zero.

4. Given the value of $\sin\alpha$ show *à priori* that $\sin\frac{3}{4}\alpha$ will have eight different values, when expressed in terms of $\sin\alpha$.

5. From the formula

$$\cot 3\alpha = \frac{3\cot\alpha - \cot^3\alpha}{1 - 3\cot^2\alpha}$$

deduce three formulæ analogous to those found in Example 4.

6. If $\sin\alpha/3$ is found from a given value of $\cot\alpha$, find how many values of $\sin\alpha/3$ may be expected *à priori*.

7. Prove *à priori* that $\sin n\alpha$, when expressed by $\sin\alpha$, will have one or two values, according as n is odd or even; and that $\cos n\alpha$, expressed by $\cos\alpha$, will have only one value: n is a positive integer.

8. Show that $\tan\alpha$, expressed in terms of $\sin 4\alpha$, will have four values; and find these values.

John's Camb': '40.

9. If $\sin\alpha$ is expressed as a function of $\sin 6\alpha$, determine by *à priori* reasoning how many values the resulting expression ought to have for a given value of $\sin 6\alpha$.

Pet' Camb': '61.

10. Form the equation for determining $\tan\alpha$ in terms of $\tan 6\alpha$; and show *à priori* that it ought to have six values.

Prove that the sum of these six values is $-6\cot 6\alpha$.

Pet' Camb': '60.

11. Prove that

$$\tan\theta + \tan(\pi/5 + \theta) + \tan(2\pi/5 + \theta) + \tan(3\pi/5 + \theta) + \tan(4\pi/5 + \theta) = 5\tan 5\theta.$$

Q. C. B. '88.

12. Prove that

$$x^3 - 3x^2\tan 3\theta - 3x + \tan 3\theta \equiv (x - \tan\theta)(x - \tan\overline{\pi/3 + \theta})(x - \tan\overline{2\pi/3 + \theta})$$

R. U. Schol': '83.

13. Prove that

$$\operatorname{cosec}\theta + \operatorname{cosec}(\pi/3 + \theta) + \operatorname{cosec}(\pi/3 - \theta) = 3\operatorname{cosec} 3\theta$$

Christ's Camb': '78.

14. If $\sin 3\theta$ is given, and thence $\tan\theta$ is found, show *à priori* that six values are generally to be expected.

Hence show that

$$\tan^2\theta\tan^2(\pi/3 + \theta) + \tan^2(\pi/3 + \theta)\tan^2(\pi/3 - \theta) + \tan^2(\pi/3 - \theta)\tan^2\theta = 6\sec^2 3\theta + 3$$

Clare, Caius and King's Camb': '80.

15. Form the equation whose roots are

$$\tan^2\pi/11,\quad \tan^2 2\pi/11,\quad \tan^2 3\pi/11,\quad \tan^2 4\pi/11,\quad \tan^2 5\pi/11$$

Christ's, Emm' and Sid' Camb': '82.

16. Prove that

$$(x - 2\cos 2\pi/7)(x - 2\cos 4\pi/7)(x - 6\pi/7) \equiv x^3 + x^2 - 2x - 1$$

17. Prove that—

$$\cos 11\pi/36 + \cos 13\pi/36 + \cos 35\pi/36 = 0$$

$$\cos 11\pi/36 \cos 13\pi/36 + \cos 13\pi/36 \cos 35\pi/36 + \cos 35\pi/36 \cos 11\pi/36 = -\tfrac{3}{4}$$

$$\cos 11\pi/36 \cos 13\pi/36 \cos 35\pi/36 = -\frac{\sqrt{3}+1}{8\sqrt{2}}$$

18. Prove that

$$\frac{1}{\cos 2\pi/7 + \cos 2\phi} + \frac{1}{\cos 4\pi/7 + \cos 2\phi} + \frac{1}{\cos 6\pi/7 + \cos 2\phi} = \frac{7\tan 7\phi - \tan\phi}{2\sin 2\phi}$$

R. U. Schol': '86.

NOTE—*Express (see Example* 5) $\tan 7\phi$ *in terms of* $\tan\phi$: *the result will give an eq'n of the 7th degree in* $\tan\phi$, *the sum of whose roots is* $7\tan 7\phi$. *These 7 roots will be*

$$\tan\phi,\ \tan(\pi/7 \pm \phi),\ \tan(2\pi/7 \pm \phi),\ \tan(3\pi/7 \pm \phi);$$

and these added will give the result.

19. Assuming that the sum of the cubes of the roots of the equation

$$x^n + p_1 x^{n-1} + p_2 x^{n-2} + \&c + p_{n-1} x + p_n = 0$$

is $-p_1^3 + 3p_1p_2 - 3p_3$

prove that $\cot^3\theta + \cot^3(\theta + \pi/n) + \cot^3(\theta + 2\pi/n)$ &c, to n terms,

$$= n\cot n\theta\,(n^2 \operatorname{cosec}^2 n\theta - 1)$$

R. U. Schol': '90.

NOTE—*If* $x = \cot\theta$, *then (Example* 5) *the expansion of* $\cot n\theta$ *in terms of* x *gives*

$$x^n - n\cot n\theta . x^{n-1} + \frac{n(n-1)}{2} x^{n-2} + \frac{n(n-1)(n-2)}{6}\cot n\theta . x^{n-3} + \&c = 0,$$

whence result easily follows.

20. Prove that

$$\cot^2 \pi/11 + \cot^2 2\pi/11 + \cot^2 3\pi/11 + \cot^2 4\pi/11 + \cot^2 5\pi/11 = 15$$

Ox' Jun' Math' Schol': '90.

CHAPTER VII

On Angles expressed as the Inverses of Trigonometrical Functions

§ **42.** If $\sin x = \alpha$, the fact implied by this direct statement could be otherwise expressed by, what may be termed, the inverse statement—

x *is the angle whose* sine *is* α

But this latter statement is ambiguous; for we know that there is an infinite number of $\wedge^s$ whose sines are each α.

Hence, if we are to use such '*inverse*' statements, we must have some convention to limit their ambiguity. The convention usually made is this—When an angle is expressed as the inverse of a T. F. *the least positive* angle for which the function has the specified value is always to be understood, unless the contrary is expressly stated.

Now inasmuch as the phrase—'*the least positive angle whose* sine *is* α'—is too tedious for general use, some brief notation to express these words is desirable. The notation usually adopted is * $\sin^{-1}\alpha$; with a similar notation for $\wedge^s$ expressed inversely by means of the other T. F.s. For example $\tan^{-1}\alpha$ means *the least positive angle whose tangent is* α.

Note (1)—The origin of this notation seems to have come from considering that, if it is supposed possible that the symbol sin could be separated from its $\wedge$, and treated as an algebraic entity, we should then be able to write the eq'n

$$\sin x = \alpha$$

as $$x = \frac{\alpha}{\sin} = \sin^{-1}\alpha$$

---

* Sometimes *arc* $\sin\alpha$.

Note (2)—Since $\sin\alpha = \sin\{n\pi + (-1)^n\alpha\}$, $\therefore$, strictly speaking,

$$\sin^{-1}(\sin\alpha) = n\pi + (-1)^n\alpha$$

so $$\cos^{-1}(\cos\alpha) = 2n\pi \pm \alpha$$

and $$\tan^{-1}(\tan\alpha) = n\pi + \alpha\text{*}$$

Def'—If an assigned trigonometrical function has a given value, then the least positive angle (expressed by means of this trigonometrical function) for which this is true, is called the corresponding **inverse trigonometrical function.**

Note—For brevity we shall use I. T. F. to denote the words *inverse trigonometrical function.*

$\wedge$s expressed as I. T. F.s are capable of combination in manner very analogous to the combinations of the corresponding T. F.s. The mode of procedure is always uniform; and, once laid hold of, the manipulation of the I. T. F.s is easy.

Thus a table similar to that on p. 28, may be formed for the I. T. F.s. An example will show how any I. T. F. may be expressed in terms of any other I. T. F.: for instance, to express $\sec^{-1}\alpha$ in terms of $\text{cosec}^{-1}\alpha$.

$$\text{If} \quad \sec^{-1}\alpha = \theta$$

$$\sec\theta = \alpha$$

$$\therefore \quad \text{cosec}\,\theta, \text{ which} = \frac{\sec\theta}{\sqrt{\sec^2\theta - 1}},$$

$$\text{also} = \frac{\alpha}{\sqrt{\alpha^2 - 1}}$$

$$\therefore \theta, \text{ i.e. } \sec^{-1}\alpha, = \text{cosec}^{-1}\frac{\alpha}{\sqrt{\alpha^2 - 1}}$$

Proceeding in this way we get the table here given: the Student should verify this table.

---

* For a full development of this idea see *Chrystal's Algebra,* vol' II, pp. 235–238; where also the graph of $\sin^{-1}x$ is drawn.

	$\sin^{-1}$	$\cos^{-1}$	$\tan^{-1}$	$\cot^{-1}$	$\sec^{-1}$	cosec^{-1}
$\sin^{-1}\alpha$	$\sin^{-1}\alpha$	$\cos^{-1}\sqrt{1-\alpha^2}$	$\tan^{-1}\frac{\alpha}{\sqrt{1-\alpha^2}}$	$\cot^{-1}\frac{\sqrt{1-\alpha^2}}{\alpha}$	$\sec^{-1}\frac{1}{\sqrt{1-\alpha^2}}$	$\mathrm{cosec}^{-1}\frac{1}{\alpha}$
$\cos^{-1}\alpha$	$\sin^{-1}\sqrt{1-\alpha^2}$	$\cos^{-1}\alpha$	$\tan^{-1}\frac{\sqrt{1-\alpha^2}}{\alpha}$	$\cot^{-1}\frac{\alpha}{\sqrt{1-\alpha^2}}$	$\sec^{-1}\frac{1}{\alpha}$	$\mathrm{cosec}^{-1}\frac{1}{\sqrt{1-\alpha^2}}$
$\tan^{-1}\alpha$	$\sin^{-1}\frac{\alpha}{\sqrt{1+\alpha^2}}$	$\cos^{-1}\frac{1}{\sqrt{1+\alpha^2}}$	$\tan^{-1}\alpha$	$\cot^{-1}\frac{1}{\alpha}$	$\sec^{-1}\sqrt{1+\alpha^2}$	$\mathrm{cosec}^{-1}\frac{\sqrt{1+\alpha^2}}{\alpha}$
$\cot^{-1}\alpha$	$\sin^{-1}\frac{1}{\sqrt{\alpha^2+1}}$	$\cos^{-1}\frac{\alpha}{\sqrt{\alpha^2+1}}$	$\tan^{-1}\frac{1}{\alpha}$	$\cot^{-1}\alpha$	$\sec^{-1}\frac{\sqrt{\alpha^2+1}}{\alpha}$	$\mathrm{cosec}^{-1}\sqrt{\alpha^2+1}$
$\sec^{-1}\alpha$	$\sin^{-1}\frac{\sqrt{\alpha^2-1}}{\alpha}$	$\cos^{-1}\frac{1}{\alpha}$	$\tan^{-1}\sqrt{\alpha^2-1}$	$\cot^{-1}\frac{1}{\sqrt{\alpha^2-1}}$	$\sec^{-1}\alpha$	$\mathrm{cosec}^{-1}\frac{\alpha}{\sqrt{\alpha^2-1}}$
$\mathrm{cosec}^{-1}\alpha$	$\sin^{-1}\frac{1}{\alpha}$	$\cos^{-1}\frac{\sqrt{\alpha^2-1}}{\alpha}$	$\tan^{-1}\frac{1}{\sqrt{\alpha^2-1}}$	$\cot^{-1}\sqrt{\alpha^2-1}$	$\sec^{-1}\frac{\alpha}{\sqrt{\alpha^2-1}}$	$\mathrm{cosec}^{-1}\alpha$

Note—When $\alpha > 1$, $\sin^{-1}\alpha$ and $\cos^{-1}\alpha$ are imaginary,
and " < 1, $\sec^{-1}\alpha$ " $\text{cosec}^{-1}\alpha$ " " ;
bnt $\tan^{-1}\alpha$ and $\cot^{-1}\alpha$ are always real.

Exercise

Fill in an 8th column and row with the corresponding values for $\text{vers}^{-1}\alpha$

So again we may combine two (or more) I. T. F.s into a single I. T. F. of the same kind; or we may prove a formula for I. T. F.s analogous to any formula for T. F.s; or we may solve an eq′n expressed in terms of I. T. F.s; or we may by means of the I. T. F. notation write a symbolical expression for the general solution of a trig′l eq′n, when we have found from it a value for a T. F. which does not give any known $\wedge$; or, finally, we may by means of I. T. F.s find symbolical expressions for the roots of many alg′l eq′ns.

Note—Obviously, from the very meaning of the co- in cosine,

$$\sin^{-1}\theta + \cos^{-1}\theta = \pi/2$$

$$\therefore \quad \sin^{-1}\theta = \pi/2 - \cos^{-1}\theta$$

$$\text{Sim'ly} \quad \tan^{-1}\theta = \pi/2 - \cot^{-1}\theta$$

$$\text{and} \quad \sec^{-1}\theta = \pi/2 - \text{cosec}^{-1}\theta$$

These transformations are sometimes useful.

Examples

1. *To express* $\tan^{-1}\alpha + \tan^{-1}\beta$ *as a single similar* I. T. F.

$$\text{If} \quad x = \tan^{-1}\alpha, \text{ and } y = \tan^{-1}\beta$$

$$\text{then} \quad \tan x = \alpha, \text{ and } \tan y = \beta$$

$$\therefore \quad \tan(x+y), \text{ which} = \frac{\tan x + \tan y}{1 - \tan x \tan y},$$

$$\text{also} = \frac{\alpha + \beta}{1 - \alpha\beta}$$

$$\therefore \quad x + y, \text{ i.e. } \tan^{-1}\alpha + \tan^{-1}\beta, = \tan^{-1}\frac{\alpha + \beta}{1 - \alpha\beta}$$

Note—This formula may be taken to mean—

any ∧ whose tan is α + *any* ∧ whose tan is β

= *some one* of the ∧s whose tan is $\dfrac{\alpha+\beta}{1-\alpha\beta}$

2. *To express* $\text{vers}^{-1}\alpha + \text{vers}^{-1}\beta$ *as a single similar* I. T. F.

If $\text{vers}^{-1}\alpha = x$, and $\text{vers}^{-1}\beta = y$

then $\text{vers}\, x = \alpha$, and $\text{vers}\, y = \beta$

$\therefore\ \cos x = 1 - \alpha$, and $\cos y = 1 - \beta$

$\therefore\ \sin x = \sqrt{2\alpha - \alpha^2}$, and $\sin y = \sqrt{2\beta - \beta^2}$

$\therefore\ \cos(x+y)$, which $= \cos x \cos y - \sin x \sin y$,

also $= (1-\alpha)(1-\beta) - \sqrt{(2\alpha-\alpha^2)(2\beta-\beta^2)}$

$\therefore\ \text{vers}(x+y) = \alpha + \beta - \alpha\beta + \sqrt{(2\alpha-\alpha^2)(2\beta-\beta^2)}$

$\therefore\ x + y$, i.e. $\text{vers}^{-1}\alpha + \text{vers}^{-1}\beta$,

$$= \text{vers}^{-1}\left\{\alpha + \beta - \alpha\beta + \sqrt{(2\alpha-\alpha^2)(2\beta-\beta^2)}\right\}$$

3. *To show that the three angles,* $\tan^{-1}\alpha$, $\tan^{-1}\beta$, $\tan^{-1}\dfrac{1-\alpha-\beta-\alpha\beta}{1+\alpha+\beta-\alpha\beta}$ *together make up half a right angle.*

Using the formula of Example 1, we get sum of ∧s

$$= \tan^{-1}\frac{\alpha+\beta}{1-\alpha\beta} + \tan^{-1}\frac{1-\alpha-\beta-\alpha\beta}{1+\alpha+\beta-\alpha\beta}$$

$$= \tan^{-1}\frac{\dfrac{\alpha+\beta}{1-\alpha\beta} + \dfrac{1-\alpha-\beta-\alpha\beta}{1+\alpha+\beta-\alpha\beta}}{1 - \dfrac{(\alpha+\beta)(1-\alpha-\beta-\alpha\beta)}{(1-\alpha\beta)(1+\alpha+\beta-\alpha\beta)}}$$

$$= \tan^{-1}\frac{\alpha+\alpha^2+\alpha\beta-\alpha^2\beta+\beta+\alpha\beta+\beta^2-\alpha\beta^2+1-\alpha-\beta-\alpha\beta-\alpha\beta+\alpha^2\beta+\alpha\beta^2+\alpha^2\beta^2}{1+\alpha+\beta-\alpha\beta-\alpha\beta-\alpha^2\beta-\alpha\beta^2+\alpha^2\beta^2-\alpha+\alpha^2+\alpha\beta+\alpha^2\beta-\beta+\alpha\beta+\beta^2+\alpha\beta^2}$$

$$= \tan^{-1}\frac{1+\alpha^2+\beta^2+\alpha^2\beta^2}{1+\alpha^2+\beta^2+\alpha^2\beta^2}$$

$= \tan^{-1} 1$

= half a right ∧

4. *To find a formula connecting* I. T. F.*s, analogous to the formula*

$$\sin 3\alpha = 3\sin\alpha - 4\sin^3\alpha$$

$$\text{If} \quad \sin\alpha = x$$

$$\text{then} \quad \alpha = \sin^{-1}x$$

$$\text{and} \quad 3\alpha = 3\sin^{-1}x$$

$$\text{But} \quad 3\alpha = \sin^{-1}(3\sin\alpha - 4\sin^3\alpha)$$

$$\therefore \quad 3\sin^{-1}x = \sin^{-1}(3x - 4x^3)$$

which is the formula req'd

5. *If* $\sec^{-1}\frac{x}{a} + \sec^{-1}a = \sec^{-1}\frac{x}{b} + \sec^{-1}b$, *to find* x *in terms of* a *and* b.

Putting $\alpha, \beta, \theta, \phi$ for $\sec^{-1}a$, $\sec^{-1}b$, $\sec^{-1}\frac{x}{a}$, $\sec^{-1}\frac{x}{b}$, respectively we get

$$\sec\alpha = a, \quad \sec\beta = b, \quad \sec\theta = \frac{x}{a}, \quad \sec\phi = \frac{x}{b};$$

$$\text{and} \quad \theta + \alpha = \phi + \beta$$

$$\therefore \cos\theta\cos\alpha - \sin\theta\sin\alpha = \cos\phi\cos\beta - \sin\phi\sin\beta$$

$$\text{and} \quad \cos\theta\cos\alpha = \frac{a}{x}\cdot\frac{1}{a} = \frac{1}{x} = \frac{b}{x}\cdot\frac{1}{b} = \cos\phi\cos\beta$$

$$\therefore \quad \sin\theta\sin\alpha = \sin\phi\sin\beta$$

$$\therefore \quad \left(1 - \frac{a^2}{x^2}\right)\left(1 - \frac{1}{a^2}\right) = \left(1 - \frac{b^2}{x^2}\right)\left(1 - \frac{1}{b^2}\right)$$

$$\therefore \quad b^2(x^2 - a^2)(a^2 - 1) = a^2(x^2 - b^2)(b^2 - 1)$$

$$\therefore \quad (a^2 - b^2)x^2 = a^2b^2(a^2 - b^2)$$

$$\therefore \quad x = \pm\sqrt{ab}$$

6. *To express, in its most general terms, the solution of the equation*

$$\sin(n\cos\theta) = \cos(n\sin\theta)$$

By the hypothesis, we have at once

$$\sin(n\cos\theta) = \sin(\pi/2 - n\sin\theta)$$

$$\therefore \quad n\cos\theta = r\pi + (-1)^r(\pi/2 - n\sin\theta)$$

$$\therefore \quad n\{\cos\theta + (-1)^r\sin\theta\} = \{2r + (-1)^r\}\frac{\pi}{2}$$

$$\therefore \quad \frac{1}{\sqrt{2}} \cos\theta + (-1)^r \frac{1}{\sqrt{2}} \sin\theta = \frac{2r + (-1)^r}{n\sqrt{2}} \cdot \frac{\pi}{2}$$

$$\therefore \quad \cos\left\{\theta - (-1)^r \frac{\pi}{4}\right\} = \left\{\frac{2r + (-1)^r}{n\sqrt{2}}\right\} \frac{\pi}{2}$$

$$\therefore \quad \theta = (-1)^r \frac{\pi}{2} + \cos^{-1}\left\{\frac{2r + (-1)^r}{n\sqrt{2}}\right\} \frac{\pi}{2}$$

where r is any integer, pos′, neg′, or zero.

7. *Solve, by the aid of trigonometry, the equations*

$$\left.\begin{array}{l} \sqrt{x(1-y)} + \sqrt{y(1-x)} = a \\ \sqrt{xy} + \sqrt{(1-x)(1-y)} = b \end{array}\right\}$$

where a *and* b *are real.*

Since a and b are real, it is clear, from the eq′ns, that x and y are pos′ fractions.

$\therefore$ we may put $\sin^2\theta$ for x, and $\sin^2\phi$ for y.

This gives
$$\left.\begin{array}{l} \sin\theta\cos\phi + \cos\theta\sin\phi = a \\ \sin\theta\sin\phi + \cos\theta\cos\phi = b \end{array}\right\}$$

whence
$$\left.\begin{array}{l} \sin(\theta + \phi) = a \\ \cos(\theta - \phi) = b \end{array}\right\}$$

$$\therefore \quad \theta = \tfrac{1}{2}\left\{\sin^{-1} a + \cos^{-1} b\right\}$$

$$\phi = \tfrac{1}{2}\left\{\sin^{-1} a - \cos^{-1} b\right\}$$

$$\therefore x = \left[\sin\left(\frac{\sin^{-1} a}{2}\right)\cos\left(\frac{\cos^{-1} b}{2}\right) + \cos\left(\frac{\sin^{-1} a}{2}\right)\sin\left(\frac{\cos^{-1} b}{2}\right)\right]^2$$

$$y = \left[\sin\left(\frac{\sin^{-1} a}{2}\right)\cos\left(\frac{\cos^{-1} b}{2}\right) - \cos\left(\frac{\sin^{-1} a}{2}\right)\sin\left(\frac{\cos^{-1} b}{2}\right)\right]^2$$

Note—If we give to a, b the numerical values of known sines and cosines, the results will be verified.

For example, if $a = \frac{\sqrt{3}}{2} = b$,

so that $\sin^{-1} a = 60°$, and $\cos^{-1} b = 30°$,

then $x = \frac{1}{2}$, and $y = \frac{2 - \sqrt{3}}{4}$

which values will be found to satisfy the eq′ns.

Exercises

Prove that—

1. $\tan^{-1}\frac{1}{2} + \tan^{-1}\frac{1}{3} = \frac{\pi}{4}$ (*Euler's formula*)

2. $\cot^{-1}\frac{3}{4} + \cot^{-1}\frac{1}{7} = 135^\circ$

3. $\sin^{-1}\frac{1}{\sqrt{5}} + \cot^{-1}3 = 45^\circ$

4. $\tan^{-1}\frac{1}{3} + \tan^{-1}\frac{1}{5} + \tan^{-1}\frac{1}{7} + \tan^{-1}\frac{1}{8} = 45^\circ$

5. $4\tan^{-1}\frac{1}{5} - \tan^{-1}\frac{1}{239} = \frac{\pi}{4}$ (*Machin's formula*)

6. $\tan^{-1}\frac{3}{5} - \cot^{-1}\frac{7}{3} = \cot^{-1}\frac{22}{3}$

7. $2\tan^{-1}\frac{1}{408} - \tan^{-1}\frac{1}{1393} = \tan^{-1}\frac{1}{239}$

8. $\cos^{-1}\frac{9}{\sqrt{82}} + \sin^{-1}\frac{4}{\sqrt{41}} = \frac{\pi}{4}$

9. $\tan\left(\tan^{-1}\frac{1}{3} + 3\tan^{-1}\frac{1}{7} + \tan^{-1}\frac{1}{26} - \frac{\pi}{4}\right) = \frac{1}{2057}$

10. $\sin^{-1}\sqrt{\frac{\theta}{\theta + \alpha}} = \tan^{-1}\sqrt{\frac{\theta}{\alpha}}$

11. $\sin^{-1}x \pm \sin^{-1}y = \sin^{-1}(x\sqrt{1 - y^2} \pm y\sqrt{(1 - x^2)})$

12. $\cos^{-1}x \pm \cos^{-1}y = \cos^{-1}(xy \mp \sqrt{(1 - x^2)(1 - y^2)})$

13. $\tan^{-1}\left(\frac{\alpha\cos\theta}{1 - \alpha\sin\theta}\right) - \tan^{-1}\left(\frac{\alpha - \sin\theta}{\cos\theta}\right) = \theta$

14. $\cos^{-1}\frac{\alpha + \beta\cos x}{\beta + \alpha\cos x} + \cos^{-1}\frac{\alpha - \beta\cos x}{\beta - \alpha\cos x}$

$$= \cos^{-1}\frac{\alpha^2\sin^2 x + \alpha^2 - \beta^2}{\alpha^2\sin^2 x - \alpha^2 + \beta^2}$$

15. $\frac{1}{2}\tan^{-1}\left[2\tan\left\{\alpha + \tan^{-1}(\tan^3\alpha)\right\}\right] = \alpha$

16. $2\tan^{-1}\left\{\sqrt{\frac{\alpha - \beta}{\alpha + \beta}}\tan\frac{\theta}{2}\right\} = \cos^{-1}\left(\frac{\beta + \alpha\cos\theta}{\alpha + \beta\cos\theta}\right)$

17. $\cos^{-1}\frac{1 - x^2}{1 + x^2} - \cos^{-1}\frac{1 - y^2}{1 + y^2} = \tan^{-1}\frac{2(x - y)(1 + xy)}{(1 + xy)^2 - (x - y)^2}$

18. $\tan^{-1}\dfrac{\alpha_1-\alpha_2}{1+\alpha_1\alpha_2}+\tan^{-1}\dfrac{\alpha_2-\alpha_3}{1+\alpha_2\alpha_3}+\&c+\tan^{-1}\dfrac{\alpha_{n-1}-\alpha_n}{1+\alpha_{n-1}\alpha_n}$

$$=\tan^{-1}\alpha_1-\tan^{-1}\alpha_n$$

19. $\sin^{-1}\dfrac{2\alpha_1\beta_1}{\alpha_1^2+\beta_1^2}+\sin^{-1}\dfrac{2\alpha_2\beta_2}{\alpha_2^2+\beta_2^2}+\&c+\sin^{-1}\dfrac{2\alpha_n\beta_n}{\alpha_n^2+\beta_n^2}$

$$=\sin^{-1}\frac{2xy}{x^2+y^2}$$

where x, y are rational functions of α_1, β_1, α_2, β_2 &c

20. $\dfrac{\alpha^3}{2}\operatorname{cosec}^2\left(\tfrac{1}{2}\tan^{-1}\dfrac{\alpha}{\beta}\right)+\dfrac{\beta^3}{2}\sec^2\left(\tfrac{1}{2}\tan^{-1}\dfrac{\beta}{\alpha}\right)=(\alpha+\beta)(\alpha^2+\beta^2)$

21. $\tan^3\left\{\dfrac{\sin^{-1}(3\sin\alpha)+\alpha}{4}\right\}=\tan\left\{\dfrac{\sin^{-1}(3\sin\alpha)-3\alpha}{4}\right\}$

22. $\sin\left(\dfrac{2\pi}{3}+\cos^{-1}\dfrac{\alpha}{\beta}\right)\sin\left(\dfrac{2\pi}{3}-\cos^{-1}\dfrac{\alpha}{\beta}\right)$

$$-\cos\left(\frac{2\pi}{3}+\cos^{-1}\frac{\alpha}{\beta}\right)\cos\left(\frac{2\pi}{3}-\cos^{-1}\frac{\alpha}{\beta}\right)=\tfrac{1}{2}$$

23. $\cos(6\tan^{-1}x)=\dfrac{1-15x^2+15x^4-x^6}{(1+x^2)^3}$

24. $\cos\sec^{-1}\sin\tan^{-1}\cos\tan^{-1}\sin\cos^{-1}\tan\sin^{-1}x$

$$=\sqrt{\frac{3-4x^2}{1-x^2}}$$

25. $\sin 2\cos^{-1}\tan 3\cot^{-1}x=\dfrac{(6x^2-2)\sqrt{x^6-15x^4+15x^2-1}}{x^2(x^2-3)^2}$

And, if $x = 5{\cdot}1761328$, calculate the value of the dexter, by finding the value of the sinister from a table book.

NOTE—'*The left side of this is merely a hard phrase to be construed from Trigonometry into Algebra.*' (*De Morgan*)

26. If $2\theta=\phi+\sin^{-1}(\alpha\sin\phi)$

then $\phi=\theta+\tan^{-1}\left(\dfrac{1-\alpha}{1+\alpha}\tan\theta\right)$

27. If $\cos^{-1}\dfrac{x}{a}=2\sin^{-1}\dfrac{y}{a}$, then $\alpha^2=\alpha x+2y^2$

28. If $x^2 = a^2 + b^2 + ab,$

and $\tan^2\phi = 2\,\text{cosec}\left(2\tan^{-1}\frac{b}{a}\right)$

then $x = \sqrt{ab}.\sec\phi$

29. If $\cos^{-1}\frac{x}{a} + \cos^{-1}\frac{y}{b} = \alpha,$

then $\left(\frac{x}{a}\right)^2 - \frac{2xy}{ab}\cos\alpha + \left(\frac{y}{b}\right)^2 = \sin^2\alpha$

Sandhurst: '88.

30. If $\tan^{-1}\frac{1}{4} + 2\tan^{-1}\frac{1}{5} + \tan^{-1}\frac{1}{6} = \frac{\pi}{2} - \cot^{-1}x$

then $x = \frac{235}{226}$

31. If $\tan^{-1}\frac{1}{x} + \tan^{-1}\frac{1}{\alpha^2 - x + 1} = \tan^{-1}\frac{1}{\alpha - 1}$

then $x = \alpha$, or $\alpha^2 - \alpha + 1$

32. The algebraic equivalent of

$$\sin^{-1}x \pm \sin^{-1}y \pm \sin^{-1}z \pm \sin^{-1}u = n\pi,$$

where n is any integer, is

$$\{4(s-x)(s-y)(s-z)(s-u) - (xy+zu)(xz+yu)(xu+yz)\}$$
$$\times\{4s(s-x-y)(s-x-z)(s-x-u) - (zu-xy)(yu-xz)(yz-xu)\} = 0$$

where $2s = x + y + z + u$

Math' Tri': '88.

33. If $a = b\cos\theta + c\sin\theta$
and $b = a\cos\theta + \partial\sin\theta$

then either $a^2 - b^2 = 0$
and $\sin\theta = 0$

or $a^2 + \partial^2 = b^2 + c^2$
and $\theta = \tan^{-1}\frac{\partial}{a} + \tan^{-1}\frac{c}{b}$

Q. C. B.: '69.

34. If $\tan^3\theta = \tan(\theta - \alpha)$

then $\theta = \frac{1}{4}\{\alpha + \sin^{-1}(3\sin\alpha)\}$

where, unless $\alpha \ngtr \sin^{-1}\frac{1}{3}$, there are no real solutions.

35. If $\tan(n\cot\theta) = \cot(n\tan\theta)$

then $\theta = \frac{m\pi}{2} + (-1)^m \frac{1}{2}\sin^{-1} 4n/(2r+1)\pi$

Pet' Camb': '60.

36. If $\cos^2\theta\cos^2\alpha + 4\cos\theta\sin(\theta-\alpha)\sin^3\alpha = \sin^2\alpha\cos^2\alpha$

then $\theta = n\pi + \tan^{-1}(2\tan\alpha \pm \cot\alpha)$

37. Find the value of x in each of the following—

(1) $\tan^{-1}x + \frac{1}{2}\sec^{-1}5x = \frac{\pi}{4}$

(2) $\cot^{-1}(x-1) - \cot^{-1}(x+1) = \frac{\pi}{12}$

(3) $\operatorname{vers}^{-1}x - \operatorname{vers}^{-1}\alpha x = \operatorname{vers}^{-1}(1-\alpha)$

(4) $\operatorname{vers}^{-1}(1+x) - \operatorname{vers}^{-1}(1-x) = \tan^{-1}(2\sqrt{1-x^2})$

(5) $\cos^{-1}\frac{1-x^2}{1+x^2} + \tan^{-1}\frac{2x}{1-x^2} = \frac{4\pi}{3}$

(6) $2\sin^{-1}\frac{1}{\sqrt{x^2-2x+2}} = \frac{\pi}{6} + \operatorname{chd}^{-1}\frac{2}{\sqrt{x^2+2x+2}}$,

where $\operatorname{chd}\alpha$ (contracted for chord α) is defined as $2\sin\frac{\alpha}{2}$.

38. The following equations require the inverse trigonometrical functions to express their complete solutions: find the solution in each case: θ, ϕ are always the unknown angles; others are supposed known—

(1) $\sqrt{2}\sin^2\theta = \cos 3\theta$

(2) $\operatorname{cosec}\theta + \operatorname{cosec}3\theta = 4\cos^2\frac{\theta}{2}\operatorname{cosec}2\theta$

(3) $\frac{\sin\alpha\cos(\beta+\theta)}{\sin\beta\cos(\alpha+\theta)} = \frac{\tan\beta}{\tan\alpha}$

Math' Tri': '62.

(4) $35\sin 3\theta + 20\cos 3\theta + 39\sin\theta - 20\cos\theta = 0$

Pet' and Pemb', Camb': '79.

(5) $\tan(\theta+\alpha)\tan(\theta+\beta) = \tan^2\theta$

(6) $(1+\cos\alpha\cos\theta)\tan(\theta+\alpha) + (\cos\alpha+\cos\theta)\sin\alpha = 0$

(7) $m\sec^2\theta\tan(\alpha-\theta) = n\tan\theta\sec^2(\alpha-\theta)$

(8) $\sin(\pi\cos\theta) = \cos(\pi\sin\theta)$

(9) $\tan(\theta - \alpha)\tan(\theta - \beta) = \tan^2\theta$

(10) $$\left.\begin{array}{l} p\sin^4\theta - q\sin^4\phi = p \\ p\cos^4\theta - q\cos^4\phi = q \end{array}\right\}$$

(11) $$\left.\begin{array}{l} 2m\cos\theta = a\sin(\theta+\phi) - b\sin(\theta-\phi) \\ 2n\cos\phi = a\sin(\theta+\phi) + b\sin(\theta-\phi) \end{array}\right\}$$

(12) $$\left.\begin{array}{l} a\tan\phi = b\tan\theta \\ \sin(\theta+\phi) = a + b \end{array}\right\}$$

(13) $$\left.\begin{array}{l} n\sin\theta - m\cos\theta = 2m\sin\phi \\ n\sin 2\theta - m\cos 2\phi = m \end{array}\right\}$$

(14) $$\left.\begin{array}{l} \sin(\theta-\phi) = \frac{1}{10} \\ \dfrac{\sin^2\theta\cos^2\phi + \cos^2\theta\sin^2\phi}{\sin^2\theta\cos^2\phi - \cos^2\theta\sin^2\phi} = \frac{5}{3} \end{array}\right\}$$

(15) $$\left.\begin{array}{l} \sin\theta + \cos\phi = \alpha \\ \sin\phi + \cos\theta = \beta \end{array}\right\}$$

(16) $\cos 7\theta + 7\cos\theta = 0$

(17) $\tan\alpha\tan\theta = \tan^2(\alpha+\theta) - \tan^2(\alpha-\theta)$

(18) $\tan^2\theta = 2\tan\alpha\tan\beta\sec\theta + \tan^2\alpha + \tan^2\beta$

(19) $\tan^2\theta + \sec^2 2\theta = \dfrac{7\sqrt{3} - 10}{\sqrt{3}}$

Show that *one* of the solutions is *nearly* $\pi/3$.

(20) $6\cos 3\theta - 10\cos 2\theta \div 5\sin 2\theta + 22\cos\theta - 5\sin\theta = 10$

Math' Tri': '84.

39. Show that 2α is one value of

$$\cos^{-1}\frac{1 - a^2\cos 2\alpha - 2a\sin\alpha}{1 + a^2 - 2a\sin\alpha} - \cos^{-1}\frac{\cos 2\alpha + 2a\sin\alpha - a^2}{1 + a^2 - 2a\sin\alpha}$$

Ox' First Public Exam': '79.

40. Show that the sum of any number of terms such as

$$a\sin(\theta+\alpha),\quad b\sin(\theta+\beta),\quad c\sin(\theta+\gamma),\quad \&c$$

can always be reduced to the form $x\sin(\theta+\phi)$

where $x^2 = \Sigma a^2 - 2\Sigma(ab\cos\overline{\alpha-\beta})$

and $\phi = \tan^{-1}\overline{\Sigma a\sin\alpha/\Sigma a\cos\alpha}$

41. Solve, by the aid of trigonometry, the following equations—

(1) $$\left.\begin{aligned}\sqrt{x(1-y)} + \sqrt{y(1-x)} &= a\\ \sqrt{x(1-x)} + \sqrt{y(1-y)} &= b\end{aligned}\right\}$$

(2) $$\left.\begin{aligned}\frac{x+y}{1-xy} + \frac{1+xy}{x-y} &= a\\ \frac{x-y}{1+xy} + \frac{1-xy}{x+y} &= b\end{aligned}\right\}$$

(3) $$\left.\begin{aligned}x &= a\sqrt{1-z^2}\\ y &= b\sqrt{1-z^2}\\ z &= \sqrt{(1-x^2)(1-y^2)}\end{aligned}\right\}$$

(4) $$\left.\begin{aligned}\frac{x+y}{\sqrt{(1+x^2)(1+y^2)}} &= a\\ \frac{1+xy}{\sqrt{(1+x^2)(1+y^2)}} &= b\end{aligned}\right\}$$

42. Two circular discs, in the same plane, have radii R, r; and the distance apart of their centres is a: an endless cord passes round the discs; prove that the length of the cord—

1°, when it *crosses* between the discs is

$$2\sqrt{a^2-(R+r)^2} + \pi(R+r) + 2(R+r)\sin^{-1}\frac{R+r}{a}$$

and, 2°, when it does *not* cross is

$$2\sqrt{a^2-(R-r)^2} + \pi(R+r) + 2(R-r)\sin^{-1}\frac{R-r}{a}$$

43. Show that the solution of the equation

$$\begin{vmatrix} 1 & \cos\theta & 0 & 0\\ \cos\theta & 1 & \cos\alpha & \cos\beta\\ 0 & \cos\alpha & 1 & \cos\gamma\\ 0 & \cos\beta & \cos\gamma & 1\end{vmatrix} = 0$$

is

$$\theta = n\pi + (-1)^n \sin^{-1}\left(\sqrt{\cos^2\alpha + \cos^2\beta - 2\cos\alpha\cos\beta\cos\gamma}\,\big/\sin\gamma\right)$$

NOTE—*Multiply 1st row by* $\cos\theta$, *and subtract result from 2nd: this will reduce the determinant to 3 rows: it can be then expanded by Sarrus' rule; and result follows at once.*

44. Construct geometrically the angles

$$\tan^{-1}(3-\sqrt{2}) \quad \text{and} \quad \sec^{-1}(\sqrt{5}-1)$$

45. If $\sin^2\theta + \sin^2\phi = \frac{1}{2}$, show that *one* value of

$$\sin^{-1}(\sin\theta + \sin\phi) + \sin^{-1}(\sin\theta - \sin\phi)$$

is a right angle.

46. Two circles (radii R, r) cut at an angle α; prove that the area common to them is

$$(R^2 - r^2)\tan^{-1}\frac{r\sin\alpha}{R + r\cos\alpha} + r^2\theta - Rr\sin\theta$$

Find also the length of their common chord.

47. If α, β, γ, &c are the roots of the equation

$$x^n - p_1 x^{n-1} + p_2 x^{n-2} - p_3 x^{n-3} + \&c = 0$$

show that

$$\tan^{-1}\alpha + \tan^{-1}\beta + \tan^{-1}\gamma + \&c = \tan^{-1}\frac{p_1 - p_3 + p_5 - \&c}{1 - p_2 + p_4 - \&c}$$

48. In a triangle ABC, if S is the circumcentre, I the incentre, O the orthocentre, M the mid point of BC, and L the mid point of the altitude of A; show that the angle made by BC with—

1°, ML is $\cot^{-1}(\cot B \sim \cot C)$

2°, IS „ $\tan^{-1}\overline{\cos B + \cos C - 1}/\overline{\sin B \sim \sin C}$

3°, SO „ $\tan^{-1}\overline{\tan B\tan C - 3}/\overline{\tan B \sim \tan C}$

4°, IO „ $\tan^{-1}\left(\cot\frac{B\sim C}{2} - \cos A\Big/\overline{2\sin\frac{B}{2}\sin\frac{C}{2}\sin\frac{B\sim C}{2}}\right)$

Wolstenholme: 533–6.

CHAPTER VIII

Elimination

§ **43**. *Def′*—As in Algebra, when from a given set of equations we deduce a new equation, not containing one (or more) of the letters originally involved, we are said to **eliminate** such letter (or letters) between the given equations, and the new equation is called the **eliminant** (or **resultant**) of the given equations with respect to the letter (or letters) got rid of; so also in Trigonometry, when from a given set of equations we deduce a new equation, not containing one (or more) of the angles originally involved, we are said to **eliminate** such angle (or angles) between the given equations, and the new equation is called the **eliminant** (or **resultant**) of the given equations with respect to the angle (or angles) got rid of.

To acquire facility in eliminating between eq′ns (whether in Alg′ or Trig′) is of the first consequence in a mathematical training; for, in almost all mathematical problems, treated analytically, the solution consists mainly of two processes—

1°, writing down eq′ns; and,

2°, eliminating between them.

Now this facility comes from observation of methods, and practice in applying them. No rules of general application can be given; but certain methods and artifices are frequently useful; and should certainly be familiar to the Student. Of these some Examples, which should be carefully studied, will now be given.

In effecting a trig′l elimination we are, of course, at liberty to make use of any identity that can be usefully pressed into our service. Thus, if we can reduce two eq′ns to the forms

$$\left.\begin{aligned}\sin\theta &= a\\ \cos\theta &= b\end{aligned}\right\}$$

then, squaring and adding, and recollecting that

$$\sin^2\theta + \cos^2\theta = 1$$

we get

$$a^2 + b^2 = 1$$

and θ is eliminated.

This artifice is of constant use.

So also, if eq'ns are reduced to the forms

$$\left.\begin{aligned}\sec\theta = a\\ \tan\theta = b\end{aligned}\right\}$$

then, squaring and subtracting, and recollecting that

$$\sec^2\theta - \tan^2\theta = 1$$

we get

$$a^2 - b^2 = 1$$

as the eliminant of the eq'ns with respect to θ.

Or suppose that we are given the eq'ns

$$\left.\begin{aligned}1 - \cos\theta = a\sin\theta\\ 1 + \cos\theta = b\sin\theta\end{aligned}\right\}$$

and it is req'd to elim' θ.

Multiplying the corresponding sides of the given eq'ns together, and recollecting that

$$1 - \cos^2\theta = \sin^2\theta$$

we get

$$ab = 1$$

as the req'd eliminant.

Again the relations between the roots and coeff's of an eq'n can often be advantageously used.

E.g. if $\left.\begin{aligned}a\sin\theta + b\cos\theta = c\\ \text{and}\quad a\sin\phi + b\cos\phi = c\end{aligned}\right\}$

then $\sin\theta$, $\sin\phi$ are the roots of the eq'n

$$b^2 - b^2x^2 = c^2 + a^2x^2 - 2acx$$

i.e. of $(a^2 + b^2)x^2 - 2acx + c^2 - b^2 = 0$

$$\left.\begin{aligned} \therefore\quad \sin\theta + \sin\phi &= 2\,ac/\overline{a^2+b^2} \\ \text{and}\quad \sin\theta\sin\phi &= \overline{c^2-b^2}/\overline{a^2+b^2} \end{aligned}\right\}$$

$$\left.\begin{aligned} \text{Sim'ly}\quad \cos\theta + \cos\phi &= 2\,bc/\overline{a^2+b^2} \\ \text{and}\quad \cos\theta\cos\phi &= \overline{c^2-a^2}/\overline{a^2+b^2} \end{aligned}\right\}$$

From which results, if any other eq'n is given connecting any of the T. F.s $\sin\theta$, $\sin\phi$, $\cos\theta$, $\cos\phi$, the elimination can be effected.

Some of the eliminations now to be given are solutions of Geometrical Problems: although these will probably be at present beyond the Student's range of reading, he may find them interesting to refer to hereafter; and the statement of the Problem will ∴ be given in connection with the eq'ns by means of which it is solved.

Examples

1. *To eliminate θ between the equations*

$$\left.\begin{aligned} x &= a\cos(\theta-\alpha) \\ y &= b\cos(\theta-\beta) \end{aligned}\right\}$$

we have

$$\left.\begin{aligned} \frac{x}{a} &= \cos\theta\cos\alpha + \sin\theta\sin\alpha \\ \text{and}\quad \frac{y}{b} &= \cos\theta\cos\beta + \sin\theta\sin\beta \end{aligned}\right\}$$

$$\left.\begin{aligned} \therefore\quad \frac{x}{a}\cos\beta - \frac{y}{b}\cos\alpha &= \sin\theta\sin(\alpha-\beta) \\ \text{and}\quad -\frac{x}{a}\sin\beta + \frac{y}{b}\sin\alpha &= \cos\theta\sin(\alpha-\beta) \end{aligned}\right\}$$

Squaring and adding, we get

$$\left(\frac{x}{a}\right)^2 - \frac{2\,xy}{ab}\cos(\alpha-\beta) + \left(\frac{y}{b}\right)^2 = \sin^2(\alpha-\beta)$$

NOTE—*The above gives a solution of this Problem—A given △ moves so that two corners are on fixed intersecting lines, find the Locus of the* 3rd *corner.*

2. *To eliminate θ and φ between the equations*

$$\left.\begin{aligned}\frac{x}{a}\cos\theta + \frac{y}{b}\sin\theta &= 1\\ \frac{x}{a}\cos\phi + \frac{y}{b}\sin\phi &= 1\\ \frac{\cos\theta\cos\phi}{a^2} + \frac{\sin\theta\sin\phi}{b^2} &= 0\end{aligned}\right\}$$

Put Z for $\sin\theta$: then $\sin\theta$ is a root of the eq'n in Z

$$\left\{\left(\frac{x}{a}\right)^2 + \left(\frac{y}{b}\right)^2\right\} Z^2 - \frac{2y}{b} Z + 1 - \left(\frac{x}{a}\right)^2 = 0$$

And, from 2nd given eq'n, $\sin\phi$ is also a root of this eq'n

$$\therefore\quad \sin\theta\sin\phi = \overline{1 - \left(\frac{x}{a}\right)^2}\Big/\overline{\left(\frac{x}{a}\right)^2 + \left(\frac{y}{b}\right)^2}$$

$$\text{Sim'ly}\quad \cos\theta\cos\phi = \overline{1 - \left(\frac{y}{b}\right)^2}\Big/\overline{\left(\frac{x}{a}\right)^2 + \left(\frac{y}{b}\right)^2}$$

∴, substituting in the 3rd given eq'n, we get

$$b^2 - y^2 + a^2 - x^2 = 0$$

i. e. $x^2 + y^2 = a^2 + b^2$ is the eliminant

NOTE—*The eliminant gives the Locus (which is a ⊙) of the cross of tangents to the ellipse* $(x/a)^2 + (y/b)^2 = 1$, *which are ⊥ to each other.*

3. *To eliminate θ and φ between the equations*

$$\left.\begin{aligned}x\cos\theta + y\sin\theta &= 2a\\ x\cos\phi + y\sin\phi &= 2a\\ 2\cos\frac{\theta}{2}\cos\frac{\phi}{2} &= 1\end{aligned}\right\}$$

From the two 1st, we have, by well-known algebraic processes,

$$\frac{x}{\cos\dfrac{\theta+\phi}{2}} = \frac{y}{\sin\dfrac{\theta+\phi}{2}} = \frac{2a}{\cos\dfrac{\theta-\phi}{2}}$$

$$\text{and}\ \therefore\ = \frac{x + 2a}{2\cos\dfrac{\theta}{2}\cos\dfrac{\phi}{2}} = x + 2a, \quad \text{by 3rd eq'n}$$

$$\therefore\ x^2 + y^2 = (x + 2a)^2$$

$$\therefore\ y^2 = 4a(x + a) \text{ is the eliminant.}$$

Otherwise—Put Z for $\cos\frac{\theta}{2}$, then $\cos\frac{\theta}{2}$, $\cos\frac{\phi}{2}$ are roots of the eq'n in Z

$$x(2Z^2 - 1) + 2yZ\sqrt{1 - Z^2} = 2a$$

i.e. of $4(x^2 + y^2)Z^4 - 4(x^2 + y^2 + 2a)Z^2 + (x + 2a)^2 = 0$

$$\therefore\ 4\cos^2\frac{\theta}{2}\cos^2\frac{\phi}{2} = \frac{(x + 2a)^2}{x^2 + y^2}$$

$\therefore$, from 3rd eq'n, $x^2 + y^2 = (x + 2a)^2$

or $y^2 = 4a(x + a)$, as before.

4. *Show that the elimination of* ϕ *between the equations*

$$\left.\begin{array}{l}\cos\alpha\cos\phi/a^2 + \sin\alpha\sin\phi/b^2 + 1 = 0\\ \cos\beta\cos\phi/a^2 + \sin\beta\sin\phi/b^2 + 1 = 0\end{array}\right\}$$

where $a^2 + b^2 = 1$, *produces*

$$\cos\alpha\cos\beta/a^2 + \sin\alpha\sin\beta/b^2 + 1 = 0.$$

Math' Tri': '79.

The following solution was given by Prof' Cayley in the *Messenger of Mathematics*, vol' v, p. 24 (old series)

$$\cos\phi/a^2 : \sin\phi/b^2 : 1$$

$$= \sin\alpha - \sin\beta : \cos\beta - \cos\alpha : \sin(\beta - \alpha)$$

$$= \cos\frac{\alpha + \beta}{2} : \sin\frac{\alpha + \beta}{2} : -\cos\frac{\alpha - \beta}{2}$$

Whence, eliminating ϕ by squaring and adding, we get

$$a^4\cos^2\frac{\alpha + \beta}{2} + b^4\sin^2\frac{\alpha + \beta}{2} = \cos^2\frac{\alpha - \beta}{2}$$

$$\therefore\ a^4(1 + \cos\overline{\alpha + \beta}) + b^4(1 - \cos\overline{\alpha + \beta}) - (1 + \cos\overline{\alpha - \beta}) = 0$$

$$\therefore\ a^4 + b^4 - 1 + (a^4 - b^4 - 1)\cos\alpha\cos\beta + (-a^4 + b^4 - 1)\sin\alpha\sin\beta = 0$$

But, since $a^2 + b^2 = 1$,

$$\therefore\ a^4 + b^4 - 1 = -2a^2b^2$$

$$a^4 - b^4 - 1 = -2b^2$$

$$-a^4 + b^4 - 1 = -2a^2$$

$$\therefore\ \cos\alpha\cos\beta/a^2 + \sin\alpha\sin\beta/b^2 + 1 = 0$$

5. *To show that, if α, β, γ are angles, unequal and less than 2π, which satisfy the equation*

$$\frac{a}{\cos\theta} + \frac{b}{\sin\theta} + c = 0$$

then $\Sigma \sin(\alpha + \beta) = 0$

Wolstenholme: 425.

Here we have to elim′ a, b, c; and ∴, at once,

$$\begin{vmatrix} \operatorname{cosec}\alpha & \sec\alpha & 1 \\ \operatorname{cosec}\beta & \sec\beta & 1 \\ \operatorname{cosec}\gamma & \sec\gamma & 1 \end{vmatrix} = 0$$

Expanding this, by Sarrus' rule, we have

$$\Sigma \operatorname{cosec}\alpha \sec\beta - \Sigma \operatorname{cosec}\beta \sec\alpha = 0$$

whence $\Sigma \sin\beta \sin\gamma \cos\alpha \cos\gamma - \Sigma \sin\alpha \sin\gamma \cos\beta \cos\gamma = 0$

$$\therefore \quad \Sigma \sin 2\gamma . \sin \overline{\beta - \alpha} = 0$$

∴ [Exercise (43) p. 88] $\Sigma \sin \overline{\alpha + \beta} . \Sigma \sin \overline{\alpha - \beta} = 0$

$$\therefore \quad \Sigma \sin \overline{\alpha + \beta} = 0,$$

the other factor being rejected by cond′ns of question.

NOTE—*The eliminant gives the cond′n that three normals to an ellipse at the p′ts whose eccentric* ∧*s are α, β, γ, may be concurrent.*

The following is another* solution

Put x for $\tan\frac{\theta}{2}$, then

$$a\frac{1 + x^2}{1 - x^2} + b\frac{1 + x^2}{2x} + c = 0$$

$$\text{whence} \quad x^4 + \frac{2}{b}(c - a)x^3 - \frac{2}{b}(c + a)x - 1 = 0$$

∴, if δ is the 4th value of θ corresponding to the 4th root of this eq′n, we have, by Theory of Eq′ns,

$$\left.\begin{aligned} \Sigma \tan\frac{\alpha}{2}\tan\frac{\beta}{2} &= 0 \\ \text{and} \quad \Pi \tan\frac{\alpha}{2} &= -1 \end{aligned}\right\}$$

where the *four* roots are taken into account.

* Kindly sent to me by Prof′ Wolstenholme. R. C. J. N.

Eliminating δ between these, we get

$$\tan\frac{\alpha}{2}\tan\frac{\beta}{2}+\tan\frac{\beta}{2}\tan\frac{\gamma}{2}+\tan\frac{\gamma}{2}\tan\frac{\alpha}{2}$$
$$-\left(\tan\frac{\alpha}{2}+\tan\frac{\beta}{2}+\tan\frac{\gamma}{2}\right)\cot\frac{\alpha}{2}\cot\frac{\beta}{2}\cot\frac{\gamma}{2}=0$$

Whence, taking account only of α, β, γ, we get

$$\Sigma\left(\cot\frac{\alpha}{2}\cot\frac{\beta}{2}-\tan\frac{\alpha}{2}\tan\frac{\beta}{2}\right)=0$$

$$\therefore\quad \Sigma\frac{\cos^2\frac{\alpha}{2}\cos^2\frac{\beta}{2}-\sin^2\frac{\alpha}{2}\sin^2\frac{\beta}{2}}{\sin\frac{\alpha}{2}\sin\frac{\beta}{2}\cos\frac{\alpha}{2}\cos\frac{\beta}{2}}=0$$

$$\therefore\quad \Sigma\cos\frac{\alpha+\beta}{2}\cos\frac{\alpha-\beta}{2}\sin\frac{\gamma}{2}\cos\frac{\gamma}{2}=0$$

$$\therefore\quad \Sigma(\cos\alpha+\cos\beta)\sin\gamma=0$$

$$\therefore\quad \Sigma\sin\overline{\alpha+\beta}=0$$

6. *To eliminate ϕ, ϕ' between*

$$\left.\begin{aligned}\frac{ax}{\cos\phi}-\frac{by}{\sin\phi}&=c^2\\ \frac{ax}{\cos\phi'}-\frac{by}{\sin\phi'}&=c^2\\ \frac{\cos\frac{\phi-\phi'}{2}}{\cos\frac{\phi+\phi'}{2}}&=\frac{c}{a}\end{aligned}\right\}\quad\text{where}\quad c^2=a^2-b^2$$

Solving for x, we get

$$ax\sin\overline{\phi-\phi'}=c^2(\sin\phi-\sin\phi')\cos\phi\cos\phi'$$

whence

$$ax\cos\frac{\phi-\phi'}{2}=c^2\cos\frac{\phi+\phi'}{2}\left\{\cos^2\frac{\phi+\phi'}{2}+\cos^2\frac{\phi-\phi'}{2}-1\right\}$$

$\therefore$, putting Ω for $\cos^2\frac{\phi-\phi'}{2}$, and using 3rd eq'n, we get

$$c(x+c)=(a^2+c^2)\,\Omega\quad\ldots\ldots\ldots\quad (A)$$

Again solving for y, from two 1st eq'ns,

$$\text{by} \sin \overline{\phi - \phi'} = + c^2 (\cos\phi - \cos\phi') \sin\phi \sin\phi'$$

$$\therefore \text{by} \cos\frac{\phi - \phi'}{2} = - c^2 \sin\frac{\phi + \phi'}{2} \left\{\cos^2\frac{\phi - \phi'}{2} - \cos^2\frac{\phi + \phi'}{2}\right\}$$

$$\therefore \text{by} \sqrt{\Omega} = - c^2 \sqrt{1 - \frac{a^2}{c^2}\,\Omega}\left(1 - \frac{a^2}{c^2}\right)\Omega$$

$$\therefore \text{by} = \frac{b^2}{c}\sqrt{c^2\,\Omega - a^2\,\Omega^2}$$

$$\therefore c^2 y^2 + a^2 b^2 \Omega^2 = b^2 c^2 \Omega \quad . \; . \; . \; . \; . \; . \; . \; . \quad \text{(B)}$$

$\therefore$, substituting for Ω from eq'n (A) in eq'n (B) we have as the eliminant

$$(a^2 + c^2)^2 y^2 + a^2 b^2 (x + c)^2 = b^2 c (a^2 + c^2)(x + c)$$

NOTE—*This gives the Locus of the cross of normals at the ends of a focal chord of the ellipse* $(x/a)^2 + (y/b)^2 = 1$, ϕ, ϕ' *being the eccentric* $\wedge^s$ *of the ends of the chord.*

The same result is found, by a quite different method, as a solution of the Problem, in Salmon's Conics, 6th ed, p. 211.

Exercises

1. Eliminate θ between the equations—

(1) $\left.\begin{aligned} \tan\theta + \sin\theta &= a \\ \tan\theta - \sin\theta &= b \end{aligned}\right\}$

(2) $\left.\begin{aligned} \sec\theta + \tan\theta &= a \\ \operatorname{cosec}\theta + \cot\theta &= b \end{aligned}\right\}$

(3) $\left.\begin{aligned} \operatorname{cosec}\theta - \sin\theta &= a \\ \sec\theta - \cos\theta &= b \end{aligned}\right\}$

(4) $\left.\begin{aligned} \sin\theta + \sin 2\theta &= a \\ \cos\theta + \cos 2\theta &= b \end{aligned}\right\}$

(5) $\left.\begin{aligned} \operatorname{cosec}\theta \tan^3\theta (\operatorname{cosec}^2\theta + 1) + \sec\theta &= a \\ \tan^3\theta (\operatorname{cosec}^2\theta + 1) - \tan\theta &= b \end{aligned}\right\}$

(6) $\left.\begin{aligned} \sin(\theta - \alpha)/\sin(\theta - \beta) &= a/b \\ \cos(\theta - \alpha)/\cos(\theta - \beta) &= a'/b' \end{aligned}\right\}$

(7) $$\left.\begin{aligned}\sin\theta\cos\theta(\cos\theta-\sin\theta)&=a\\ \sin\theta\cos\theta(\cos\theta+\sin\theta)&=b\end{aligned}\right\}$$

(8) $$\left.\begin{aligned}3\cos(\theta+\alpha)-2\sin(\theta+\alpha)&=\cos(\theta-\alpha)\\ 3\cos(\theta+\beta)+4\sin(\theta+\beta)&=\cos(\theta-\beta)\end{aligned}\right\}$$

(9) $$\left.\begin{aligned}\tan\alpha&=\overline{(1+m)\tan\theta}\big/\overline{1-m\tan^2\theta}\\ \tan\beta&=\overline{(1-m)\tan\theta}\big/\overline{1+m\tan^2\theta}\end{aligned}\right\}$$

(10) $$\sin\theta\cot x=\sin(\theta+\alpha)\cot y=\sin(\theta+\alpha+\beta)\cot z$$

(11) $$\left.\begin{aligned}(a-b)\sin(\theta+\alpha)&=(a+b)\sin(\theta-\alpha)\\ a\tan\frac{\theta}{2}&=b\tan\frac{\alpha}{2}+c\end{aligned}\right\}$$

(12) $$a\sin\theta+b\cos\theta=c=a\operatorname{cosec}\theta+b\sec\theta$$

(13) $$\left.\begin{aligned}x&=a(1+\sin^2\theta\cos 2\theta)\\ y&=a\sin^2\theta\sin 2\theta\end{aligned}\right\}$$

(14) $$\left.\begin{aligned}a\sin^2\theta+b\cos^2\theta&=c\\ a\operatorname{cosec}^2\theta+b\sec^2\theta&=\partial\end{aligned}\right\}$$

(15) $$\left.\begin{aligned}\sin(\theta+\alpha)&=k\sin 2\theta\\ \sin(\theta+\beta)&=k\sin 2\theta\end{aligned}\right\}$$

(16) $$\cos(\theta-\alpha-\beta)\cos(\theta-\beta)=m=\cos(\theta-\alpha+\beta)\cos(\theta+\beta)$$

(17) $$a\cos\theta+b\sin\theta=c=a\cos 3\theta+b\sin 3\theta$$

Ox' 2nd Public Exam': '90.

(18) $$x\sin\theta+y\cos\theta=a\sqrt{1+\sin 2\theta\cos\omega}$$

$$x\cos\theta-y\sin\theta=a\cos 2\theta\cos\omega\big/\sqrt{1+\sin 2\theta\cos\omega}$$

Ox' 1st Public Exam': '77.

2. Eliminate θ and ϕ between the equations—

(1) $$\left.\begin{aligned}x&=a\cos^m\theta\cos^m\phi\\ y&=b\cos^m\theta\sin^m\phi\\ z&=c\sin^m\theta\end{aligned}\right\}$$

(2)
$$\left.\begin{aligned} \cos^2\theta &= \cos\alpha/\cos\beta \\ \cos^2\phi &= \cos\gamma/\cos\beta \\ \tan\theta/\tan\phi &= \tan\alpha/\tan\gamma \end{aligned}\right\}$$

(3)
$$\left.\begin{aligned} a\cos^2\theta + b\sin^2\theta &= m\cos^2\phi \\ a\sin^2\theta + b\cos^2\theta &= n\sin^2\phi \\ m\tan^2\theta &= n\tan^2\phi \end{aligned}\right\}$$

(4)
$$\left.\begin{aligned} a\cos\theta + b\sin\theta &= c \\ a\cos\phi + b\sin\phi &= c \\ \tan\theta\tan\phi &= m \end{aligned}\right\}$$

(5)
$$\left.\begin{aligned} a\sin\phi &= b\sin\theta \\ a\sin(\theta+\phi) &= c\sin\theta \\ \cos\theta - \cos\phi &= 2m \end{aligned}\right\}$$

(6)
$$\left.\begin{aligned} \sec\theta - \sec\phi &= a \\ \tan\frac{\theta}{2}\Big/\tan\frac{\phi}{2} &= b \\ \tan\theta\tan\phi &= c \end{aligned}\right\}$$

(7)
$$\left.\begin{aligned} a\cos\theta + b\cos\phi &= c \\ a\sin\theta + b\sin\phi &= c \\ \theta + \phi &= n\pi \end{aligned}\right\}$$

where n is any integer

Caius Camb': '78.

(8)
$$\left.\begin{aligned} x\cos\theta + y\sin\theta &= 1 \\ x\cos\phi + y\sin\phi &= 1 \\ a\cos\theta\cos\phi + b\sin\theta\sin\phi + c(\cos\theta + \cos\phi) & \\ + \partial(\sin\theta + \sin\phi) &= 0 \end{aligned}\right\}$$

(9)
$$\left.\begin{aligned} \sin\theta + \cos\phi &= 4mn \\ \cos\theta + \sin\phi &= 2(m^2 - n^2) \\ \cos\frac{\theta+\phi}{2} - \sin\frac{\theta+\phi}{2} &= \sqrt{2}(m^2 - n^2) \end{aligned}\right\}$$

Ox' Jun' Math' Schol': '78.

3. Eliminate ϕ from the equations

$$\left.\begin{aligned} \frac{x}{a}\cos\phi + \frac{y}{b}\sin\phi &= 1 \\ \frac{ax}{\cos\phi} - \frac{by}{\sin\phi} &= a^2 - b^2 \end{aligned}\right\}$$

NOTE—*These are the eq'ns to the tangent and normal to an ellipse at the p't whose eccentric* $\wedge$ *is* ϕ : *the elimination gives the Locus of their cross, which is of course the ellipse itself.*

4. Eliminate ϕ from the equations

$$\left.\begin{aligned} ax/\cos\phi - by/\sin\phi &= a^2 - b^2 \\ ax\sin\phi/\cos^2\phi + by\cos\phi/\sin^2\phi &= 0 \end{aligned}\right\}$$

NOTE—*The elimination gives the Envelope of the normals (i.e. the evolute) of the ellipse* $(x/a)^2 + (y/b)^2 = 1$

5. Eliminate θ between the equations

$$\left.\begin{aligned} -bA^2x\sec\theta + aB^2y\,\mathrm{cosec}\,\theta &= ab(A^2 - B^2) \\ bA^2x\sec^2\theta\sin\theta + aB^2y\,\mathrm{cosec}^2\theta\cos\theta &= 0 \end{aligned}\right\}$$

NOTE—*The elimination gives the Envelope of the* $\perp$ *from any p't on the ellipse* $(x/a)^2 + (y/b)^2 = 1$, *to the polar of that p't with respect to the ellipse* $(x/A)^2 + (y/B)^2 = 1$

6. If
$$\tan\phi = \frac{1 + 2e^2}{1 - e^2}\tan\theta$$

and
$$\tan\left(\frac{\pi}{4} + \frac{\phi}{2}\right) = \frac{1 + e}{1 - e}\tan\left(\frac{\pi}{4} + \frac{\theta}{2}\right)$$

show that
$$\sin\theta = \frac{2}{e}$$

7. Eliminate θ between the equations

$$\left.\begin{aligned} a\cos 2\theta + b\sin 2\theta &= c \\ a'\cos 3\theta + b'\sin 3\theta &= 0 \end{aligned}\right\}$$

John's Camb': '87.

NOTE—*Put* x *for* $\tan\theta$, *and elim' between*

$$\left.\begin{aligned} b'x^3 + 3a'x^2 - 3b'x - a' &= 0 \\ \textit{and}\quad (a + c)x^2 - 2bx + c - a &= 0 \end{aligned}\right\}$$

8. Show that if θ is eliminated between the equations

$$6 \tan(\theta + \alpha) = 3 \tan(\theta + \beta) = 2 \tan(\theta + \gamma)$$

the eliminant can be arranged as

$$3 \sin^2(\alpha - \beta) + 5 \sin^2(\beta - \gamma) - 2 \sin^2(\gamma - \alpha) = 0$$

Pet' Camb': '72.

9. Eliminate θ from the equations

$$\left.\begin{array}{lll} x(1 + \sin^2\theta - \cos\theta) & - y \sin\theta (1 + \cos\theta) & = a(1 + \cos\theta) \\ y(1 + \cos^2\theta) & - x \sin\theta \cos\theta & = a \sin\theta \end{array}\right\}$$

NOTE—*Solve for* x *and* y.

Wolstenholme: 416.

10. If
$$\left.\begin{array}{ll} a \cos 3\theta + b \cos 2\theta + c \cos\theta & = 0 \\ \text{and} \quad a \cos 2\theta + b \cos\theta + c & = 0 \end{array}\right\}$$

eliminate θ; and show that $\theta = m\pi$, if a and c are unequal.

11. Show that if x, y, z are eliminated between the equations

$$\left.\begin{array}{l} x = y \cos\gamma + z \cos\beta \\ y = z \cos\alpha + x \cos\gamma \\ z = x \cos\beta + y \cos\alpha \end{array}\right\}$$

the eliminant can be arranged as

$$\Pi \sin^2\alpha = \Pi (\cos\alpha + \cos\beta \cos\gamma)$$

Clare Camb': '74.

NOTE—*Use Sarrus' rule: then mult' by* $\cos\alpha \cos\beta \cos\gamma$, *and re-arrange.*

12. Eliminate θ and ϕ between

$$\left.\begin{array}{r} \sin(\phi - \theta)/\sin(\phi + \theta) = a/b \\ \sin\theta / \sin\phi = b/x \\ \cos(\theta - \phi) = c/x \end{array}\right\}$$

13. Eliminate θ and ϕ between

$$\frac{x}{a}\cos\theta + \frac{y}{b}\cos\phi = 1 = \frac{x}{a}\sin\theta + \frac{y}{b}\sin\phi$$

$$\text{and} \quad \left(\frac{x}{a}\right)^2 - \left(\frac{y}{b}\right)^2 = \mu^2(\sin 2\theta - \sin 2\phi)$$

E. T. xxiii.

14. Show that the elimination of β between

$$\left.\begin{aligned} x^2 \cos\alpha \cos\beta + x(\sin\alpha + \sin\beta) + 1 &= 0 \\ \text{and}\quad x^2 \cos\beta \cos\gamma + x(\sin\beta + \sin\gamma) + 1 &= 0 \end{aligned}\right\}$$

produces $x^2 \cos\gamma \cos\alpha + x(\sin\gamma + \sin\alpha) + 1 = 0$

E. T. xlv. *Math' Tri'*: '69. *John's Camb'*: '81.

15. Show that an eliminant of θ, ϕ between

$$\left.\begin{aligned} b\tan\phi - c\tan\theta &= a \\ \partial\sin\phi - e\sin\theta &= 0 \\ be\sec^3\phi - c\partial\sec^3\theta &= 0 \end{aligned}\right\}$$

is $(c\partial)^{\frac{2}{3}} - (be)^{\frac{2}{3}} = (e^2 - \partial^2)^{\frac{1}{3}}\, a^{\frac{2}{3}}$

Christ's Camb': '49.

16. Show that the elimination of θ, ϕ between

$$\left.\begin{aligned} \frac{x}{a}\cos\theta + \frac{y}{b}\sin\theta &= 1 \\ \frac{x}{a}\cos\phi + \frac{y}{b}\sin\phi &= 1 \\ 4\cos\frac{\theta-\phi}{2}\cos\frac{\alpha-\theta}{2}\cos\frac{\alpha-\phi}{2} &= 1 \end{aligned}\right\}$$

produces $\left(\frac{x}{a} - \cos\alpha\right)^2 + \left(\frac{y}{b} - \sin\alpha\right)^2 = 3$

Wolstenholme: 460.

17. Eliminate ϕ, i from

$$\left.\begin{aligned} A &= \left(\frac{\sin^2\phi}{a^2} + \frac{\cos^2\phi}{b^2}\right)\cos^2 i + \frac{\sin^2 i}{c^2} \\ B &= \frac{\cos^2\phi}{a^2} + \frac{\sin^2\phi}{b^2} \\ C &= \left(\frac{1}{b^2} - \frac{1}{a^2}\right)\sin\phi\cos\phi\cos i \end{aligned}\right\}$$

Math' Tri': '40.

18. If

$$\left.\begin{aligned} \sin(\theta+\alpha) = \sin(\phi+\alpha) &= \sin\beta \\ \text{and}\quad a\sin(\theta+\phi) + b\sin(\theta-\phi) &= c \end{aligned}\right\}$$

show that either $a\sin(2\alpha \pm 2\beta) = -c$

or $a\sin 2\alpha \pm b\sin 2\beta = c$

Math' Tri': '71.

19. If
$$\left.\begin{aligned}\tan y/\tan\beta &= \sin(x-\alpha)/\sin\alpha\\ \text{and}\quad \tan y/\tan 2\beta &= \sin(x-2\alpha)/\sin 2\alpha\end{aligned}\right\}$$
show that $\tan x/\sin 2\alpha = 1/(\cos 2\alpha - \cos 2\beta)$

Math' Tri': '74.

20. Eliminate ϕ between
$$\left.\begin{aligned}a\cos x\cos\phi + b\sin x\sin\phi &= c\\ a\cos y\cos\phi + b\sin y\sin\phi &= c\end{aligned}\right\}$$

21. If $x = \lambda a\cos\phi,\quad y = \lambda b\sin\phi$

and $\frac{x}{a}\cos\phi + \frac{y}{b}\sin\phi - 1 = \lambda(a^2\sin^2\phi + b^2\cos^2\phi)/(a^2+b^2)$

show that $\left(\frac{x^2+y^2}{a^2+b^2}\right)^2 = \left(\frac{x}{a}\right)^2 + \left(\frac{y}{b}\right)^2$

NOTE—*This is a solution of the Problem to find the Locus of the centres of all rectangular hyperbolas having contact of the 3rd order with the ellipse*
$$(x/a)^2 + (y/b)^2 = 1.$$

22. Eliminate θ between
$$\left.\begin{aligned}x\cos\theta + y\sin\theta &= 2a\sin 2\theta\\ y\cos\theta - x\sin\theta &= a\cos 2\theta\end{aligned}\right\}$$

NOTE—*The elimination gives the Envelope of normals (i.e. the evolute) of the hypocycloid* $x^{\frac{2}{3}} + y^{\frac{2}{3}} = a^{\frac{2}{3}}$.

23. Eliminate ϕ from
$$\left.\begin{aligned}\frac{x}{a}\cos\phi - \frac{y}{b}\sin\phi &= \cos 2\phi\\ \frac{x}{a}\sin\phi + \frac{y}{b}\cos\phi &= 2\sin 2\phi\end{aligned}\right\}$$

NOTE—*The elimination gives the Envelope of the chords of curvature of the ellipse* $(x/a)^2 + (y/b)^2 = 1$.

24. Eliminate θ and ϕ between
$$\left.\begin{aligned}a\tan 3\theta + b\tan\theta &= c\\ a\tan 3\phi + b\tan\phi &= c\\ \theta + \phi &= \pi/4\end{aligned}\right\}$$

R. U. Schol': '90.

25. Eliminate θ between

$$\left.\begin{aligned} a\cos^3\theta + b\cos^2\theta + c\cos\theta + d &= 0 \\ a\cos^3\phi + b\cos^2\theta + c\cos\phi + d &= 0 \\ \cos\theta + \cos\phi &= m \end{aligned}\right\}$$

26. Eliminate θ between

$$\left.\begin{aligned} x\cos(\theta+\alpha) + y\sin(\theta+\alpha) &= a\sin 2\theta \\ y\cos(\theta+\alpha) - x\sin(\theta+\alpha) &= 2a\cos 2\theta \end{aligned}\right\}$$

R. U. Schol': '87.

27. Show that if x, y, a, b are eliminated between the equations

$$\left.\begin{aligned} \left(\frac{x}{a}\right)^2 + \left(\frac{y}{b}\right)^2 &= 1 \\ x = r\cos\theta, \quad y &= r\sin\theta \\ a^2 + b^2 = r^2 + r'^2, \quad ab &= rr'\sin\alpha \end{aligned}\right\}$$

the eliminant may be arranged either as

$$\pm\cot 2\theta = \cot 2\alpha + \left(\frac{r}{r'}\right)^2 \operatorname{cosec} 2\alpha$$

or $$\left(\frac{r}{r'}\right)^2 \sin 2\theta = \sin 2(\alpha \pm \theta)$$

28. Eliminate θ from the eq'ns

$$\left.\begin{aligned} \rho\sin(\alpha - 3\theta) &= 2a\sin^3\theta \\ \rho\cos(\alpha - 3\theta) &= 2a\cos^3\theta \end{aligned}\right\}$$

T. C. D.: '75.

29. Show that, unless $\sin 2\alpha = 0$, the result of eliminating θ and ϕ from the simultaneous equations

$$\left.\begin{aligned} x\operatorname{cosec}\theta + y\sec\theta = 1 &= x\operatorname{cosec}\phi + y\sec\phi \\ \text{and}\quad \theta - \phi &= 2\alpha \end{aligned}\right\}$$

is $(1 - 4\cos^2\alpha)^3 x^2 y^2 = (x^2 + y^2 - \cos^2 2\alpha)^2 (x^2 + y^2 - \cos^2\alpha)\cos^2\alpha$

Ox' Jun' Math' Schol': '80.

30. Eliminate θ from

$$\left.\begin{aligned} x &= a\cos\overline{2\theta + \alpha} \\ y &= b\sin\overline{3\theta + \beta} \end{aligned}\right\}$$

Ox' Jun' Math' Schol': '75.

31. Eliminate θ from the equations

$$\left.\begin{aligned} x\cos(\theta - A) + y\cos\theta &= 2R\sin(\theta + C)\cos\theta\cos(\theta - A) \\ x\sin(\theta - A) + y\sin\theta &= 2R\{\sin(\theta + C)\sin\theta\cos(\theta - A) - \cos B\cos\theta\} \end{aligned}\right]$$

where A, B, C are angles of a triangle.

Tucker: *E. T.* vii.

32. Show that if θ, ϕ, χ are eliminated between

$$\left.\begin{aligned} a\cos(\theta + \phi) + b\cos(\theta - \phi) + c &= 0 \\ a\cos(\phi + \chi) + b\cos(\phi - \chi) + c &= 0 \\ a\cos(\chi + \theta) + b\cos(\chi - \theta) + c &= 0 \end{aligned}\right\}$$

where θ, ϕ, χ are unequal, the resultant is $a^2 - b^2 + 2bc = 0$.

Math' Tri': '79.

NOTE—*From the 1st two eq'ns we see that* $\cos\theta$, $\cos\chi$ *are the two values of* x *got from the eq'n*

$$\{(a + b)^2\cos^2\phi + (a - b)^2\sin^2\phi\}x^2 + 2(a + b)c\cos\phi . x + c^2 - (a - b)^2\sin^2\phi = 0$$

$\therefore$ $\cos\theta\cos\chi$ (*and sim'ly* $\sin\theta\sin\chi$) *are known: substitute in 3rd eq'n.*

33. If

$$\left.\begin{aligned} \frac{\cos^2\theta}{a^2} + \frac{\sin^2\theta}{b^2} &= \frac{1}{r^2} \\ r\sin\theta &= \rho\sin\phi \\ \text{and } r^2 + \rho^2 - 2r\rho\cos(\theta + \phi) &= a^2 - b^2 \end{aligned}\right\}$$

prove that $\rho\left\{1 \pm \sqrt{1 - \frac{b^2}{a^2}} . \cos\phi\right\} = \pm\frac{b^2}{a}$

34. Eliminate θ between

$$\left.\begin{aligned} a\sec\theta + b\operatorname{cosec}\theta &= c \\ a\sec\theta - b\operatorname{cosec}\theta &= \cos 2\theta \end{aligned}\right\}$$

Joh' Camb': '42.

35. Eliminate ϕ between

$$\left.\begin{aligned} x\cos 3\phi + y\sin 3\phi &= b\cos\phi \\ x\sin 3\phi + y\cos 3\phi &= b\cos\left(\phi + \frac{\pi}{6}\right) \end{aligned}\right\}$$

Christ's Emm' and Sid', Camb': '81.

36. Prove that the eliminant of θ between

$$4(\cos\alpha\cos\theta+\cos\phi)(\cos\alpha\sin\theta+\sin\phi)$$
$$=4(\cos\alpha\cos\theta+\cos\chi)(\cos\alpha\sin\theta+\sin\chi)$$
$$=(\cos\phi-\cos\chi)(\sin\phi-\sin\chi)$$

can be written $\{\cos(\phi-\chi)-\cos 2\alpha\}\{\cos(\phi-\chi)\pm 1\}=0$

Math' Tri': '73. *Caius Camb'*: '74.

37. Show that the elimination of x, y, z between the equations

$$a/\sin\overline{x+y}+b/\cos\overline{x+y}=c/\cos\overline{x-y}$$
$$a/\sin\overline{y+z}+b/\cos\overline{y+z}=c/\cos\overline{y-z}$$
$$a/\sin\overline{z+x}+b/\cos\overline{z+x}=c/\cos\overline{z-x}$$

where no two of x, y, z are equal, or differ by a multiple of π, produces

$$a^2+b^2=c^2$$

R. U. Schol': '88.

38. Eliminate θ between

$$\left.\begin{array}{l}\sin 3\theta+m\sin\overline{\alpha+\theta}+2\sin\overline{\beta-\theta}\sin\overline{\gamma-\theta}\sin\overline{\beta+\gamma+\theta}=0\\ \cos 3\theta+m\cos\overline{\alpha+\theta}+2\cos\overline{\beta-\theta}\cos\overline{\gamma-\theta}\cos\overline{\beta+\gamma+\theta}=0\end{array}\right\}$$

Math' Tri': '87.

NOTE—*Mult' 1st eq'n by* $\sin\theta$, *2nd eq'n by* $\cos\theta$, *and add; then mult' 1st by* $\cos\theta$, *2nd by* $\sin\theta$, *and subtract: this will give* $\cos 2\theta\,(1+\cos 2\theta)$ *and* $\sin 2\theta\,(1-\cos 2\theta)$ *in terms of* α, β, γ, m.

39. ABC is a triangle, whose sides respectively opposite A, B, C are denoted by a, b, c: the coordinates (see p. 20) of A, with respect to rectangular axes through C as origin, are $e-\cos\alpha$, $\sqrt{1-e^2}\,.\sin\alpha$; and the coordinates of B with respect to the same axes are $e-\cos\beta$, $\sqrt{1-e^2}\,.\sin\beta$:

$$\text{if } a+b+c=2(1-\cos\phi)$$
$$\text{and } a+b-c=2(1-\cos\theta)$$

prove that $\phi-\theta=\alpha-\beta$, and $\cos\dfrac{\phi+\theta}{2}=e\cos\dfrac{\alpha+\beta}{2}$

Joh' Camb': '40.

CHAPTER IX

Some Important Limits

§ **44**. **Theorem**—*If* θ *radians is a positive angle less than* $\pi/2$, *then* $\sin\theta$, θ, $\tan\theta$ *are in ascending order of magnitude; and vanish in a ratio of equality.*

Ax'—If two tangents are drawn to a circle from the same point, then the arc intercepted between these tangents is less than the sum of the tangents, but is greater than the chord joining the points of contact.

Note—Such an arc must be less than a semi-⊙; and ∴ the ∧ subtended by it at the centre must be less than 2 r't ∧s.

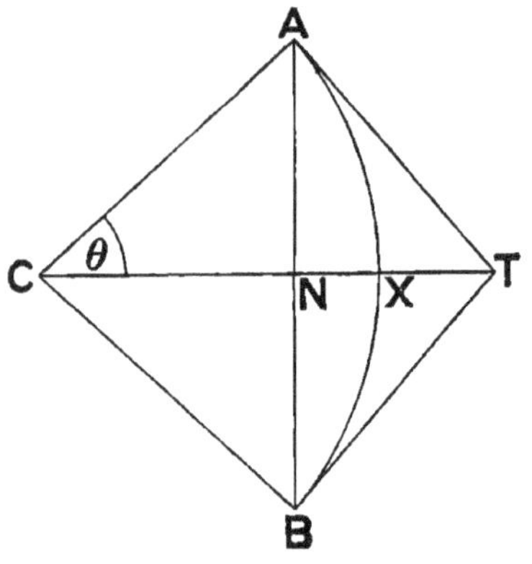

Let $A\hat{C}B$ be 2θ.

Describe a ⊙, with C as centre, cutting the arms of ACB in A, B.

Draw AT, BT tang's at A, B; and let CT cut ch'd AB in N, and arc AB in X.

Then $A\hat{C}T = \theta$; and, by the above axiom,

$$\text{arc } AXB > \text{ch'd } ANB, \text{ but} < AT + BT$$

$$\therefore \text{ arc } AX > NA, \text{ but} < AT$$

$$\therefore \frac{AX}{CA} > \frac{NA}{CA}, \text{ but} < \frac{AT}{CA}$$

$$\therefore \quad \theta > \sin\theta, \text{ but } < \tan\theta$$

i.e. $\sin\theta$, θ, $\tan\theta$ are in ascending order of magnitude

$\therefore$ also $\dfrac{\sin\theta}{\theta}$, 1, $\dfrac{\tan\theta}{\theta}$ " " " "

or $\dfrac{\sin\theta}{\theta}$, 1, $\dfrac{\sin\theta}{\theta}\cdot\sec\theta$ " " "

Now, as θ diminishes, $\sec\theta$ continually approaches (and ultimately has) unity for its Limit.

$\therefore$ $\dfrac{\sin\theta}{\theta}$, 1, $\dfrac{\tan\theta}{\theta}$ approach equality, as θ is diminished:

$$\text{i.e.} \quad \text{Lim}'_{\theta=0}\left(\frac{\sin\theta}{\theta}\right) = 1$$

$$\text{and} \quad \text{Lim}'_{\theta=0}\left(\frac{\tan\theta}{\theta}\right) = 1$$

Cor. (1)—If $x^\circ = {}^r\theta$

$$\text{Lim}'_{x=0}\left(\frac{\sin x}{x}\right) = \text{Lim}'_{\theta=0}\left(\frac{\sin\theta}{\frac{180}{\pi}.\theta}\right) = \text{Lim}'_{\theta=0}\left(\frac{\sin\theta}{\theta}\right)\frac{\pi}{180} = \frac{\pi}{180}$$

So also, if $y'' = {}^r\phi$

$$\text{Lim}'_{y=0}\left(\frac{\sin y}{y}\right) = \text{Lim}'_{\phi=0}\left(\frac{\sin\phi}{\frac{180\times 60\times 60}{\pi}.\phi}\right)$$

$$= \text{Lim}'_{\phi=0}\left(\frac{\sin\phi}{\phi}\right)\frac{\pi}{64800} = \frac{\pi}{64800}$$

Cor. (2)—By the formula at end of p. 64, we have

$$\cos\frac{\alpha}{2}\cos\frac{\alpha}{4}\cos\frac{\alpha}{8}\;\&\text{c}\;\cos\frac{\pi}{2^n} = \frac{\sin\alpha}{2^n\sin\frac{\alpha}{2^n}}$$

$$\text{Now} \quad \frac{\sin\alpha}{2^n\sin\frac{\alpha}{2^n}} = \left(\frac{\sin\alpha}{\alpha}\right)\left(\frac{\frac{\alpha}{2^n}}{\sin\frac{\alpha}{2^n}}\right)$$

And, by the Theorem, if α is in radian measure, and n is increased indefinitely,

$$\text{Lim}'_{n=\infty}\left(\frac{\frac{\alpha}{2^n}}{\sin\frac{\alpha}{2^n}}\right)=1$$

$$\therefore\quad \text{Lim}'_{n=\infty}\left(\frac{\sin\alpha}{2^n\sin\frac{\alpha}{2^n}}\right)=\frac{\sin\alpha}{\alpha}$$

$$\therefore\quad \text{Lim}'_{n=\infty}\left(\cos\frac{\alpha}{2}\cos\frac{\alpha}{4}\cos\frac{\alpha}{8}\ \&c\ \cos\frac{\alpha}{2^n}\right)=\frac{\sin\alpha}{\alpha}$$

(*Euler's Formula*)

Put $\pi/2$ for α, in this formula, and we get

$$\frac{2}{\pi}=\frac{\sqrt{2}}{2}\cdot\frac{\sqrt{2+\sqrt{2}}}{2}\cdot\frac{\sqrt{2+\sqrt{2+\sqrt{2}}}}{2}\ \&c\ \textit{ad infin}'$$

(*Vieta's Formula*)

This gives another method for approximating to π.

§ **45. Theorem**—*If θ radians is a positive angle less than $\pi/2$, then* $\tan\theta-\theta>\theta-\sin\theta$.

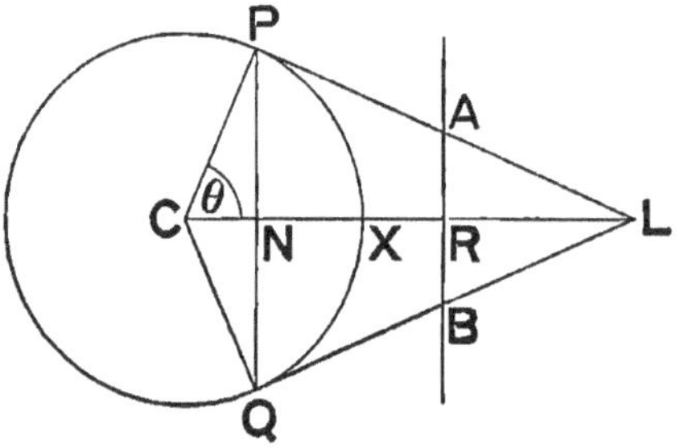

Let C be the centre of a ⊙; 2θ the inclination of two of its radii CP, CQ; PL, QL the tang's at P, Q; and A, B the respective mid' p'ts of PL, QL.

Let CL cut PQ in N, AB in R, and the ⊙ in X.

Then RA is the radical axis of a system of which the ⊙ is one, and L is a limiting p't *.

* See *Euclid Revised*, pp. 349, 350.

$\therefore$ AB will not meet the $\odot$

$\therefore$ PA + AB + BQ > arc PXQ

$$\therefore \frac{PL}{CP} + \frac{NP}{CP} > 2\left(\frac{\text{arc } PX}{CP}\right)$$

i.e. $\tan\theta + \sin\theta > 2\theta$

i.e. $\tan\theta - \theta > \theta - \sin\theta$, if θ is pos' and $< \pi/2$ *

§ **46.** Though, when θ is a small positive angle, $\sin\theta$, θ, $\tan\theta$ are nearly equal; yet, since then $\tan\theta - \theta > \sin\theta - \theta$, the approximations $\tan\theta = \theta$, $\cos\theta = 1$, are not so close as $\sin\theta = \theta$.

But, since $1 - \cos\theta = \dfrac{\sin^2\theta}{1 + \cos\theta}$,

$$\therefore \text{Lim}'_{\theta=0}(1 - \cos\theta) = \text{Lim}'_{\theta=0}\left(\frac{\sin^2\theta}{1+\cos\theta}\right) = \frac{\theta^2}{1+1}$$

$$\therefore \cos\theta = 1 - \frac{\theta^2}{2}, \text{ nearly}$$

From this $\tan\theta = \dfrac{\theta}{1 - \dfrac{\theta^2}{2}} = \theta\left(1 - \dfrac{\theta^2}{2}\right)^{-1}$

$$\text{i.e. } \tan\theta = \theta + \frac{\theta^3}{2}, \text{ nearly}$$

$\therefore \theta$, $1 - \dfrac{\theta^2}{2}$, $\theta + \dfrac{\theta^3}{2}$ are respectively approximations to $\sin\theta$, $\cos\theta$, $\tan\theta$, when θ is a small positive angle.

We proceed to get still closer approximations.

* This proof is due to Prof' Genese: *E. T.* xlii.

§ 47. Theorem—*If* θ *radians is a positive angle less than* $\pi/2$, *then* $\sin\theta$ *lies between* θ *and* $\theta - \theta^3/4$; $\cos\theta$ *lies between* $1 - \theta^3/2$ *and* $1 - \theta^2/2 + \theta^4/16$; *and* $\tan\theta$ *lies between* $\theta + \theta^3/2$ *and* $\theta + \theta^3/4$.

For $$\sin\theta = 2\tan\frac{\theta}{2}\cos^2\frac{\theta}{2}$$

and ∴ $$> \theta\cos^2\frac{\theta}{2},\quad \text{when } \theta < \pi/2 \text{ and pos}'$$

i. e. $$> \theta\left(1 - \sin^2\frac{\theta}{2}\right)$$

∴, *à fortiori*, $$> \theta\left\{1 - \left(\frac{\theta}{2}\right)^2\right\}$$

i. e. $$> \theta - \theta^3/4$$

And it has been shown (§ 44) that $\sin\theta < \theta$, ∴ $\sin\theta$ lies between θ and $\theta - \theta^3/4$, when θ is pos′ and $< \pi/2$.

Again $$\cos\theta = 1 - 2\sin^2\frac{\theta}{2}$$

∴ $$> 1 - 2\left(\frac{\theta}{2}\right)^2$$

i. e. $$> 1 - \theta^2/2$$

Also $$\cos\theta < 1 - 2\left\{\frac{\theta}{2} - \frac{1}{4}\left(\frac{\theta}{2}\right)^3\right\}^2$$

∴ $$< 1 - \theta^2/2 + \theta^4/16$$

∴ $\cos\theta$ lies between $1 - \theta^2/2$ and $1 - \theta^2/2 + \theta^4/16$, when θ is pos′ and $< \pi/2$.

Since, 1°, $\left.\begin{array}{l}\sin\theta < \theta \\ \cos\theta > 1 - \dfrac{\theta^2}{2}\end{array}\right\}$ and, 2°, $\left.\begin{array}{l}\sin\theta > \theta - \dfrac{\theta^3}{4} \\ \cos\theta < 1 - \dfrac{\theta^2}{2} + \dfrac{\theta^4}{16}\end{array}\right\}$

$$\therefore,\ 1^\circ,\ \tan\theta < \frac{\theta}{1-\dfrac{\theta^2}{2}};\ \text{and, } 2^\circ,\ \tan\theta > \frac{\theta-\dfrac{\theta^3}{4}}{1-\dfrac{\theta^2}{2}+\dfrac{\theta^4}{16}}$$

$$\text{i.e. } \tan\theta < \theta\left(1-\frac{\theta^2}{2}\right)^{-1}$$

$$\text{but } \tan\theta > \left(\theta-\frac{\theta^3}{4}\right)\left(1-\frac{\theta^2}{2}+\frac{\theta^4}{16}\right)^{-1}$$

$$\text{i.e. } \tan\theta < \theta+\frac{\theta^3}{2};\ \text{ but } \tan\theta > \theta+\frac{\theta^3}{4}$$

So that $\tan\theta = \theta + \dfrac{\theta^3}{3}$ is a good approximation.

Hence, for small values of θ, we have that

$$\sin\theta = \theta - \frac{\theta^3}{4}$$

$$\cos\theta = 1 - \frac{\theta^2}{2}$$

$$\tan\theta = \theta + \frac{\theta^3}{3}$$

are good approximations, if powers of θ higher than θ^3 are of no consequence.

We can however go closer still to $\sin\theta$ without taking account of powers of θ higher than θ^3.

§ **48. Theorem**—*If* θ *radians is a positive angle less than* $\pi/2$ *then* $\sin\theta$ *is greater than* $\theta - \theta^3/6$; $\cos\theta$ *is less than* $1 - \theta^2/2 + \theta^4/24$; *and* $\tan\theta$ *is greater than* $\theta + \theta^3/3$ *.

* Rawson: *Messenger of Mathematics*: *III.* 101.

Since $\sin\theta = 3\sin\frac{\theta}{3} - 4\sin^3\frac{\theta}{3}$

$$\therefore\quad 3\sin\frac{\theta}{3} = 3^2\sin\frac{\theta}{3^2} - 3\times 4\sin^3\frac{\theta}{3^2}$$

$$3^2\sin\frac{\theta}{3^2} = 3^3\sin\frac{\theta}{3^3} - 3^2\times 4\sin^3\frac{\theta}{3^3}$$

&c &c

$$3^{n-1}\sin\frac{\theta}{3^{n-1}} = 3^n\sin\frac{\theta}{3^n} - 3^{n-1}\times 4\sin^3\frac{\theta}{3^n}$$

∴, adding corresponding sides, and omitting terms common,

$$\sin\theta = 3^n\sin\frac{\theta}{3^n} - 4\left\{\sin^3\frac{\theta}{3} + 3\sin^3\frac{\theta}{3^2} + 3^2\sin^3\frac{\theta}{3^3} + \&c + 3^{n-1}\sin^3\frac{\theta}{3^n}\right\}$$

But $\sin({}^r\phi) < {}^r\phi$

$$\sin\theta > 3^n\sin\frac{\theta}{3^n} - 4\left\{\left(\frac{\theta}{3}\right)^3 + 3\left(\frac{\theta}{3^2}\right)^3 + 3^2\left(\frac{\theta}{3^3}\right)^3 + \&c + 3^{n-1}\left(\frac{\theta}{3^n}\right)^3\right\}$$

$$\therefore\quad \sin\theta > \theta\left(\frac{\sin\frac{\theta}{3^n}}{\frac{\theta}{3^n}}\right) - \frac{4\theta^3}{27}\left(1 + \frac{1}{3^2} + \frac{1}{3^4} + \&c + \frac{1}{3^{2n-2}}\right)$$

∴, passing to the Limit when $n = \infty$, we get

$$\sin\theta > \theta - \frac{\frac{4\theta^3}{27}}{1-\frac{1}{9}}$$

$$\text{i.e.}\quad > \theta - \theta^3/6$$

Hence $\sin\frac{\theta}{2} > \frac{\theta}{2} - \frac{\theta^3}{48}$

$$\therefore \quad \cos\theta \left(\text{i.e. } 1 - 2\sin^2\frac{\theta}{2}\right) < 1 - 2\left(\frac{\theta}{2} - \frac{\theta^3}{48}\right)^2$$

$$\text{i.e.} \quad < 1 - \frac{\theta^2}{2} + \frac{\theta^4}{24}$$

$\therefore$, on account of both the foregoing inequalities,

$$\tan\theta > \frac{\theta - \frac{\theta^3}{6}}{1 - \frac{\theta^2}{2} + \frac{\theta^4}{24}}$$

$$\text{i.e.} \quad > \left(\theta - \frac{\theta^3}{6}\right)\left\{1 + \left(\frac{\theta^2}{2} - \frac{\theta^4}{24}\right)\right\}$$

$$> \theta + \frac{\theta^3}{3} - \frac{\theta^5}{8}$$

$$\therefore \quad > \theta + \frac{\theta^3}{3}$$

Hence we have the following very good approximations—

When θ is so small an angle that θ^4 may be neglected,

$$\sin\theta = \theta - \frac{\theta^3}{6}$$

$$\cos\theta = 1 - \frac{\theta^2}{2}$$

$$\tan\theta = \theta + \frac{\theta^3}{3}$$

where, in each case, the dexter expression is slightly less than the true value of the corresponding T. F.

These are the approximations which the Student should recollect for use.

Note—Since $1^\circ = \cdot 01745$ of a radian; and that $(\cdot 01745)^4/24$ has no significant figure in the 1st six places of decimals; we see that the above approximations are true if $\theta \not> 1^\circ$, at least as far as the sixth decimal place.

§ **49.** Theorem—*If θ radians is a positive angle less than $\pi/2$, then*

$$\frac{3\sin\theta}{2+\cos\theta},\quad \theta,\quad \text{and}\quad \frac{\tan\theta+2\sin\theta}{3}$$

are in ascending order of magnitude.

$$\text{For}\quad \frac{\sin\theta}{2+\cos\theta} = \frac{2\sin\frac{\theta}{2}\cos\frac{\theta}{2}}{1+2\cos^2\frac{\theta}{2}} = \frac{2\sin\frac{\theta}{2}}{\sec\frac{\theta}{2}+2\cos\frac{\theta}{2}}$$

But $\sec\frac{\theta}{2}+\cos\frac{\theta}{2} > 2$, since θ is pos′ and $< \pi/2$

$$\therefore\quad \frac{\sin\theta}{2+\cos\theta} < 2\left\{\frac{\sin\frac{\theta}{2}}{2+\cos\frac{\theta}{2}}\right\}$$

$$\therefore \text{ also } < 2^2\left\{\frac{\sin\frac{\theta}{4}}{2+\cos\frac{\theta}{4}}\right\}$$

$$\therefore, \text{ by n repetitions of this process, } < 2^n\left\{\frac{\sin\frac{\theta}{2^n}}{2+\cos\frac{\theta}{2^n}}\right\}$$

$$\text{i.e. } < \theta\left\{\frac{\sin\frac{\theta}{2^n}}{\frac{\theta}{2^n}}\right\}\left\{\frac{1}{2+\cos\frac{\theta}{2^n}}\right\}$$

$$\therefore\quad < \frac{\theta}{3}, \text{ by increasing n to } \infty.$$

$$\text{Again } \tan\theta - 2\tan\frac{\theta}{2} > 2\left(2\sin\frac{\theta}{2} - \sin\theta\right)$$

$$\text{if } 2\tan\frac{\theta}{2}\left\{\frac{1}{1-\tan^2\frac{\theta}{2}} - 1\right\} > 4\sin\frac{\theta}{2}\left(1 - \cos\frac{\theta}{2}\right)$$

$$\text{i. e. if } \frac{1}{\cos\frac{\theta}{2}}\left\{\frac{\sin^2\frac{\theta}{2}}{\cos^2\frac{\theta}{2} - \sin^2\frac{\theta}{2}}\right\} > 4\sin^2\frac{\theta}{4}$$

$$\text{or if } \frac{4\sin^2\frac{\theta}{4}\cos^2\frac{\theta}{4}}{\cos\theta\cos\frac{\theta}{2}} > 4\sin^2\frac{\theta}{4}$$

$$\text{or if } \cos^2\frac{\theta}{4} > \cos\theta\cos\frac{\theta}{2}$$

But, as $\theta < \pi/2$, $\therefore \cos\frac{\theta}{4} >$ both $\cos\theta$ and $\cos\frac{\theta}{2}$

$\therefore$ the above inequality is true

$$\therefore \tan\theta + 2\sin\theta > 2\left(\tan\frac{\theta}{2} + 2\sin\frac{\theta}{2}\right)$$

$$\therefore \text{ also } > 2^2\left(\tan\frac{\theta}{4} + 2\sin\frac{\theta}{4}\right)$$

$\therefore$, by n repetitions of this process, $> 2^n\left(\tan\frac{\theta}{2^n} + 2\sin\frac{\theta}{2^n}\right)$

$$\text{i.e. } > \theta\left\{\left(\frac{\tan\frac{\theta}{2^n}}{\frac{\theta}{2^n}}\right) + 2\left(\frac{\sin\frac{\theta}{2^n}}{\frac{\theta}{2^n}}\right)\right\}$$

$\therefore > 3\theta$, by increasing n to ∞.

$$\left.\begin{aligned}\therefore \theta &> \frac{3\sin\theta}{2+\cos\theta}\\ \text{and} &< \frac{\tan\theta + 2\sin\theta}{3}\end{aligned}\right\}$$

when θ is pos′ and $< \pi/2$

Note—By the above investigations, we see that the smaller θ is taken the more nearly will these limits approximate to each other. Hence, if we can find the numerical value of $\sin\theta$, $\cos\theta$, $\tan\theta$ (where θ is small), we shall get Limits between which π must lie. Thus from Chapter V we have that

$$\sin\frac{\pi}{12} = \cdot259, \quad \cos\frac{\pi}{12} = \cdot966, \quad \tan\frac{\pi}{12} = \cdot268$$

$$\therefore \frac{\pi}{12} < \frac{\cdot268 + \cdot518}{3} < \frac{\cdot786}{3} < \cdot262$$

$$\text{and } \frac{\pi}{12} > \frac{\cdot777}{2\cdot966} > \cdot261$$

$\therefore$ π lies between 3·13 and 3·14

Of course by taking θ smaller we shall get closer Limits.

Examples

1. *To find approximate values for* $\sin 1'$ and $\cos 1'$.

By an easy numerical calculation it will be found that, if ${}^{r}x = 1'$, then $x^3/6$ and $x^4/24$ have no significant figure in the 1st ten decimal places.

$\therefore$, to that order of approximation,

$$\sin 1' = {}^{r}x = \frac{3\cdot1416}{180 \times 60} = \cdot00029088$$

$$\text{And} \quad \cos 1' = 1 - \tfrac{1}{2}\left(\frac{3\cdot1416}{180 \times 60}\right)^2$$

$$= 1 - \tfrac{1}{2}(\cdot0002909)^2$$

$$= \cdot99999995$$

2. *To find* $\sin 10''$ approximately.

$$10'' < \frac{10 \times 3\cdot1416}{180 \times 60 \times 60} \text{ of a radian}$$

i. e. $< \cdot000048481$,, ,, ,,

$\therefore \sin 10'' < \cdot000048481$

but $> \cdot000048481 - \frac{1}{6}(\cdot00005)^3$

$\therefore \sin 10'' = \cdot000048481$ correct to 9 dec'l places.

Cor'—When $n \not> 10$, $\sin n'' =$ radian measure of n''

$= n \times$ radian measure of $1''$

$= n \sin 1''$

This result will be needed hereafter.

3. *To find the Limit, when* $\theta = 0$, *of*

$$\frac{m \sin \theta - \sin m\theta}{\theta(\cos\theta - \cos m\theta)}$$

Putting $\theta - \frac{\theta^3}{6}$ for $\sin\theta$, and $1 - \frac{\theta^2}{2}$ for $\cos\theta$, the given fraction becomes

$$\frac{m\left(\theta - \frac{\theta^3}{6}\right) - \left(m\theta - \frac{m^3\theta^3}{6}\right)}{\theta\left\{\left(1 - \frac{\theta^2}{2}\right) - \left(1 - \frac{m^2\theta^2}{6}\right)\right\}}$$

$$\text{which} = \frac{m\theta - \frac{m\theta^3}{6} - m\theta + \frac{m^3\theta^3}{6}}{\theta - \frac{\theta^3}{2} - \theta + \frac{m^2\theta^3}{2}}$$

$$= \frac{m^3 - m}{3(m^2 - 1)} = \frac{m}{3}, \text{ the req'd Limit}$$

4. *To find the Limit, when* $n = \infty$, *of* $\left(\cos\frac{\theta}{n}\right)^{n^2}$.

Put x for it, and take logarithms; then

$$\log x = n^2 \log\cos\frac{\theta}{n}$$

$$= n^2 \log\left(1 - \frac{\theta^2}{2n^2}\right), \text{ since } \frac{\theta}{n} \text{ is very small ultimately}$$

$$= -n^2\left(\frac{\theta^2}{2n^2} + \frac{\theta^4}{8n^4} + \&c\right)$$

$$= -\theta^2/2, \text{ ultimately}$$

$$\therefore x = e^{-\theta^2/2}$$

5. *To find the mean centre of a circular arc.*

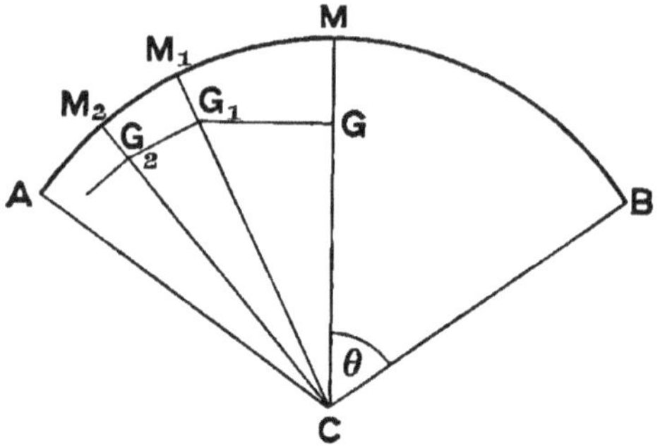

Let AB be an arc of a ⊙, whose centre is C, and rad′ r; M the mid p′t, and G the mean centre of this arc.

Obviously CGM is a st′ line.

Also, if M_1, G_1 are respectively the mid p′t and mean centre of arc AM, and, if M_2, G_2 ,, ,, ,, ,, ,, ,, ,, ,, AM_1, and sim′r notation is continued, then

$$G_1 G \text{ is } \perp \text{ to } CM,\ G_2 G_1 \text{ is } \perp \text{ to } CM_2, \&c$$

$$\therefore\ CG = CG_1 \cos\frac{\theta}{2}, \text{ if } A\hat{C}B = 2\theta$$

$$= CG_2 \cos\frac{\theta}{2} \cos\frac{\theta}{4}$$

and, after n substitutions $= CG_n \cos\frac{\theta}{2} \cos\frac{\theta}{4} \cos\frac{\theta}{8} \ldots \cos\frac{\theta}{2^n}$

∴, using *Euler's formula,* and noting that $\text{Lim}'_{n=\infty} CG_n$ is r,

we get that $CG = r\left(\frac{\sin\theta}{\theta}\right)$*

Exercises

1. If $6^\circ = {}^r\theta$, find which is the first place of decimals affected by using $1 - \theta^2/2$ for cos 6°, instead of $1 - \theta^2/2 + \theta^4/24$; and verify your result by a table-book of natural cosines.

2. Find the Limit, when ${}^r\theta = \pi/2$, of each of the following—

(1) $\sec\theta - \tan\theta$

(2) $\operatorname{cosec}\theta - \cot\theta$

* The above proof is due to Prof′ Crofton : *E. T.* xiii.

3. Find the Limit, when $x = 0$, of each of the following—

(1) $(\sin x'')/x''$

(2) $(\tan x' + \tan x'')/x''$

(3) $(\sin \alpha x^\circ)/(\sin \beta x^\circ)$

(4) $(\text{vers } \alpha x^\circ)/(\text{vers } \beta x^\circ)$

4. Evaluate $\text{Lim}'_{\theta=0} (\tan 2\theta - 2 \tan \theta)/\theta^3$

Clare and Caius Camb' : 74.

5. When ${}^r\theta$ is small, show that

$$3\theta = 2(\sin 2\theta - \sin \theta) + \tan 2\theta - \tan \theta, \text{ nearly}$$

6. By putting $\pi/24$ for θ in the results of § 49, find Limits for π.

7. Apply *Euler's formula* to show that the Limit of

$$\left(1 - \tan^2\frac{\theta}{2}\right)\left(1 - \tan^2\frac{\theta}{4}\right)\left(1 - \tan^2\frac{\theta}{8}\right)\ldots \textit{ad infin'} \text{ is } \theta/\tan\theta$$

8. Knowing $\tan 45^\circ$ find $\tan 46^\circ$, without tables, and correct to four decimal places.

R. U. Schol': '82.

9. If $\sin \theta = 2165/2166$, show that θ is very nearly $/19^{th}$ of a radian.

10. If $\sin \theta/\theta = 19493/19494$, show that θ has very nearly the magnitude of 1°.

11. If $\cos \theta/\theta + \theta/\cos \theta$ has its least possible positive value, show that

$$\theta > \sqrt{3} - 1$$

Math' Tri': '64.

12. Evaluate $\text{Lim}'_{x=0} \operatorname{cosec}^2\beta x^\circ . \log_e \cos \alpha x^\circ$.

13. Defining $\text{chd } \theta$ as $2 \sin \frac{\theta}{2}$, show that, for a small value of θ

$$3\theta = 8 \text{ chd} \frac{\theta}{2} - \text{chd } \theta$$

14. If $\cos(\theta + x) \cosh \cos h' + \sin h \sin h' = \cos \theta$, where x^2, h^3, h'^3 may be neglected; and if $h + h' = p$, $h - h' = q$, show that the number of seconds in x is, approximately,

$$\frac{1}{4 \sin 1''}\left\{p^2 \tan \frac{\theta}{2} - q^2 \cot \frac{\theta}{2}\right\}$$

Enclopædia Metropolitana : *Trigonometry*.

15. If θ is measured by the radian, find the Limit, when θ is indefinitely diminished, of each of the following—

(1) $(1 - \cos\theta)/\theta^2$

(2) $(\theta - \sin\theta)/\theta^3$

(3) $(\tan\theta - \sin\theta)/\sin^3\theta$

(4) $\sin n\theta \cos(n - 1)\theta/\sin\theta$

(5) $(\theta + \sin\theta - \sin 2\theta)/(2\theta + \tan\theta - \tan 3\theta)$

(6) $(\theta + \tan\theta - \tan 2\theta)/(2\theta + \tan\theta - \tan 3\theta)$

(7) $(2\sin\theta - \sin 2\theta)^2/(\sec\theta - \cos 2\theta)^3$

(8) $\left(\theta + \sin 2\theta - 6\sin\frac{\theta}{2}\right)^2 \bigg/ \left(4 + \cos\theta - 5\cos\frac{\theta}{2}\right)^3$

(9) $(\tan\sin\theta - \sin\tan\theta)/\theta^7$

(10) $(\tan\theta/\theta)^{1/\theta^2}$

NOTE—*Some of these are given in Williamson's, and some in Greenhill's Differential Calculus, as Exercises to be done by the Calculus: they will all come easily, by using the approximations at the end of* § 48.

16. AB, BC, CD are equal consecutive arcs of a circle: TA, TD are tangents: AB, DC meet in Y: XB, XC are tangents, and are produced to meet AD produced in E, F: if B, C, D move up to A, show that, ultimately,

$$\triangle YAD : \triangle TAD : \triangle XEF = 18 : 27 : 25$$

W. H. H. Hudson: *E. T.* xxxiv.

§ **50.** If a trig'l eq'n involves a small ∧, we may approximate to the value of that ∧, by means of some of the preceding formulæ. The method of proceeding is this—Substitute for the T. F.s of the small ∧ the values given in § 46, or in § 48, according to the smallness of the ∧, and the degree of approximation required. In the eq'n thus obtained neglect all powers of the ∧ except the 1st; and hence find a rough 1st approx' to the value of the ∧. Substitute this value in the terms involving the 2nd power of the ∧; this will give a 2nd approx' to the ∧. This 2nd approx' may then be substituted in the 3rd powers of the ∧, giving a 3rd approx'; &c.

Example

If $\cot\alpha\sin(C+\theta) = \cot\alpha\sin C\cos\phi - \cos C\sin\phi$, *and* θ *and* ϕ *are so small that their cubes may be neglected, show that, approximately,*

$$\phi = -\theta\cot\alpha + \frac{\theta^2}{2}\cot\alpha\tan C\,(1-\cot^2\alpha)$$

Ox' 1st Public Exam': '74.

From the conditions of the question we have

$$\cot\alpha\left\{\theta\cot C + \left(1-\frac{\theta^2}{2}\right)\sin C\right\} = \left(1-\frac{\phi^2}{2}\right)\cot\alpha\sin C - \phi\cos C$$

$$\therefore\ \phi = -\theta\cot\alpha + \left\{\left(1-\frac{\phi^2}{2}\right)-\left(1-\frac{\theta^2}{2}\right)\right\}\cot\alpha\tan C$$

$$= -\theta\cot\alpha + \tfrac{1}{2}(\theta^2-\phi^2)\cot\alpha\tan C$$

$\therefore$, as a 1st approximation, we have

$$\phi = -\theta\cot\alpha + \frac{\theta^2}{2}\cot\alpha\tan C$$

Substituting this for ϕ^2, and omitting terms in θ^3, θ^4, we get

$$\phi = -\theta\cot\alpha + \tfrac{1}{2}(\theta^2 - \theta^2\cot^2\alpha)\cot\alpha\tan C$$

$$= -\theta\cot\alpha + \frac{\theta^2}{2}\cot\alpha\tan C\,(1-\cot^2\alpha)$$

Exercises

1. If θ is so small that its cube may be neglected, and if

$\cos(\alpha-\theta) = \cos\alpha\cos\beta$, show that

$$\theta = -2\cot\alpha\sin^2\frac{\beta}{2}\left(1-\cot^2\alpha\sin^2\frac{\beta}{2}\right), \text{ approximately.}$$

Joh' Camb': '35.

2. If θ is so small that its cube may be neglected, and if

$$\sin(\alpha-\theta) = \sin\alpha\cos\beta,$$

show that the number of seconds in θ is, approximately,

$$2\left(\frac{\tan\alpha}{\sin 1''}\sin^2\frac{\beta}{2} - \frac{\tan^3\alpha}{\sin 1''}\sin^4\frac{\beta}{2}\right)$$

3. If θ and ϕ are so small that their cubes may be neglected, and if

$$\cos(\alpha + \theta) = \sin\phi \sin\alpha \cos\beta + \cos\phi \cos\alpha$$

show that the number of seconds in θ is, approximately,

$$-\frac{\phi \cos\beta}{\sin 1''} + \frac{\phi^2 \sin^2\beta}{\sin 2''} \cot\alpha$$

Hymers' Astronomy.

4. If α and δ are small, and

$$\cos(\theta + \delta) = \sin^2\alpha + \cos^2\alpha \cos\theta,$$

find 1st, 2nd, and 3rd approximations for δ.

5. If $\cos\theta = {}^{r}\theta$, approximate to the value of θ.

NOTE—*Since* $\pi/4 = \cdot 7854$, *and* $\cos 45^\circ = \cdot 7071$, $\therefore\ \theta = \pi/4 - \phi$, *where* ϕ *is very small*: *hence* $\cos(\pi/4 - \phi) = \pi/4 - \phi$. *Now expand the* cosine *and use the approx's of* § 48.

6. If ${}^{r}x = \cot x$, show that ${}^{r}x$ is nearly equal to $49^\circ\ 17'$.

CHAPTER X

The Theory of Proportional Differences

§ **51.** We assume, in this part of our subject, that the Student has a logarithmic * table-book, and is familiar with its use: we also assume that he is acquainted with the theory of logarithms, as given in any good Algebra.†

Now it is shown in treatises on Algebra that, if n is an integer of five figures, and d a decimal, it is approximately (viz. to seven decimal places) true that

$$\log(n + d) - \log n : \log(n + 1) - \log n = d : 1$$

Or, expressed in words—*The change in the logarithm of a number is approximately proportional to the change in the number, when the change in the number is less than unity.*

This Law is called **the principle of proportional differences** (or **proportional parts**).

The principle (some special cases excepted) holds *mutatis mutandis* for the T. F.s, and also for their logarithms. This we now proceed to prove.

Def'—By an **increment** of a variable is meant a *small* increase or decrease of that variable.

Note—In the case of an ∧ an increment means an increase or decrease of not more than 1′

In what follows all ∧s are supposed to be measured by the radian, unless it is otherwise stated.

---

* *Chambers' Mathematical Tables* is a good and handy book.

† See, for example, *C. Smith's Treatise on Algebra*, Chapter XXIV; or *Chrystal's Algebra*, Chapters XXI and XXVIII.

Def'—If $f(\theta)$ is any function of θ, and $f(\theta + h)$ the *same* function of $\theta + h$, where h is an increment of θ, then $\Delta f(\theta)$ denotes $f(\theta + h) \sim f(\theta)$.

The principle of proportional differences will be proved for $f(\theta)$ if we can show that

$$\Delta f(\theta) \propto h$$

Note—It is to be recollected that h may be pos' or neg'.

The Student wlll find that the numerical values of the **sines, cosines**, &c of successive $\wedge^s$, as registered in table-books, are there called **natural sines, cosines**, &c; while their logarithms, *increased by* 10, are called **logarithmic sines, cosines**, &c. He should be careful to notice that the *tabular* (i. e. registered) logarithm of a T. F. is its *true* logarithm plus 10.

Note—The customary notation $L \sin \alpha$, to denote the tabular logarithm of $\sin \alpha$, i. e. the true logarithm of $\sin \alpha$ increased by 10, will be used. So that $L \sin \alpha = \log \sin \alpha + 10$; with sim'r notation for the other T. F.s.

To prove the principle of proportional differences for each of the natural T. F.s

§ **52.** Since $1' = {}^r(\pi/10800) = {}^r(\cdot00029)$ very nearly

$$\therefore, \text{ if } h \not> 1', \text{ then } \frac{h^2}{2} \not> \cdot000000042$$

$$\left.\begin{aligned} \text{But then } \sin h \sim h &< \frac{h^3}{6} \\ \text{and } \cos h \sim \left(1 - \frac{h^2}{2}\right) &< \frac{h^4}{24} \end{aligned}\right\} \text{See § 48}$$

$$\left.\begin{aligned} \therefore \quad \sin h &= h \\ \text{and } \cos h &= 1 - \frac{h^2}{2} \end{aligned}\right\} \begin{matrix}\text{to more than 7 dec'l places} \\ \text{when } h \not> 1'\end{matrix}$$

$$\text{Now } \sin(\theta + h) = \sin\theta \cos h + \cos\theta \sin h$$

$$= \left(1 - \frac{h^2}{2}\right) \sin\theta + h\cos\theta, \text{ to 7 places}$$

$$\therefore \quad \sin(\theta + h) - \sin\theta = h\cos\theta - \frac{h^2}{2}\sin\theta, \text{ to 7 places}$$

Unless θ is nearly $\pi/2$, the term $\frac{h^2}{2}\sin\theta$ will not affect the 7th dec'l place.

$\therefore$, in general, $\Delta \sin\theta = h\cos\theta$, and $\therefore \propto h$

which proves the principle for $\sin\theta$, unless θ is near $\pi/2$.

But if $\theta = \pi/2$ nearly, then—

1°, $h\cos\theta$ is so small that it will not affect 7 figure tables at all; and, for such tables, the differences are said to be *insensible*: also,

2°, the term $\frac{h^2}{2}\sin\theta$ would need to be taken into account; but the diff's will not occur at regular intervals; and they are said to be *irregular*; and the principle does not hold.

Note—The practical application of the principle can be seen thus—

The T. F. of an $\wedge$ has the same value no matter in what unit the $\wedge$ is measured.

Also the ratio of 2 $\wedge^s$, measured in any unit = their ratio measured in any other unit.

$\therefore$, if A is an $\wedge$ (not near 90°) measured in degrees and minutes,

$$\sin(A + n'') - \sin A : \sin(A + 1') - \sin A = n'' : 60''$$

i. e. diff' of sines for an increment of $n'' = \frac{n}{60} \times$ diff' for an increment of 1'

But diff' for 1', corresponding to A, is registered in the tables.

Hence the diff' for n'' can be reckoned.

A precisely similar investigation (which the Student should go through) will show that

$$\Delta \cos\theta = -h\sin\theta, \quad \text{and} \quad \therefore \propto h$$

unless θ is very small, when the differences become *insensible* and *irregular*.

Hence, with that exception, the principle of proportional parts holds for cosines.

Note (1)—The diff's for sines and cosines may be deduced, each from the other, by taking the complementary $\wedge$.

Note (2)—A consideration of the graphs of the sine and cosine (see § 11) will remind us when their diff's become insensible.

§ **53.** Again $\Delta \tan\theta = \sin h/\overline{\cos\theta\cos(\theta+h)}$

$$= \tan h/\overline{\cos^2\theta\,(1 - \tan h\,.\,\tan\theta)}$$

But, since $\sin h = h$, $\cos h = 1 - \frac{h^2}{2}$ } to more than 7 dec'l places when $h \not> 1'$

$$\therefore \text{ then } \tan h = h, \text{ to at least 7 places}$$

$$\therefore\ \Delta\tan\theta = h\sec^2\theta/(1 - h\tan\theta)$$

$$= h\sec^2\theta + h^2\sec^2\theta\tan\theta$$

If $\theta < \pi/4$ the 2nd term may be neglected.

And then $\Delta\tan\theta = h\sec^2\theta$, and $\therefore \propto h$

So that the principle is true for tangents.

But if $\theta > \pi/4$ the 2nd term may not be neglected.

And then the diff's are *irregular*.

The diff's are never *insensible*.

In a precisely similar manner it can be shown that

$$\Delta\cot\theta = -h\,\text{cosec}^2\,\theta, \text{ and } \therefore \propto h$$

unless $\theta < \pi/4$, when the diff's become *irregular*.

Hence, with that exception, the principle of proportional parts holds for cotangents.

Note—The diff's for tangents and cotangents may be deduced, each from the other, by taking the complementary $\wedge$.

§ **54.** Next, since

$$\sec(\theta + h) = /\overline{\cos\theta\cos h - \sin\theta\sin h}$$

$$= /\overline{\cos\theta\left(1 - \frac{h^2}{2} - h\tan\theta\right)}$$

$$= \sec\theta\left(1 + h\tan\theta + \frac{h^2}{2} + h^2\tan^2\theta\right)$$

$$\therefore\ \Delta\sec\theta = h\sin\theta\sec^2\theta + \frac{h^2}{2}\sec\theta\,(1 + 2\tan^2\theta)$$

When θ is neither near 0 or $\pi/2$ the 2nd term may be neglected, and then

$$\Delta\sec\theta = h\sin\theta\sec^2\theta, \text{ and } \therefore \propto h$$

so that the principle holds for **secants.**

But when θ is nearly 0 the diff's are *irregular* and *insensible*; and when θ is nearly $\pi/2$ they are *irregular* though sensible.

Sim'ly it may be shown that

$$\Delta\operatorname{cosec}\theta = -h\cos\theta\operatorname{cosec}^2\theta, \text{ and } \therefore \propto h;$$

unless θ is nearly 0, when diff's are *irregular* though sensible; or unless θ is nearly $\pi/2$ when they are *irregular* and *insensible.*

§ **55.** Hence if $f(\theta)$ is any T. F., and δ an increment of θ, then

$$f(\theta + \delta) - f(\theta) \propto \delta$$

excepting in particular cases: cf. § 60.

Note (1)—The particular cases will be best recollected by thinking of the *graph* of each function.

Note (2)—The two terms got in each case for $\Delta f(\theta)$ can be at once recovered, by anyone who knows *Taylor's Theorem* in the Differential Calculus, from the consideration that

$$f(\theta + h) = f(\theta) + hf'(\theta) + \frac{h^2}{2}f''(\theta) + \&c$$

Examples

1. *Given that* sin $18^\circ\ 1' = \cdot 3092936$, *find* sin $18^\circ\ 0'\ 23''$, *without the aid of a table-book.*

$$\sin 18^\circ = \frac{\sqrt{5}-1}{4} = \frac{1\cdot 2360679}{4} = \cdot 3090169$$

$$\therefore \text{ diff}' \text{ for } 60'' = \cdot 0002767$$

$$\therefore \quad ,, \quad ,, \quad 23'' = \frac{23}{60} \times \cdot 0002767$$

$$= \cdot 0001060$$

$$\therefore \sin 18^\circ\ 0'\ 23'' = \cdot 3091229$$

2. *Given that* $\cos \alpha = \cdot 3711999$, *find* α *by the aid of a table-book.*

From the tables $\cos 68^\circ\ 13' = \cdot 3710977$

and diff′ for $60'' = \cdot 0002701$

Also if $\alpha = 68^\circ\ 13' - \delta''$

diff′ for $\delta'' = \cdot 0001022$

$$\therefore \delta'' = \frac{1022}{2701} \times 60''$$

$$= 22''\cdot 7$$

$$\therefore \alpha = 68^\circ\ 12'\ 37''\cdot 3$$

Note—Notice that for $\wedge^s$ between 0° and 90°, the diff′s are to be *added* for the sine, tangent, and secant; but *subtracted* for the cosine, cotangent, and cosecant. This is evident from the consideration that, as the generator of an $\wedge$ passes thro′ the 1st quadrant, the T. F.s beginning with co *decrease*, while the others *increase*.

Exercises

1. Find sin $37^\circ\ 23'\ 47''$, by the aid of a table-book.

2. If $\cos \alpha = \cdot 8241657$, find α by the aid of a table-book.

3. Given that sin $30^\circ\ 1' = \cdot 5002519$, find sin $30^\circ\ 1'\ 17''$ *without* using tables.

4. Given that $\cos \alpha = \cdot 4996532$, and that $\cdot 0002519$ is the difference for $1'$, find α *without* the aid of tables.

To prove the principle of proportional differences for each of the logarithmic T. F.s.

§ **56.** $\sin(\theta + h)/\sin\theta = 1 - \frac{h^2}{2} + h\cot\theta$, when $h \not> 1'$

$\therefore \Delta L \sin\theta$ (which $= \Delta \log \sin\theta$)

$$= \log\left(1 + h\cot\theta - \frac{h^2}{2}\right)$$

$$= M\left\{\left(h\cot\theta - \frac{h^2}{2}\right) - \tfrac{1}{2}\left(h\cot\theta - \frac{h^2}{2}\right)^2 \&c\right\}$$

$$= M h \cot\theta - \tfrac{1}{2} M h^2 \operatorname{cosec}^2\theta$$

where $M = /\log_e 10$, and $\therefore < \frac{1}{2}$,

and powers of h higher than h^2 are neglected.

When θ is very small $\operatorname{cosec}\theta$ is very large, and the diff's are sensible but *irregular*.

When θ is nearly $\pi/2$ $\cot\theta$ is very small, and $\operatorname{cosec}\theta$ nearly unity, so that then diff's are *insensible* and *irregular*.

But, with these exceptions, we have from above that

$$\Delta L \sin\theta = M h \cot\theta, \text{ and } \therefore \propto h$$

So that the principle holds for $L \sin\theta$.

§ **57.** $\cos(\theta + h)/\cos\theta = 1 - \frac{h^2}{2} - h\tan\theta$, when $h \not> 1'$

$\therefore \Delta L \cos\theta$ (which $= \Delta \log \sin\theta$)

$$= \log\left\{1 - \left(h\tan\theta + \frac{h^2}{2}\right)\right\}$$

$$= -M\left\{\left(h\tan\theta + \frac{h^2}{2}\right) + \tfrac{1}{2}\left(h\tan\theta + \frac{h^2}{2}\right)^2 \&c\right\}$$

$$= -M h \tan\theta - \frac{M h^2}{2}\sec^2\theta$$

When θ is very small $\tan\theta$ is very small, and $\sec\theta$ is nearly unity, so that then the diff's are *insensible* and *irregular*.

When θ is nearly $\pi/2$, $\sec\theta$ is very large, and then diff's are sensible but *irregular*.

But, with these exceptions, we have from above

$$\Delta\, \text{L}\cos\theta = -\text{M h}\tan\theta, \text{ and } \therefore \propto \text{h}$$

$\therefore$ the principle holds for $\text{L}\cos\theta$

§ **58**. $\tan(\theta+\text{h})/\tan\theta = \left(1+\dfrac{\tan \text{h}}{\tan\theta}\right)\Big/(1-\tan \text{h}\tan\theta)$

$$= (1+\text{h}\cot\theta)/(1-\text{h}\tan\theta)$$

$$\therefore \Delta\,\text{L}\tan\theta \text{ (which} = \Delta\log\tan\theta)$$

$$= \text{M}\log_e(1+\text{h}\cot\theta) - \text{M}\log_e(1-\text{h}\tan\theta)$$

$$= \text{M}\left\{\text{h}\cot\theta - \frac{(\text{h}\cot\theta)^2}{2}\right\} + \text{M}\left\{\text{h}\tan\theta + \frac{(\text{h}\tan\theta)^2}{2}\right\}$$

$$= 2\,\text{M h}\operatorname{cosec} 2\theta + \frac{\text{M h}^2}{2}(\tan^2\theta - \cot^2\theta)$$

When θ is small $\cot\theta$ is large, and the diff's *irregular*.

When θ is nearly $\pi/2$ $\tan\theta$ is large, and the diff's *irregular*.

But, with these exceptions, we have

$$\Delta\,\text{L}\tan\theta = 2\,\text{M h}\operatorname{cosec} 2\theta, \text{ and } \therefore \propto \text{h}$$

$\therefore$ the principle holds for $\text{L}\tan\theta$.

§ **59**. The principle, for the remaining three T. F.s, can now easily be deduced from the former three, by considering that

$$\text{L}\operatorname{cosec}\theta = 20 - \text{L}\sin\theta$$

$$\text{L}\sec\theta = 20 - \text{L}\cos\theta$$

$$\text{L}\operatorname{cotan}\theta = 20 - \text{L}\tan\theta$$

Or they can be deduced by putting for θ its complement.

$\therefore$, unless θ is small or near $\pi/2$, if $f(\theta)$ is any T. F., and δ an increment of θ,

$$\text{L}f(\theta+\delta) - \text{L}f(\theta) \propto \delta$$

Examples

1. *Find* L cos $25° 36' 19''$.

In the table-book we find that

$$\text{L}\cos 25° 37' = 9\cdot9550653$$
$$\text{diff}' \text{ for } 1' = \cdot0000606$$
$$\therefore \text{ if L}\cos 25° 36' 19'' = 9\cdot9550653 + x$$
$$\text{then } x = \frac{19}{60} \times \cdot0000606$$
$$= \cdot00000191$$
$$\therefore \text{ the answer is } 9\cdot9550844$$

2. *Given that* L sin $\alpha = 9\cdot6448213$, *find* α.

In the table-book we find that

$$\text{L}\sin 26° 11' = 9\cdot6446796$$
$$\text{diff}' \text{ for } 1' = \cdot0002569$$
$$\therefore \text{ if } \alpha = 26° 11' + x''$$
$$\text{then } x = \frac{1417}{2569} \times 60 = 33$$
$$\therefore \alpha = 26° 11' 33''$$

Exercises

1. Given that L sin $32° 28' = 9\cdot7298197$, and L sin $32° 29' = 9\cdot7300182$, find L sin $32° 28' 36''$.

2. Given that L tan $21° 17' = 9\cdot5905617$, and L tan $21° 18' = 9\cdot590951$, find L tan $21° 17' 12''$.

3. Given that L cot $72° 15' = 9\cdot5052819$, and L cot $72° 16' = 9\cdot5048538$, find L cot $72° 15' 35''$.

4. Given that L cos $22° 28' 20'' = 9\cdot9657025$, L cos $22° 28' 10'' = 9\cdot9657112$, and L cos $\alpha = 9\cdot657056$, find α.

5. Given that L cos $20° 35' 20'' = 9\cdot9713351$, difference for $10'' = \cdot0000079$, and L cos $\alpha = 9\cdot9713383$, find α.

6. Given that L cot $44° 59' = 10\cdot0002527$, difference for $1' = \cdot0002527$, and L cot $\alpha = 10\cdot0001234$, find α.

7. Show that, by taking logarithms of the particular case of *Euler's formula* when the angle is a right angle, viz.

$$\cos\frac{\pi}{4}\cos\frac{\pi}{8}\cos\frac{\pi}{16}\,\&c = \frac{2}{\pi},$$

retaining ten of the cosines, and using a seven figure table-book, we shall get π correct to five decimal places.

§ **60.** If $f(\theta)$ denote any T. F.—

$\Delta f(\theta)$ is *irregular* or *insensible* according to this table

θ	$\Delta \sin\theta$	$\Delta \cos\theta$	$\Delta \tan\theta$	$\Delta \cot\theta$	$\Delta \sec\theta$	$\Delta \operatorname{cosec}\theta$
0	neither	both	neither	irreg'	both	irreg'
$\pi/2$	both	neither	irreg'	neither	irreg'	both

$\Delta\, Lf(\theta)$ is *irregular* or *insensible* according to this table

θ	$\Delta L\sin\theta$	$\Delta L\cos\theta$	$\Delta L\tan\theta$	$\Delta L\cot\theta$	$\Delta L\sec\theta$	$\Delta L\operatorname{cosec}\theta$
0	irreg'	both	irreg'	irreg'	both	irreg'
$\pi/2$	both	irreg'	irreg'	irreg'	irreg'	both

Note—In the above tables the 0 and $\pi/2$, under θ, are not intended to indicate that θ is absolutely 0 or $\pi/2$, but *near* those values.

Hence when an $\wedge$ is small, it should not be found from its cosine, or cotangent, or cosecant; when it is near a right $\wedge$, it should not be found from its sine, or tangent, or secant.

We may however get round the difficulty by artifices.

For if α is very small, so that

$$\cos \alpha = a, \text{ where } a \text{ is very nearly unity,}$$

$$\text{then } \sin \frac{\alpha}{2} = \sqrt{(1 - a)/2}$$

from which $\frac{\alpha}{2}$, and $\therefore$ α, can be easily found.

Again if α is nearly a right $\wedge$, so that

$$\sin \alpha = a, \text{ where } a \text{ is nearly unity,}$$

$$\text{then } \sin \left(45^\circ - \frac{\alpha}{2}\right) = \sqrt{(1 - a)/2}$$

from which $45^\circ - \frac{\alpha}{2}$, and $\therefore$ α, can be found.

So also when α is nearly a right $\wedge$, so that

$$\tan \alpha = a, \text{ where } a \text{ is very large,}$$

$$\text{then } \tan (\alpha - 45^\circ) = \overline{a - 1} \big/ \overline{a + 1}$$

from which $\alpha - 45^\circ$, and $\therefore$ α, can be found.

Besides the above we have seen (in § 53) that the diff's are *irregular* for $\tan \alpha$, when $\alpha > \pi/4$, and for $\cot \alpha$, when $\alpha < \pi/4$; so that these exceptional cases must be added to the above.

From the 2nd table, we see that when an $\wedge$ is small, or near a right $\wedge$, it cannot be accurately found from any one of its logarithmic T. F.s.

§ **61.** As the case when θ is small is one of frequent occurrence in practical work, various methods to overcome the difficulty of calculating $L \sin \theta$, $L \cos \theta$, and $L \tan \theta$, when θ is small, have been devised. Such methods are of the greater value because in practice $\wedge$s are more frequently reckoned from their logarithmic than from their natural T. F.s.

Method 1—For ordinary tables the following method (*Maskelyne's*) is available.

$$\left.\begin{aligned}\text{Since } \sin\theta &= \theta - \frac{\theta^3}{6}\\ \text{and } \cos\theta &= 1 - \frac{\theta^2}{2}\end{aligned}\right\} \text{for small } \wedge^s\text{, i.e. when } \theta \ngtr 3^\circ$$

$$\therefore \frac{\sin\theta}{\theta} = 1 - \frac{\theta^2}{6} = \left(1 - \frac{\theta^2}{2}\right)^{\frac{1}{3}} \text{ nearly}$$

$$= (\cos\theta)^{\frac{1}{3}}$$

$$\therefore \log\sin\theta = \log\theta + \tfrac{1}{3}\log\cos\theta$$

Now for small $\wedge^s$ diff's of $\log\cos\theta$ are insensible.

$\therefore$ from this eq'n, given $\log\theta$ we can find $\log\sin\theta$.

The practical rule (given in *Chambers' Tables*, p. xviii) is

"*To the* logarithm *of the angle, reduced to seconds, add* 4·6855749, *and from this sum subtract one-third of its* L-secant, *the index of the latter* logarithm *being previously diminished by* 10: *the remainder is the required* L sin θ."

The rationale of the Rule is easily seen to follow from the formula thus—

$$\text{If } {}^r\theta = n''$$

then $\theta = \pi n/(180 \times 60 \times 60) = 3{\cdot}141592 \times n/648000$

$\therefore \log\theta = \log n + \log 3{\cdot}141592 - \log 648000$

Now, from table-book, $\log 3{\cdot}1415$	=	·4971371
diff' for 9		125
" " 2		28
$\therefore \log 3{\cdot}141592$	=	·49714988
But $\log 648000$	=	5·8115750
Subtracting, we get		$\bar{6}{\cdot}68557488$

$$\therefore \log \theta = \log n + \bar{6}\cdot6855749$$

$\therefore$, by Maskelyne's formula,

$$L \sin \theta = \log n + 4\cdot6855749 - \tfrac{1}{3}(L \sec \theta - 10)$$

which is Chambers' Rule.

Example

To find L sin 1° 44′ 36″·8

Here n = 6276″·8, and its log = 3·7977383

Here n = 6276″·8, and its log	= 3·7977383
Constant number is	4·6855749
	8·4833132
$\frac{1}{3}$ (L sec n − 10) = $\frac{1}{3}$ × ·0002011 =	·0000670
∴ L sin 1° 44′ 36″. 8 =	8·4832462

Conversely—given L sin θ to find θ—for the given value of L sin θ find the nearest value of θ given in the tables, and use this to get L cos θ; which, since the diff′s of L cos θ for small $\wedge^s$ are insensible, will be a sufficient approx′: then, from the formula $\log \theta = \log \sin \theta - \frac{1}{3} \log \cos \theta$, we get a closer approx′ to θ.

The practical rule (given in *Chambers' Tables*, p. xix) is

"*To the given* L sin θ *add* $\bar{5}\cdot3144251$, *and one-third of the* L sec θ, *the index of the latter* logarithm *being previously diminished by* 10, *and the sum will be the* logarithm *of the number of seconds in the angle*".

To show that this Rule comes from the formula, we have, as before,

$$\log n = \log \theta - \bar{6}\cdot6855749$$

$\therefore$, by Maskelyne's formula,

$$= L \sin \theta + \bar{5}\cdot3144251 + \tfrac{1}{3}(L \sec \theta - 10)$$

which is Chambers' Rule.

Example

To find θ, *if* $\mathsf{L}\sin\theta = 8{\cdot}4832462$

The 1st approx′ to θ is $1^\circ\,44'\,36''$

$$\text{and } \mathsf{L}\sec 1^\circ\,44'\,36'' = 10{\cdot}0002011$$

$$\therefore\ \tfrac{1}{3}(\mathsf{L}\sec\theta - 10) = {\cdot}0000670$$

$$\mathsf{L}\sin\theta = 8{\cdot}4832462$$

$$\text{constant number is } \bar{5}{\cdot}3144251$$

$$\text{sum} = 3{\cdot}7977383$$

$$\text{which, from tables, } = \log 6276{\cdot}8$$

$$\therefore\ n'' = 6276''{\cdot}8, \text{ where } n'' = {}^r\theta$$

$$= 1^\circ\,44'\,36''{\cdot}8$$

For the **L-tangents**, we get

$$\log\tan\theta = \log\sin\theta - \log\cos\theta$$

$$= \log\theta - \tfrac{2}{3}\log\cos\theta$$

Exercises

Deduce, as above, the following rules, given in *Chambers' Tables.*

1. To find $\mathsf{L}\tan\theta$, when θ is small—

"*To the* logarithm *of* θ, *reduced to seconds, add* 4·6855749, *and two-thirds of* $\mathsf{L}\sec\theta$, *the index of the latter* logarithm *being previously diminished by* 10 : *the sum is the required* $\mathsf{L}\tan\theta$."

As a particular case find $\mathsf{L}\tan 1^\circ\,44'\,36''{\cdot}8$.

2. If $\mathsf{L}\tan\theta$ is given, θ being small, to find θ—

"*To the given* $\mathsf{L}\tan\theta$ *add* $\bar{5}{\cdot}3144251$, *and from this sum subtract two-thirds of the corresponding* $\mathsf{L}\sec\theta$, *previously diminishing its index by* 10 : *then the remainder is the* logarithm *of the number of seconds in the angle* θ."

As a particular case find θ, if $\mathsf{L}\tan\theta = 8{\cdot}4834473$.

Method 2—*Delambre* has given a method founded on the construction of a table giving the value of

$$\log\frac{\sin\theta}{\theta} + \mathsf{L}\sin 1''$$

for every second, for values of θ from 0 to 3°.

The mode of using this table is seen thus—

$$\text{If } n'' = {}^{r}\theta$$

$$\text{then } \theta = n \sin 1'' \quad (\text{Cor' to Example 2, Chap' IX})$$

$$\therefore \log \frac{\sin \theta}{\theta} = \log \frac{\sin n''}{n \sin 1''}$$

$$= L \sin n'' - L \sin 1'' - \log n$$

$$\therefore L \sin n'' = \log n + \left(\log \frac{\sin \theta}{\theta} + L \sin 1'' \right)$$

which, by the special tables, gives $L \sin n''$ when θ is known.

Conversely—given $L \sin n''$, to find n.

First find an approx' value of θ, and then from the special tables get

$$\log \frac{\sin \theta}{\theta} + L \sin 1''$$

Delambre's formula then gives $\log n$, and n is known from the ordinary tables.

Note—The error produced by using the approx' value of θ in $\log \frac{\sin \theta}{\theta}$, instead of the true value, will be insensible for 7 dec'l places.

$$\text{For if } f(\theta) = \log \frac{\sin \theta}{\theta}$$

$$\Delta f(\theta) = \log \frac{\sin (\theta + h)}{\theta + h} - \log \frac{\sin \theta}{\theta}$$

$$= \log \left\{ 1 - \frac{(\theta + h)^2}{6} \right\} - \log \left(1 - \frac{\theta^2}{6} \right)$$

$$= \left(\frac{\theta^2}{6} + \frac{\theta^4}{72} \right) - \left\{ \frac{(\theta + h)^2}{6} + \frac{(\theta + h)^4}{72} \right\}$$

$$= -\frac{\theta h}{3} + \text{terms not affecting 7th place.}$$

Method 3—Finally the difficulty may be got over by a table giving values of $L \sin \theta$ for every second for values of θ from 0 to 3°: in such a table the term $\frac{M h^2}{2} \operatorname{cosec}^2 \theta$ may be neglected.

Exercises

1. Show that $L\cos x - L\cos(x+h)$

$$= \left(h\tan x + \frac{h^2}{2}\sec^2 x + \&c\right)\Big/\log_e 10.$$

2. Prove the truth of the formula

$$\log n = L\sin n'' + (10 - L\cos n'')\big/3 - L\sin 1''$$

NOTE—*This is an amalgamation of Delambre's and Maskelyne's formulæ.*

3. An angle, which is known to be about two-thirds of a right angle, has to be found from its natural tangent, by means of a table-book which goes to seven decimal places: how nearly can the angle be calculated?

4. Show that a large error may be expected in determining α from the formula

$$\cot\alpha - \tan\alpha = 2\sqrt{1+\cos 4\alpha}\Big/\sqrt{1-\cos 4\alpha}$$

when α is small, or near 45°, or near 90°.

Also show that, if $\alpha = 30^\circ$, and m is the error in the registered value of $\cos 4\alpha$, the value deduced for $\tan\alpha$ will be less than the correct value by $2\sqrt{3}\,m/9$.

Joh' Camb': '37.

5. When an angle is nearly $64^\circ\ 36'$ show that it can be determined from its $L\sin$ within about one-tenth of a second, if the tables go to 7 decimal places, and $\log_e 10 \,.\, \tan 64^\circ\ 36' = 4{\cdot}8492$.

Joh' Camb': '50.

6. In finding an angle θ (which is nearly a multiple of a right angle) from a table of log-sines calculated to n decimal places, show that the greatest error which can be committed, measured in seconds, is

$$\tan\theta\big/(10^n \,.\, 2\sin 1'')$$

very nearly.

Joh' Camb': '50.

CHAPTER XI

Relations between the Sides and Angles of a Triangle

§ **62.** *Def's*—In the trigonometrical treatment of a triangle the following notation will be invariably used, unless the contrary is expressly stated—

A, B, C for either the *positions* of the corners; or the *magnitudes* of the angles at those corners:

a, b, c for the sides respectively opposite A, B, C:

s for the *semi-perimeter*; so that $2s = a + b + c$:

Δ (the ordinary symbol for the word 'triangle'), when used alone, for the *area* of the triangle:

s_1, s_2, s_3, respectively for $s - a$, $s - b$, $s - c$; so that

$2s_1 = b + c - a$, $2s_2 = c + a - b$, $2s_3 = a + b - c$:

D for the *diameter* of the circum-circle.

Note—$s_1 + s_2 + s_3 = s$.

§ **63. Theorem**—*The sides of a triangle are respectively proportional to the* sines *of the respective opposite angles.*

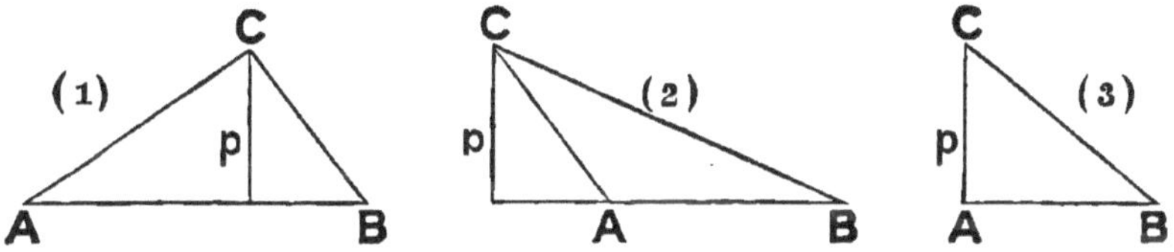

Let ABC be a Δ; then of any two of its $\wedge^s$ A, B,

either both are acute, as in fig' (1)

or one is acute, and one obtuse, as in fig' (2)

or one is acute, and one right, as in fig' (3)

∴, if p is the ⊥ from C on AB, in all cases

$$\sin A = \frac{p}{CA} = \frac{p}{b}$$

$$\sin B = \frac{p}{CB} = \frac{p}{a}$$

$$\therefore \frac{a}{\sin A} = \frac{b}{\sin B}; \text{ and, by analogy, } = \frac{c}{\sin C}$$

Cor′ (1)—A geometrical interpretation can be given to the ratios

$$a/\sin A,\ b/\sin B,\ c/\sin C.$$

For, if D is the diam′ of the circum-⊙, then, by a well-known geometrical theorem (see *Euclid Revised*, p. 288)

$$D \cdot p = ab$$

$$\therefore D = \frac{ab}{p} = \frac{ab}{b \sin A} = \frac{a}{\sin A}$$

i. e. each of the above ratios = circum-diam′ of ABC

Cor′ (2)—In any ratio, whose terms involve the sides of a △ *homogeneously*, each side may be replaced by the sine of the ∧ opposite; and *vice versâ.*

For example, the ratios

$$\sin^3 A + \sin^2 B \sin C : \sin A \sin B \sin C - \sin A \sin^2 C$$

$$\text{and} \quad a^3 + b^2c : abc - ac^2$$

are equal, and ∴ interchangeable.

§ **64. Theorem**—*Any side of a triangle can be expressed in terms of the* cosines *of the angles of which it is a common arm, and the remaining sides.*

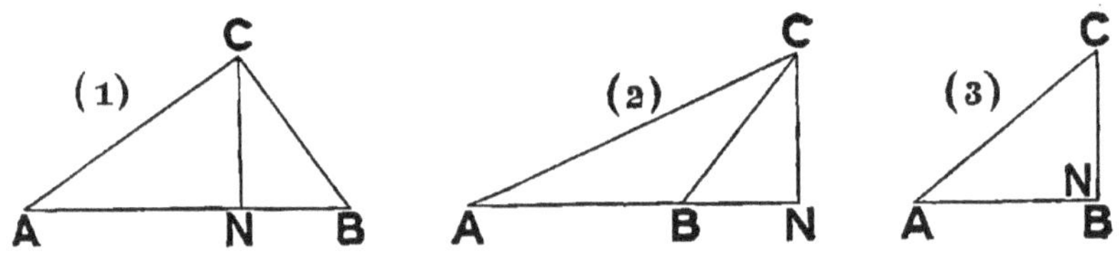

Take any side c of a $\triangle$ ABC: then the $\angle^s$ A, B, of which it is a common arm, may be

either both acute, as in fig' (1)

or one acute, and one obtuse, as in fig' (2)

or one acute, and one right, as in fig' (3)

$\therefore$, if CN is the $\perp$ from C on AB,

$$c = AN + BN, \text{ in fig}' (1)$$
$$= b \cos A + a \cos B$$
$$c = AN - BN, \text{ in fig}' (2)$$
$$= b \cos A - a \cos (180° - B)$$
$$= b \cos A + a \cos B$$
$$c = AN, \text{ in fig}' (3)$$
$$= b \cos A + a \cos B, \text{ since then } \cos B = 0,$$

$\therefore$, in all cases, $c = a \cos B + b \cos A$

and, by analogy, $b = c \cos A + a \cos C$

$$a = b \cos C + c \cos B$$

Cor' (1)—These results may be written

$$\left.\begin{array}{l} -a + b\cos C + c\cos B = 0 \\ a\cos C - b + c\cos A = 0 \\ a\cos B + b\cos A - c = 0 \end{array}\right\}$$

The eliminant of which, with respect to a, b, c, is

$$\begin{vmatrix} -1 & \cos C & \cos B \\ \cos C & -1 & \cos A \\ \cos B & \cos A & -1 \end{vmatrix} = 0$$

which, expanded by Sarrus' rule, gives

$$\Sigma \cos^2 A + 2\Pi \cos A = 1$$

a result already found otherwise (see p. 79).

Cor' (2)—Again, the results may be written

$$\left.\begin{aligned} a - b\cos C - c\cos B &= 0 \\ (b - c\cos A) - a\cos C - 0.\cos B &= 0 \\ (c - b\cos A) - 0.\cos C - a\cos B &= 0 \end{aligned}\right\}$$

The eliminant of which with respect to $-\cos C$, $-\cos B$, is

$$\begin{vmatrix} a & b & c \\ (b - c\cos A) & a & 0 \\ (c - b\cos A) & 0 & a \end{vmatrix} = 0$$

Whence, at once, $a^2 = b^2 + c^2 - 2\,bc\cos A$

Note—This last result may also be got by multiplying the eq'ns (as originally found) by c, b, a respectively, and then adding the 1st two results so found, and subtracting the 3rd; but, on account of its importance, an independent investigation of it, on first principles, will be given.

§ **65. Problem**—*To express the* sine *of an angle of a triangle in terms of the sides of that triangle.*

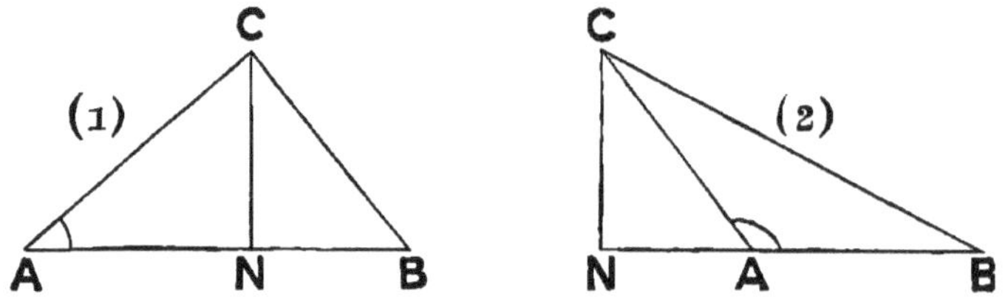

Let $\hat{A}$, of $\triangle$ ABC, be acute in fig' (1), and obtuse in fig' (2)
Drop CN $\perp$ on AB, or BA produced.

Then, in fig' (1) $a^2 = b^2 + c^2 - 2\,c\,.\,AN$, by *Euc'* ii. 13
and, in fig' (2) $a^2 = b^2 + c^2 + 2\,c\,.\,AN$, by *Euc'* ii. 12

$\therefore$, in both cases, $4\,c^2\,.\,AN^2 = (b^2 + c^2 - a^2)^2$

Also $AN^2 = b^2 - CN^2$

$$\therefore (2\,c\,.\,CN)^2 = 4\,b^2c^2 - (b^2 + c^2 - a^2)^2$$
$$= (2\,bc + b^2 + c^2 - a^2)$$
$$\times (2\,bc - b^2 - c^2 + a^2)$$

$$= (b + c + a)(b + c - a)(a - b + c)(a + b - c)$$

$$\therefore (2c \cdot b \sin A)^2 = 16\, s\, s_1\, s_2\, s_3$$

$$\therefore \sin A = \frac{2}{bc}\sqrt{s\, s_1\, s_2\, s_3}$$

In the case when $A = 90°$, $a^2 = b^2 + c^2$, and this expression for $\sin A$ reduces to unity, as it should.

Note—The expression $\sqrt{s\, s_1\, s_2\, s_3}$ is usually written

$$\sqrt{s(s-a)(s-b)(s-c)};$$

but the shorter form is very convenient.

§ **66. Problem**—*To get expressions for the area of a triangle.*

Taking the figs′ of the preceding §, we have

$$\Delta = \tfrac{1}{2} c \cdot CN = \tfrac{1}{2} bc \sin A$$

Whence also $\Delta = \sqrt{s\, s_1\, s_2\, s_3}$

But $16\, ss_1\, s_2\, s_3 \equiv 2b^2c^2 + 2c^2a^2 + 2a^2b^2 - a^4 - b^4 - c^4$

$$\therefore \Delta = \tfrac{1}{4}\sqrt{2\Sigma b^2c^2 - \Sigma a^4}$$

Again $\Delta = \tfrac{1}{2} bc \sin A$

$$= \tfrac{1}{2}\frac{a \sin B}{\sin A} \cdot \frac{a \sin C}{\sin A} \cdot \sin A$$

$$= \frac{a^2 \sin B \cdot \sin C}{2 \sin A}$$

§ **67. Problem**—*To express the* cosine *of an angle of a triangle in terms of the sides of that triangle; and thence to find the trigonometrical functions of half that angle in terms of the sides.*

Taking the figs′ of the preceding §, we have as before

$a^2 = b^2 + c^2 - 2c \cdot AN$, in fig′ (1)

$= b^2 + c^2 - 2c \cdot b \cos A$

and $a^2 = b^2 + c^2 + 2c \cdot AN$, in fig′ (2)

$= b^2 + c^2 - 2c \cdot b \cos A$

∴, in both cases $a^2 = b^2 + c^2 - 2\,bc\cos A$

$$\text{or } \cos A = \frac{b^2 + c^2 - a^2}{2\,bc}$$

When $A = 90°$, $a^2 = b^2 + c^2$, and this expression for $\cos A$ reduces to zero, as it should.

Next, to deduce $\sin\frac{A}{2}$, $\cos\frac{A}{2}$, $\tan\frac{A}{2}$, &c.

$$\text{Since} \quad 2\sin^2\frac{A}{2} = 1 - \cos A$$

$$\therefore \sin^2\frac{A}{2} = \frac{2\,bc - (b^2 + c^2 - a^2)}{4\,bc}$$

$$= \frac{(a + b - c)(a - b + c)}{4\,bc}$$

$$\text{or } \sin\frac{A}{2} = \sqrt{\frac{s_2 s_3}{bc}}$$

$$\text{Also} \quad 2\cos^2\frac{A}{2} = 1 + \cos A$$

$$\therefore \cos^2\frac{A}{2} = \frac{2\,bc + (b^2 + c^2 - a^2)}{4\,bc}$$

$$= \frac{(b + c + a)(b + c - a)}{4\,bc}$$

$$\therefore \cos\frac{A}{2} = \sqrt{\frac{ss_1}{bc}}$$

$$\text{By division } \tan\frac{A}{2} = \sqrt{\frac{s_2 s_3}{ss_1}}$$

Whence, of course, sim'r expressions follow for the reciprocal functions.

Note (1)—From these expressions

$$\sin\frac{A}{2}\cos\frac{A}{2} = \sqrt{\frac{s_2\,s_3}{bc}}\cdot\sqrt{\frac{s\,s_1}{bc}}$$

$$\therefore \quad \sin A = \frac{2}{bc}\sqrt{s\,s_1\,s_2\,s_3} \quad \text{as before}$$

Note (2)—Of course the expressions for $\sin A$ and $\cos A$, in terms of the sides, are deducible each from the other.

§ 68. Problem—*To find formulæ connecting two sides of a triangle, and their included angle.*

$$\text{Since } \frac{\sin B - \sin C}{\sin B + \sin C} = \frac{b - c}{b + c}$$

$$\therefore \frac{2\cos\frac{B+C}{2}\sin\frac{B-C}{2}}{2\sin\frac{B+C}{2}\cos\frac{B-C}{2}} = \frac{b-c}{b+c}$$

$$\therefore \cot\frac{B+C}{2}\tan\frac{B-C}{2} = \frac{b-c}{b+c}$$

$$\therefore \tan\frac{B-C}{2} = \frac{b-c}{b+c}\cot\frac{A}{2} \quad . \quad . \quad (1)$$

$$\text{Again, since } \frac{a}{b-c} = \frac{\sin A}{\sin B - \sin C}$$

$$\therefore a = (b-c)\frac{2\sin\frac{A}{2}\cos\frac{A}{2}}{2\cos\frac{B+C}{2}\sin\frac{B-C}{2}}$$

$$\therefore a\sin\frac{B-C}{2} = (b-c)\cos\frac{A}{2} \quad . \quad . \quad . \quad . \quad . \quad (2)$$

$$\text{Sim'ly } a\cos\frac{B-C}{2} = (b+c)\sin\frac{A}{2} \quad . \quad . \quad . \quad . \quad (3)$$

Formulæ (1) (2) (3) above give relations as req'd.

N. B. In all the Examples and Exercises following it is to be understood that the notation defined in § 62 is adopted, unless some other is expressly indicated.

Examples

1. *If* $a \tan A + b \tan B = (a+b) \tan \frac{A+B}{2}$, prove that $a = b$.

Math' Tri': '51.

From given cond'n we get

$$a\left(\tan A - \tan \frac{A+B}{2}\right) = b\left(\tan \frac{A+B}{2} - \tan B\right)$$

$$\therefore \quad \frac{a \sin \frac{A-B}{2}}{\cos A \cos \frac{A+B}{2}} = \frac{b \sin \frac{A-B}{2}}{\cos \frac{A+B}{2} \cos B}$$

$$\therefore \text{ either } \sin \frac{A-B}{2} = 0$$

$$\text{or } \frac{\sin A}{\cos A} = \frac{\sin B}{\cos B}$$

But either alternative gives $A = B$

$$\therefore \quad a = b$$

2. *Show that the sides and* cotangents *of the semi-angles of a triangle are so related that, if the one set are in arithmetical progression, so also are the other.*

Math' Tri': '64.

$$\text{We have } \cot \frac{A}{2} = \sqrt{\frac{s s_1}{s_2 s_3}} = \frac{s s_1}{\Delta} = \lambda s_1 \text{ say}$$

$$\text{Sim'ly } \cot \frac{B}{2} = \lambda s_2, \text{ and } \cot \frac{C}{2} = \lambda s_3$$

But if a, b, c are in A. P. so also are $s-a$, $s-b$, $s-c$;

i. e. so also are s_1, s_2, s_3;

and ∴ ,, $\cot \frac{A}{2}$, $\cot \frac{B}{2}$, $\cot \frac{C}{2}$

The converse is also evidently true.

3. *If* $\tan A, \tan B, \tan C$ *are in harmonical progression, show that* a^2, b^2, c^2 *are in arithmetical progression.*

From given cond'n we get that

$$\frac{\cos A}{\sin A} + \frac{\cos C}{\sin C} = \frac{2\cos B}{\sin B}$$

$$\therefore \quad \frac{b^2 + c^2 - a^2}{2abc} + \frac{b^2 + a^2 - c^2}{2abc} = \frac{2(a^2 + c^2 - b^2)}{2abc}$$

$$\therefore \quad 2b^2 = 2(a^2 + c^2 - b^2)$$

$$\therefore \quad 2b^2 = a^2 + c^2$$

i.e. a^2, b^2, c^2 are in A. P.

4. *If* ABC *is a triangle, and* x, y, z *connected with its angles by the relations*

$$\Sigma(y - z)\cot\frac{A}{2} = 0, \qquad \Sigma(y^2 - z^2)\cot A = 0$$

prove that

$$\frac{y^2 + z^2 - 2yz\cos A}{\sin^2 A} = \frac{z^2 + x^2 - 2zx\cos B}{\sin^2 B} = \frac{x^2 + y^2 - 2xy\cos C}{\sin^2 C}$$

Leudesdorf: E. T. xli.

We have $\Sigma(b - c)\cot\frac{A}{2} \propto \Sigma(\sin B - \sin C)\cot\frac{A}{2}$

$$\propto \Sigma \sin\frac{B - C}{2}\cos\frac{A}{2}$$

$$\propto \Sigma \sin\frac{B - C}{2}\sin\frac{B + C}{2}$$

$$\propto \Sigma(\cos C - \cos B)$$

and $\therefore \quad = 0$

Also $\Sigma(b^2 - c^2)\cot A \propto \Sigma \sin(B + C)\sin(B - C)\cot A$

$$\propto \Sigma \sin(B - C)\cos(B + C)$$

$$\propto \Sigma(\sin 2B - \sin 2C)$$

and $\therefore \quad = 0$

$\therefore$, if XYZ is the Δ whose sides are x, y, z,

then Δ XYZ is sim'r to Δ ABC

$$\therefore \quad \frac{x^2}{\sin^2 A} = \frac{y^2}{\sin^2 B} = \frac{z^2}{\sin^2 C}$$

whence the req'd relations follow at once.

5. *Given the distances of each corner of a triangle* PQR, *from the sides of a known triangle* ABC, *to find the area of* PQR.

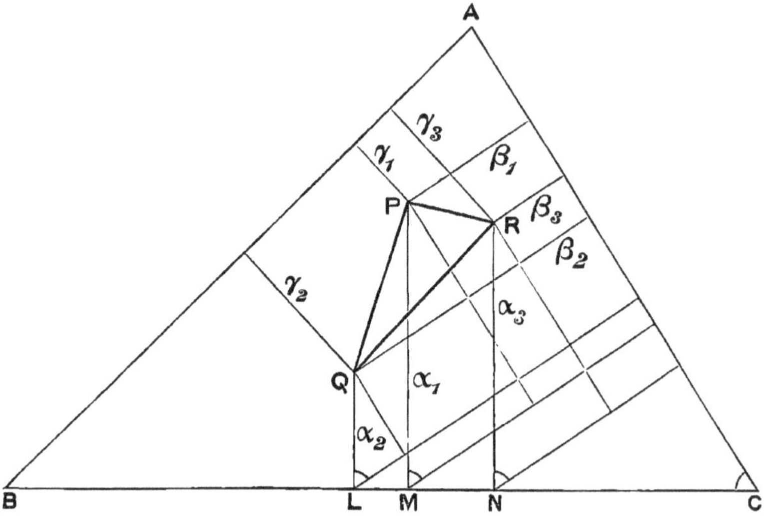

For brevity the fig′ is left to explain itself: note that all the construction lines are ⊥s.

$$\left.\begin{aligned} LC &= (\beta_2 + \alpha_2 \cos C) \operatorname{cosec} C \\ MC &= (\beta_1 + \alpha_1 \cos C) \operatorname{cosec} C \\ NC &= (\beta_3 + \alpha_3 \cos C) \operatorname{cosec} C \end{aligned}\right]$$

$$\left.\begin{aligned} \therefore\quad LM &= \{\beta_2 - \beta_1 + (\alpha_2 - \alpha_1) \cos C\} \operatorname{cosec} C \\ LN &= \{\beta_2 - \beta_3 + (\alpha_2 - \alpha_3) \cos C\} \operatorname{cosec} C \\ MN &= \{\beta_1 - \beta_3 + (\alpha_1 - \alpha_3) \cos C\} \operatorname{cosec} C \end{aligned}\right]$$

Now $\quad 2 \text{ area } PQR = (\alpha_1 + \alpha_2)\, LM + (\alpha_1 + \alpha_3)\, MN - (\alpha_2 + \alpha_3)\, LN$

$$\begin{aligned} \text{and } \therefore \quad &= (\alpha_1 + \alpha_2) \{\beta_2 - \beta_1 + (\alpha_2 - \alpha_1) \cos C\} \operatorname{cosec} C \\ &\quad + (\alpha_1 + \alpha_3) \{\beta_1 - \beta_3 + (\alpha_1 - \alpha_3) \cos C\} \operatorname{cosec} C \\ &\quad + (\alpha_2 + \alpha_3) \{\beta_3 - \beta_2 + (\alpha_3 - \alpha_2) \cos C\} \operatorname{cosec} C \end{aligned}$$

$$= \operatorname{cosec} C \{(\alpha_2 + \alpha_1)(\beta_2 - \beta_1) + (\alpha_1 + \alpha_3)(\beta_1 - \beta_3) + (\alpha_3 + \alpha_2)(\beta_3 - \beta_2)\}$$

$$= \operatorname{cosec} C \{\alpha_1 (\beta_2 - \beta_3) + \alpha_2 (\beta_3 - \beta_1) + \alpha_3 (\beta_1 - \beta_2)\}$$

$$\therefore \text{ area } PQR = \frac{\operatorname{cosec} C}{2}\begin{vmatrix} \alpha_1 & \alpha_2 & \alpha_3 \\ \beta_1 & \beta_2 & \beta_3 \\ 1 & 1 & 1 \end{vmatrix} = \frac{\operatorname{cosec} C}{4ab\Delta}\begin{vmatrix} a\alpha_1 & a\alpha_2 & a\alpha_3 \\ b\beta_1 & b\beta_2 & b\beta_3 \\ 2\Delta & 2\Delta & 2\Delta \end{vmatrix}$$

But $a\alpha_1 + b\beta_1 + c\gamma_1 = 2\Delta$, and two sim'r eq'ns.

$$\therefore \text{ area } PQR = \frac{abc}{8(\Delta)^2}\begin{vmatrix} \alpha_1 & \alpha_2 & \alpha_3 \\ \beta_1 & \beta_2 & \beta_3 \\ \gamma_1 & \gamma_2 & \gamma_3 \end{vmatrix}$$

NOTE—*This is the area of a* Δ PQR *in terms of the trilinear coordinates of its corners with respect to* ABC *as* Δ *of reference.*

6. *If* x, y, z *are the respective distances of a point from the corners* A, B, C *of a triangle; and* θ, ϕ, χ *the angles subtended at the point by* a, b, c *respectively; then*

$$ax/\sin(\theta - A) = by/\sin(\phi - B) = cz/\sin(\chi - C)$$
$$= abc/\overline{x\sin\theta + y\sin\phi + z\sin\chi}$$

Wolstenholme: 582. *Brill*: *E. T.* xlii.

We have $a^2 = y^2 + z^2 - 2yz\cos\theta$ . . . . (1)

$b^2 = z^2 + x^2 - 2zx\cos\phi$ . . . . (2)

$c^2 = x^2 + y^2 - 2xy\cos\chi$ . . . . (3)

$bc\sin A = yz\sin\theta + zx\sin\phi + xy\sin\chi$ . . . . (4)

(2) + (3) − (1) gives

$bc\cos A = x^2 - xy\cos\chi - zx\cos\phi + yz\cos\theta$ . . (5)

Elim'g yz from (4) and (5), we get, since $\theta + \phi + \chi = 2\pi$,

$$bc\sin(\theta - A) = x^2\sin\theta - xy\sin(\theta + \chi) - xz\sin(\theta + \phi)$$

Whence $abc/\Sigma(x\sin\theta) = ax/\sin(\theta - A)$

and sim'ly $= by/\sin(\phi - B)$

or $= cz/\sin(\chi - C)$

NOTE—*This is a very useful theorem; for, by means of it we can find* x, y, z, *when given* θ, ϕ, χ; *or conversely. It is* ∴ *applicable for such p'ts as the centroid, orthocentre, Lemoine point. Stated as the Problem — Given* a, b, c, θ, ϕ, χ, *to find* x, y, z—*it is known as* Hyparchus' Problem; *and is a familiar application of Trigonometry to practical surveying.*

P 2

Exercises

1. Prove that—

(1) If $\left.\begin{aligned}\sin^2 A &= \sin^2 B + \sin^2 C \\ \text{and}\quad \sin A &= 2 \sin B \cos C\end{aligned}\right\}$

the triangle is right-angled, *and* isosceles.

(2) If $\tan A/\tan B = \sin^2 A/\sin^2 B$

the triangle is right-angled, *or* isosceles.

(3) $b^2 \sin 2C - 2bc \sin(B - C) - c^2 \sin 2B = 0$

(4) $\tan A(\sin^2 B + \sin^2 C - \sin^2 A) = \tan B(\sin^2 C + \sin^2 A - \sin^2 B)$

(5) If $b = c$, $\text{vers } A = a^2/(2b^2)$

(6) If $A/2 = B/3 = C/4$, $\cos \frac{A}{2} = (a + c)/(2b)$

(7) $(a + b) \sin \frac{C}{2} = c \cos \frac{A - B}{2}$

(8) $(b + c)^2 \sin^2 \frac{A}{2} + (b - c)^2 \cos^2 \frac{A}{2} = a^2$

(9) $b^2 \cos 2C + 2bc \cos(B - C) + c^2 \cos 2B = a^2$

(10) $\Sigma\left(bc \cos^2 \frac{A}{2}\right) = s^2$

(11) $a \cos \frac{B}{2} \cos \frac{C}{2} \operatorname{cosec} \frac{A}{2} = s$

(12) $a^2 \sin 2B + b^2 \sin 2A = 2ab \sin C$

(13) $1 - \tan \frac{A}{2} \tan \frac{B}{2} = 2c/\overline{a + b + c}$

(14) $\Sigma(a) \times \Sigma \cos A = 2\Sigma\left(a \cos^2 \frac{A}{2}\right)$

(15) $\Sigma\left(\cos^2 \frac{A}{2}\Big/a\right) = s^2/(abc)$

(16) $D^2 . \Sigma\,(\overline{\cos A \cos B/ab}) = 1$

(17) $\Sigma\left(a^2 \cos \frac{B - C}{2}\Big/\cos \frac{B + C}{2}\right) = 2\Sigma(ab)$

(18) $2 \sin^2 A = a^2 \Sigma \{\cos A/(bc)\}$

(19) $\text{vers} A/\text{vers} B = \overline{a(a+c-b)}/\overline{b(b+c-a)}$

(20) $\left(\cot\frac{B}{2} + \cot\frac{C}{2}\right)\Big/\cot\frac{A}{2} = a/s_1$

(21) $\Sigma \{\sin(A-B)/(ab)\} = 0$

(22) $\Sigma \{a^2 \sin(B-C)/\sin A\} = 0$

(23) $\sqrt{c + (a-b)\cos\frac{C}{2}} + \sqrt{c - (a-b)\cos\frac{C}{2}}$
$= 2\sqrt{c}\cos\frac{A-B}{4}$

(24) If $C = 90°$, $\cos(2A - B) = a(3c^2 - 4a^2)/c^3$

2. Show that each of the following expressions is an equivalent for Δ—

(1) $a^2(\cos\overline{B-C} + \cos A)/(4\sin A)$

(2) $s^2/\Sigma\cot\frac{A}{2}$

(3) $\frac{1}{12}\Sigma\{(a^2+b^2)\sin 2C\}$

(4) $\Sigma(a^2/\sin A) \times \Pi \sin\frac{A}{2}$

(5) $(a^2 - b^2)\sin A \sin B/\overline{2\sin(A-B)}$

(6) $a^2/\overline{2(\cot B + \cot C)}$

(7) $\frac{1}{8}D\Sigma\{a\cos(B-C)\}$

(8) $Ds\Sigma\sin 2A\Big/\left(16\,\Pi\cos\frac{A}{2}\right)$

3. If p_1, p_2, p_3 are the respective altitudes from A, B, C, prove that—

(1) $p_1 p_2 p_3 = (abc)^2/D^3$

(2) $\Sigma\{p_1^2/(p_2 p_3)\} = \Sigma(ab/c^2)$

(3) $\Sigma(\sin A/p_1) = 2\Sigma(\cos A/a)$

(4) $2p_3\Sigma(ab\cos C) = ab\Sigma(a\sin A)$

(5) $p_3(a+b) = a^2\sin B + b^2\sin A$

(6) $\Sigma(p_1\sin A)^2 = 6\Delta\Pi\sin A$

4. Prove that—

(1) $2D\,\Pi \sin A = \Sigma(a \cos A)$

(2) $D\sqrt[3]{\Sigma \sin^3 A} = \sqrt[3]{\Sigma a^3}$

(3) $8\Delta^2(\Sigma \cot^2 A + 1) = \Sigma a^4$

(4) $s_1 \tan \frac{A}{2} = s_2 \tan \frac{B}{2} = s_3 \tan \frac{C}{2}$

(5) $\Sigma\left(s_1 a^2 \sec^2 \frac{A}{2}\right) = 2abc$

(6) $\Sigma\left\{(b^2 - c^2)\cot^2\frac{A}{2}\right\} + 2s^3(a - b)(b - c)(c - a)/\Delta^2 = 0$

(7) $\frac{\sin 2A}{a^2(b^2 + c^2 - a^2)} = \frac{\sin 2B}{b^2(c^2 + a^2 - b^2)} = \frac{\sin 2C}{c^2(a^2 + b^2 - c^2)}$

(8) $\Sigma\frac{a^2 - b^2}{ab\sin^2\frac{C}{2}} = -\frac{8\sin\frac{1}{2}(A - B)\sin\frac{B - C}{2}\sin\frac{C - A}{2}}{\sin A \sin B \sin C}$

(9) $\Sigma\left\{\frac{bc}{(b - a)(c - a)}\tan^2\frac{B}{2}\tan^2\frac{C}{2}\right\} = 1$

(10) $\Sigma\left(\frac{1}{a}\cos^4\frac{A}{2}\right) = \frac{\Delta^3}{abc\,s_1 s_2 s_3}\left(\frac{s}{\Delta} - \frac{1}{D}\right)$

(11) $\Sigma\frac{a\cos 2(B - C)}{\cos B \cos C} = 8\Sigma(a\cos A)$

(12) $a\left(b\cos^2\frac{C}{2} - c\cos^2\frac{B}{2}\right)^2 = (b - c)\left(b^2\cos^2\frac{C}{2} - c^2\cos^2\frac{B}{2}\right)$

(13) $\Sigma\left\{a^{\frac{1}{2}}(b^{\frac{3}{2}} + c^{\frac{3}{2}})\cos A\right\} = a^{\frac{1}{2}}b^{\frac{1}{2}}c^{\frac{1}{2}}\Sigma a^{\frac{1}{2}}$

(14) $\left[\frac{2(a^2 - b^2)}{\cos 2B - \cos 2A}\right]^{\frac{3}{2}} = \frac{abc}{\sin A \sin B \sin C}$

(15) $\begin{vmatrix} a & a^2 & \cos^2\frac{A}{2} \\ b & b^2 & \cos^2\frac{B}{2} \\ c & c^2 & \cos^2\frac{C}{2} \end{vmatrix} = 0$

(16)
$$\frac{2\Delta^3}{s^3}\begin{vmatrix} \cot\frac{A}{2} & \cot\frac{B}{2} & \cot\frac{C}{2} \\ \cot\frac{B}{2} & \cot\frac{C}{2} & \cot\frac{A}{2} \\ \cot\frac{C}{2} & \cot\frac{A}{2} & \cot\frac{B}{2} \end{vmatrix} = 3abc - \Sigma a^3$$

NOTE—*All of this group are taken from the Oxford University First Public Examinations.*

5. If O is the orthocentre of a triangle ABC, prove that

$$\Sigma(a/OA) = \Pi(a/OA)$$

6. Prove that

$$a^3\cos 3B + b^3\cos 3A = c^3 - 3abc\cos(A - B)$$

Ox' Jun' Math' Schol': '89.

7. If A, B, C are in arithmetical progression, prove that

$$2\cos\frac{A - C}{2} = \overline{a^3 + c^3}/b^3 = \overline{a + c}/\sqrt{a^2 - ac + c^2}$$

8. If a, b, c are in harmonical progression, prove that

$$\cos\frac{B}{2} = \sqrt{\sin A \sin C/(\cos A + \cos C)}$$

9. A triangle is formed by a parallel through A to BC, and perpendiculars at B, C to BA, CA respectively; show that its area is

$$\overline{a^2\cos^2(B - C)}/\overline{2\sin A\cos B\cos C}$$

10. If α is the length of the line bisecting a triangle, and making equal angles with the sides terminating it; β, γ similar lengths; prove that

$$\Sigma(\alpha^2\beta^2) = 4\Delta^2$$

Q. C. B. '69.

11. If d is the distance between the centres of two circles whose radii are R, r; show that the distance between the two points on their line of centres from which they appear equally large is

$$2dRr/(R^2 - r^2).$$

12. Show also that they appear equally large from *all* points on the circle whose diameter is the join of these points. (See *Euclid Revised*, p. 344, Ex' 6)

13. The sides of a triangle are 141, 100, 53 feet; find the length of the perpendicular on the longest side from the opposite corner.

14. In the sides BC, CA, AB respectively of a triangle are taken points P, Q, R, so that

$$BP/a = CQ/b = AR/c = \lambda$$

show that $\Sigma PQ^2 = (1 - 3\lambda + 3\lambda^2)\Sigma a^2$

15. If $C = 60°$, prove that

$$\frac{1}{a+c} + \frac{1}{b+c} = \frac{3}{a+b+c}$$

R. U. Schol': '82.

16. If the angles of a triangle are in arithmetical progression, and δ is their common difference, prove that

$$\tan^2\delta = \overline{4b^2 - (a+c)^2}/(a+c)^2$$

R. U. Schol': '84.

17. Show that the equations

$$\Sigma (ax\sqrt{x^2 - a^2}) = 2abc$$

$$2\sqrt{(x^2 - a^2)(x^2 - b^2)(x^2 - c^2)} = x(a^2 + b^2 + c^2 - 2x^2)$$

have a common root D.

18. If $\cos\theta = a/\overline{b+c}$, $\cos\phi = b/\overline{c+a}$, $\cos\chi = c/\overline{a+b}$, where a, b, c are sides of a triangle, prove that

$$\Sigma \tan^2\frac{\theta}{2} = 1, \text{ and } \Pi \tan\frac{\theta}{2} = \pm \Pi \tan\frac{A}{2}$$

19. If G is the centroid of an equilateral triangle, whose side is a; and if a line through G terminated by two sides, is divided by G into parts x, y; prove that

$$\frac{1}{x^2} - \frac{1}{xy} + \frac{1}{y^2} = \frac{9}{a^2}$$

I. I. '82.

20. Two sides of a triangle contain the vertical angle α, and are in the ratio of m to 1 (where $m > 1$) and on the lesser of them as base a similar triangle is described having again α as vertical angle : this process is continued for ever : show that the sum of all the triangles is

$$\left(1 + \frac{1}{m^2 - 2m\cos\alpha}\right) \text{original } \Delta.$$

Pet' Camb': '66.

21. Prove that $\Sigma \sqrt{(4b^2c^2 - \Delta^2)(4c^2a^2 - \Delta^2)} = \Delta^2$

22. The centres of three circles, which touch each other externally, form a triangle whose sides are 3, 4, 5 inches: prove that the area included between the circles is slightly more than half a square inch.

Assume that $\sin \frac{3\pi}{10} = .8$

23. If $\sin A$, $\sin B$, $\sin C$ are in harmonical progression, prove that so also are $1 - \cos A$, $1 - \cos B$, $1 - \cos C$.

24. If $\cos A$, $\cos B$, $\cos C$ are in arithmetical progression, prove that s_1, s_2, s_3 are in harmonical progression.

25. A pyramid is cut off from a cube; if α, β, γ are the distances of the corner of the cube from the corners of the plane of section, show that the area of the section is $\sqrt{\Sigma(\alpha^2\beta^2)}/2$.

26. If P, P′ are points in side BC; Q, Q′ points in side CA; R, R′ points in side AB; of a triangle ABC; prove that, if perpendiculars to the sides at P, Q, R concur; and perpendiculars to the sides at P′, Q′, R′ concur; then $\Sigma(PP' \sin A) = 0$; where PP′ is to be considered positive when in cyclic order ABC.

Q. C. B. '67.

27. If D is the point in side AB of triangle ABC, such that

$$CA : CB = DB : DA$$

and if ϕ denotes the angle BCD; prove that

$$(a^2 + b^2)\cot\frac{C}{2} + (a^2 - b^2)\tan\frac{C}{2} = 2b^2\cot\phi$$

28. Perpendiculars are drawn outwardly at the mid points of the sides of a triangle ABC, so that each is half the side to which it is perpendicular: the outward extremities of these are joined, forming a triangle PQR: prove that

$$\Sigma PQ^2 = \Sigma a^2 + 6\,\Delta ABC$$

29. If $b - a = mc$, prove that—

(1) $\cos\left(A + \frac{C}{2}\right) = m\cos\frac{C}{2}$

(2) $\cot\frac{B - A}{2} = \frac{1 + m\cos B}{m\sin B}$

30. In triangle ABC, AX is drawn to divide BC so that AX is a mean proportional between BX, CX: prove that

$$BC \tan AXC = (BX - CX) \tan BAC$$

31. If p_1, p_2 are the perpendiculars from the mid point of the base of a triangle on the bisectors of its vertical angle; and p_3 the perpendicular from the vertex A on the perpendicular to the base at its mid point; prove that

$$4 p_1 p_2 = a p_3 \sin A.$$

32. The distances of a point within a triangle from the corners being x, y, z; and from the sides α, β, γ; prove that

$$\Sigma (x^2 \alpha \sin A) = \alpha\beta\gamma \, \Sigma (a/\alpha)$$

33. In triangle ABC, XY is drawn parallel to BC so that XY = BX + CY; prove that

$$XY = \frac{a}{2}\left(1 + \tan\frac{B}{2} \tan\frac{C}{2}\right), \text{ or } \frac{a}{2}\left(1 + \cot\frac{B}{2} \cot\frac{C}{2}\right)$$

according as XY cuts sides, or sides produced.

R. U. Schol': '87.

34. If PQ, a chord of a circle centre A, radius R, touches a circle centre B, radius r—the second circle lying wholly within the first—and if the angle PAB is ϕ, the angle QAB is θ, and AB is δ; prove that

$$r = (R + \delta)\sin\frac{\theta}{2} \sin\frac{\phi}{2} + (R - \delta)\cos\frac{\theta}{2} \cos\frac{\phi}{2}$$

Q. C. B. '71.

35. P, Q are points on the circumcircle of ABC, such that the distance of either from A is a mean proportional to its distances from B and C: prove that

$$2(\widehat{PAB} \sim \widehat{QAC}) = \hat{B} \sim \hat{C}$$

36. The sides of a triangle, whose area is Δ_1, are b, c, d; those of Δ_2 are c, d, a; of Δ_3 are a, b, d; and of Δ_4 are a, b, c: prove that

$$\frac{\Delta_1^2 - \Delta_2^2}{a^2 - b^2} + \frac{\Delta_3^2 - \Delta_4^2}{c^2 - d^2} = \frac{\Delta_2^2 - \Delta_3^2}{b^2 - c^2} + \frac{\Delta_4^2 - \Delta_1^2}{d^2 - a^2}$$

Math' Tri': '70.

NOTE—*Use formula* $16\Delta^2 = 2\Sigma b^2c^2 - \Sigma a^4$.

37. If $y \sin^2 C + z \sin^2 B = z \sin^2 A + x \sin^2 C = x \sin^2 B + y \sin^2 A$, where A, B, C are angles of a triangle,

prove that $x/\sin 2A = y/\sin 2B = z/\sin 2C$

Wolstenholme: 545.

38. If $x \sin^2 A \cos B - y \sin^2 B \cos A + z(\cos^2 A - \cos^3 B) = 0$
and $z \sin^2 C \cos A - x \sin^2 A \cos C + y(\cos^2 C - \cos^2 A) = 0$
where A, B, C are angles of a triangle,

prove that $ax = by = cz$

Math' Tri': '63.

39. Apply Example 6 (p. 211) to show that if the cotangents of the angles of a triangle are in arithmetical progression, so also are the cotangents of the angles subtended by the sides at the centroid.

Edwardes: E. T. xxx; *but only a particular case of Wolstenholme*: 583.

40. Any point P is taken in the side AB of a triangle ABC, and Q is the point in AC for which AQ = AP; if the median AM cuts PQ in X, prove that

$$PX : QX = AC : AB$$

41. If a, b, c, Δ are in arithmetical progression, find their values in rational numbers; and show that there is only one solution in integers.

42. P is a point on the boundary of a circular field, whose centre is C: an animal, tethered by a rope fastened at P, can graze over $/$mth of the field: if X, Y are the most distant points on the boundary which the animal can reach; and if 2ϕ is the radian measure of the angle XCY; prove that

$$\sin\phi + (\pi - \phi)\cos\phi = \left(1 - \frac{1}{m}\right)\pi$$

If the field is an acre, find the length of the rope that the animal may be able to graze over half of it.

43. In the triangle ABC, AB and AC are equal; AM perpendicular to BC, and BN perpendicular to CA, cut in X; CYZ, bisecting the angle at C, crosses BN, AM in Y, Z respectively; prove that

$$\text{area } XYZ = \tfrac{1}{2}(b \sim a)^2 \sin\frac{A}{2} \tan\frac{A}{2}$$

44. The distances of a point within a square from three of its corners are l, m, n; find an equation to determine the area of the square in terms of l, m, n.

Show that in the particular case when l, m, n, are 2, 3, 4, respectively, this area is

$$\tfrac{1}{2}(25 + \sqrt{287})$$

45. If two circles whose radii are r, r′, cut at an angle θ, show that either of their common tangents is $2\sqrt{rr'} \cdot \cos\frac{\theta}{2}$

46. If $A + B + C = 180° = \alpha + \beta + \gamma$, prove that

$$\left(\frac{\sin\beta}{\sin B}\right)^2 + \left(\frac{\sin\gamma}{\sin C}\right)^2 - 2\,\frac{\sin\beta\sin\gamma}{\sin B\sin C}\cos(A+\alpha)$$

$$:\left(\frac{\sin\gamma}{\sin C}\right)^2 + \left(\frac{\sin\alpha}{\sin A}\right)^2 - 2\,\frac{\sin\gamma\sin\alpha}{\sin C\sin\alpha}\cos(B+\beta)$$

$$= \sin^2 A : \sin^2 B$$

E. T xlvi.

47. ABC is a scalene triangle; L, M, N the mid points of its sides; LX, MY, NZ outwardly drawn perpendiculars, such that

$$LX/a = MY/b = NZ/c = \tan\theta/2$$

show that there is only one value of θ for which the triangle XYZ is equilateral; and account for the second value which arises in the solution.

Miss Alice Gordon : *E. T.* xlviii.

48. Show that, if $\Sigma \cot A = \sqrt{3}$, the triangle is equilateral.

49. If G is the centroid, K the Lemoine * point, and ω the Brocard * angle of a triangle ABC, prove that—

(1) $\Sigma \cot GAB = 3\cot\omega = \Sigma\cot GBA$

(2) $\Sigma \cot AGB + \cot\omega = 0$

(3) $\Sigma \cot AKB + \frac{1}{2}\cot\omega + \frac{3}{2}\tan\omega = 0$

Joh' Camb': '88.

50. If ABC, A′B′C′ are any triangles, prove that

$$\frac{b^2}{\sin^2 B} + \frac{c^2}{\sin^2 C} - \frac{2\,b'c'}{\sin B\sin C}\cos(A+A')$$

$$= \frac{2\Delta'}{c^2\sin^2 B}(4\Delta + a^2\cot A' + b^2\cot B' + c^2\cot C')$$

Hence show that, if from any point P inside an acute-angled triangle, PA′, PB′, PC′ are dropped on the sides, and their feet joined, forming a triangle A′B′C′, then

$$\frac{1}{\Delta'} = \frac{2}{\Delta} + \frac{a^2\cot A' + b^2\cot B' + c^2\cot c'}{2\Delta'}$$

R. U. Schol': '90.

* See *Euclid Revised*, pp. 380, 395.

CHAPTER XII

On the adaptation of formulae to Logarithmic Calculation

§ **69.** When an expression consists of terms connected by the symbols + and −, it is generally unsuited for numerical calculation by means of logarithms; but, by the artifice of introducing a T. F. of a new ∧, such an expression may be thrown into a form adapted for the use of logarithms.

Def′—When an expression is modified by the introduction of a trigonometrical function of a new angle, such an angle is termed a **subsidiary angle.**

Suppose, for example, that we wish to calculate by logarithms the value of the expression

$$a \cos \alpha + b \sin \alpha$$

where a and b are numbers, and α a known ∧.

Put x for the expression; so that

$$x \equiv a\left(\cos \alpha + \frac{b}{a} \sin \alpha\right)$$

Now b/a being expressible as a number to any assigned number of dec′l places, there will be some ∧ (ϕ say) to be found in a table-book of logarithms, such that, to as many decimal places as the tables are calculated, $\tan \phi = b/a$, where ϕ is given by

$$L \tan \phi = \log b - \log a + 10$$

And then $x \equiv a(\cos \alpha + \sin \alpha \tan \phi)$

$$\equiv a \sec \phi \cos(\alpha - \phi)$$

So that x can be found from

$$\log x = \log a + L \sec \phi + L \sec(\alpha - \phi) - 20$$

The expression might also have been thrown into the form

$$\sqrt{a^2 + b^2} \cos(\alpha - \phi)$$

where ϕ has the same value as before.

The two forms will be seen to amount to the same, if we consider that

$$\sqrt{a^2 + b^2} = a\sqrt{1 + \left(\frac{b}{a}\right)^2}$$

$$= a\sqrt{1 + \tan^2\phi}$$

$$= a \sec\phi$$

Some of the formulæ in the preceding Chapter are unsuitable for numerical calculation with logarithmic tables; but may be modified into forms which are suitable.

Examples

1. *To adapt the formula of* § 68 *to logarithmic calculation.*

If $c < b$, there must be some $\cos^{-1}\frac{c}{b}$, less than 180°, call it ϕ, so that $\cos\phi = \frac{c}{b}$, where ϕ is given by

$$L\cos\phi = \log c - \log b - 10$$

Then $$\frac{b - c}{b + c} = \tan^2\frac{\phi}{2}$$

So that the formula of § 68 may be written

$$\tan\frac{B - C}{2} = \tan^2\frac{\phi}{2}\cot\frac{A}{2}$$

which is in a form adapted to log's

The result may be written as

$$B - C = 2\tan^{-1}\left\{\tan^2\tfrac{1}{2}\left(\cos^{-1}\frac{c}{b}\right)\cot\frac{A}{2}\right\}$$

2. *To adapt the first formula of* § 67 *to logarithmic calculation.*

The formula gives $a^2 = \sqrt{\{(b + c)^2 - 2bc(1 + \cos A)\}}$

$$= (b + c)\sqrt{1 - 4bc\cos^2\frac{A}{2}\Big/(b + c)^2}$$

Now, since $4\,bc\cos^2\frac{A}{2} < (b+c)^2$, there must be some

$$\sin^{-1}\left\{2\sqrt{bc}\cos\frac{A}{2}\Big/(b+c)\right\}$$

less than 180°, call it θ, so that

$$\sin\theta = 2\sqrt{bc}\cos\frac{A}{2}\Big/(b+c),$$

where θ is given by

$$L\sin\theta = \log b + \log c - \log(b+c) + L\cos\frac{A}{2}$$

So that the first formula of § 67 may be written

$$a = (b+c)\cos\theta$$

which is in a form adapted to log's.

The result may be written as

$$a = (b+c)\cos\sin^{-1}\left\{2\sqrt{bc}\cos\frac{A}{2}\Big/(b+c)\right\}$$

Another modification of this formula may be got thus—

$$a = \sqrt{\{(b-c)^2 + 2\,bc\,(1-\cos A)\}}$$

$$= (b-c)\sqrt{1 + 4\,bc\sin^2\frac{A}{2}\Big/(b-c)^2}$$

Now there must be some $\tan^{-1}\left\{2\sqrt{bc}\sin\frac{A}{2}\Big/(b-c)\right\}$ less than 180°,

call it χ, so that

$$\tan\chi = 2\sqrt{bc}\sin\frac{A}{2}\Big/(b-c)$$

where χ is given by

$$L\tan\chi = \log b + \log c - \log(b-c) + L\sin\frac{A}{2}$$

So that the first formula of § 67 may be written

$$a = (b-c)\sec\chi$$

which is in a form adapted to log's.

The result may be written as

$$a = (b-c)\sec\tan^{-1}\left\{2\sqrt{bc}\sin\frac{A}{2}\Big/(b-c)\right\}$$

A third modification of this formula may be made; for

$$a^2 = (b^2 + c^2)\left(\cos^2\frac{A}{2} + \sin^2\frac{A}{2}\right) - 2\,bc\left(\cos^2\frac{A}{2} - \sin^2\frac{A}{2}\right)$$

$$= (b - c)^2\cos^2\frac{A}{2} + (b + c)^2\sin^2\frac{A}{2}$$

$$\therefore\ a = (b - c)\cos\frac{A}{2}\sqrt{1 + \left(\frac{b + c}{b - c}\tan\frac{A}{2}\right)^2}$$

Now there must be some $\tan^{-1}\left(\frac{b + c}{b - c}\tan\frac{A}{2}\right)$ less than 180°, call it ω, so that

$$\tan\omega = \frac{b + c}{b - c}\tan\frac{A}{2}$$

where ω is given by

$$L\tan\omega = \log(b + c) - \log(b - c) + L\tan\frac{A}{2}$$

So that the first formula of § 67 may be written

$$a = (b - c)\cos\frac{A}{2}\sec\omega$$

which is in a form adapted to log's.

The result may be written as

$$a = (b - c)\cos\frac{A}{2}\sec\tan^{-1}\left(\frac{b + c}{b - c}\tan\frac{A}{2}\right)$$

Exercises

1. Adapt to the use of logarithms—

(1) $\sec\alpha - \sec\beta$

(2) $\cos A - \cos(A + \alpha) - \cos(A + \beta) + \cos(A + \alpha + \beta)$

(3) $\tan\theta = a\sin\phi/(b - a\cos\phi)$

1°, when $a > b$; and, 2°, when $a < b$.

2. If ABC is a triangle, prove that

$$\left.\begin{aligned} a &= (a + b)\cos^2\frac{\phi}{2} \\ b &= (a + b)\sin^2\frac{\phi}{2} \end{aligned}\right\}$$

where $\sin\phi = \frac{2\sqrt{ss_3}}{a + b}\sec\frac{C}{2}$

3. Prove the truth of the following transformations

(1) $\sqrt{a^2 + b^2} = a \sec \tan^{-1} b/a$

(2) $\sqrt{a^2 - b^2} = a \cos \sin^{-1} b/a$

4. Show that we may express the roots of the equation

$$ax^2 + bx + c = o$$

as $-\frac{b}{a} \cos^2 \frac{1}{2} \sin^{-1} \frac{2\sqrt{ac}}{b}$, and $-\frac{b}{a} \sin^2 \frac{1}{2} \sin^{-1} \frac{2 \wedge \overline{ac}}{b}$

5. If ABC is a triangle, show that—

(1) $\sin \frac{A + B}{2} = \frac{a + b}{2\sqrt{ab}} \sin \cos^{-1} \frac{c}{a + b}$

(2) $\cos \frac{A + B}{2} = \frac{c}{2\sqrt{ab}} \sin \cos^{-1} \frac{a - b}{c}$

(3) $\sin \frac{A - B}{2} = \frac{a^2 - b^2}{2c\sqrt{ab}} \sin \cos^{-1} \frac{c}{a + b}$

(4) $\cos \frac{A - B}{2} = \frac{a + b}{2\sqrt{ab}} \sin \cos^{-1} \frac{a - b}{c}$

6. If $n < m$, show that $m \sin^2 A - n$ can be put into the form

$$m \sin (A + \theta) \sin (A - \theta)$$

7. Show that $m \sin \theta \pm n \cos \theta$

$$\equiv m \sec \tan^{-1} \frac{n}{m} \sin \left(\theta \pm \tan^{-1} \frac{n}{m}\right)$$

8. If the roots of $x^2 + px + q = o$ are real, show that they are expressible as

$$p \sin^2 \frac{\theta}{2}, \text{ and } p \cos^2 \frac{\theta}{2}$$

9. In a triangle ABC, prove that

$$\tan A = -\frac{2a \sec 2\phi}{a - b} \tan \frac{C}{2} \cos^2 \phi$$

$$\tan B = \frac{2b}{a - b} \tan \frac{C}{2} \cos^2 \phi$$

where ϕ is given by $\tan \phi = \sqrt{\frac{a + b}{a - b}} \tan \frac{C}{2}$

10. Adapt to logarithmic calculation the expression

$$\sqrt{\frac{a-b}{a+b}} + \sqrt{\frac{a+b}{a-b}}$$

11. If $x^2 = a^2 + b^2 + ab$, show that logarithmic tables may be used to calculate x from the equations

$$\left.\begin{aligned} x &= \sqrt{ab}\sec\phi \\ \tan^2\phi &= 2\,\mathrm{cosec}\,(2\tan^{-1} b/a) \end{aligned}\right\}$$

12. Adapt to logarithmic calculation the equation

$$\cos x = \cos y \,.\, \sin A \sin B - \cos A \cos B$$

13. Show that the following transformations, for adapting the left-hand members to the use of logarithms, are justifiable—

(1) $a + a' + a'' + \&c = a\sec^2\theta\sec^2\theta'\sec^2\theta''\ \&c$

(2) $a - a' + a'' - \&c = \dfrac{a\cos^2\theta\cos^2\theta''\ \&c}{\cos^2\theta'\cos^2\theta'''\ \&c}$

where a', a', a'', $\&c$ are in descending magnitude.

(3) $(a+b)(a'+b')(a''+b'')\ \&c = aa'a''\ \&c\sec^2\theta\sec^2\theta'\sec^2\theta''\ \&c$

(4) $(a-b)(a'-b')(a''-b'')\ \&c = aa'a''\ \&c\cos^2\theta\cos^2\theta'\cos^2\theta''\ \&c$

where $a > b,\ a' > b',\ a'' > b'',\ \&c$

14. Show that $\cos(\alpha + x) = \cos\alpha\sin x + \sin\beta$ may be reduced to

$$\cos(\phi + x) = \frac{\sin\beta}{\sin\alpha}\cos\phi$$

$$\tan\phi = \frac{\sin(45^\circ + \alpha)}{\cos 45^\circ\cos\alpha}$$

15. Adapt to logarithmic computation

$$\sin^2 x - 2\sin x\sin y\cos(x+y) + \sin^2 y$$

Catalan : *E. T.* 1891.

CHAPTER XIII

Solution of Triangles

§ **70**. *Def'*—The Problem to be solved is this—*Given three parts of a triangle* (*not the three angles*) *to find the other parts.* Or we may put it thus—*Given enough* (*and only enough*) *about a triangle to distinguish it from all other triangles, it will be said to be 'solved' when its remaining details are determined.*

On consideration of the Problem in particular cases it is readily seen that its solution—

1°, for right-angled Δ^s is simple and immediate ; and,

2°, for Δ^s not right-angled has to be effected by reducing such to right-angled cases.

Note—What is called 'the solution of a triangle' is, "in truth, a reduction of it to the solution of a right-angled triangle ; and the *maker of the tables* it is who solves the right-angled triangle." (*De Morgan*)

We have ∴ to consider two classes of Δ^s—those which are *right-angled*; and those which are *not* right-angled—and there will be various cases in each class.

Class I—When a Δ ABC is right-angled—say at C.

1°, *given* a *and* b.

$$\tan A = a/b, \text{ whence } A \text{ is known}$$

$$\text{then} \quad B = 90^\circ - A$$

$$\text{and} \quad c = a/\sin A, \text{ or } = b/\sin B$$

2°, *given* b *and* c.

$$\sin B = b/c, \text{ whence } B \text{ is known}$$

$$\text{then} \quad A = 90^\circ - B$$

$$\text{Also} \quad a = \sqrt{(c+b)(c-b)}$$

3°, *given* c *and* A.

$$B = 90^\circ - A$$
$$b = c \cos A$$
$$a = c \sin A$$

4°, *given* a *and* A.

$$B = 90^\circ - A$$
$$b = a \cot A$$
$$c = a/\sin A, \text{ or } = b/\sin B$$

5°, *given* b *and* A.

$$B = 90^\circ - A$$
$$a = b \tan A$$
$$c = b/\cos A, \text{ or } = a/\sin A$$

Exercises

Verify *each* of the above methods, by using a table book, in the case of the following triangle.

$$A = 59^\circ\, 0'\, 21''.25, \quad B = 30^\circ\, 59'\, 38''.75,$$
$$a = 110.0951, \quad b = 66.1364, \quad c = 128.4327$$

NOTE—*The verification will be accomplished by assuming as data the two parts, corresponding to those of each the* 5 *cases, and finding the rem'g three parts as shown above.*

Class II—When a Δ ABC is not right-angled.

1°, *given the three sides* a, b, c.

We may use any of the formulæ such as

$$\sin \frac{A}{2} = \sqrt{\frac{s_2 s_3}{bc}}, \quad \cos \frac{A}{2} = \sqrt{\frac{s s_1}{bc}}, \quad \tan \frac{A}{2} = \sqrt{\frac{s_2 s_3}{s s_1}}$$

Note—If only *one* $\wedge$ is wanted, then by using $\sin \frac{A}{2}$ or $\cos \frac{A}{2}$ we shall only need two of the expressions s, s_1, s_2, s_3. Also if $\frac{A}{2}$ is between 45° and

135° it is better to use $\cos\frac{A}{2}$, but otherwise $\sin\frac{A}{2}$; the reason for this choice depending on the principle of the *irregularity and insensibility of differences*, summarised in § 60.

If *two* $\wedge^s$ are wanted use $\tan\frac{A}{2}$: in this case if $\frac{A}{2}$, $\frac{B}{2}$, $\frac{C}{2}$ are thus calculated, then the relation $A/2 + B/2 + C/2 = 90°$ will give a verification.

Or we may use such formulæ as

$$\cos A = (b^2 + c^2 - a^2)/(2\,bc)$$

and, if the aid of logarithms is desired, we have the adaptation effected in § 69, Example 2.

Note—The last formula will do well without logarithms when a, b, c are *small*.

Or again the solution may be effected thus—

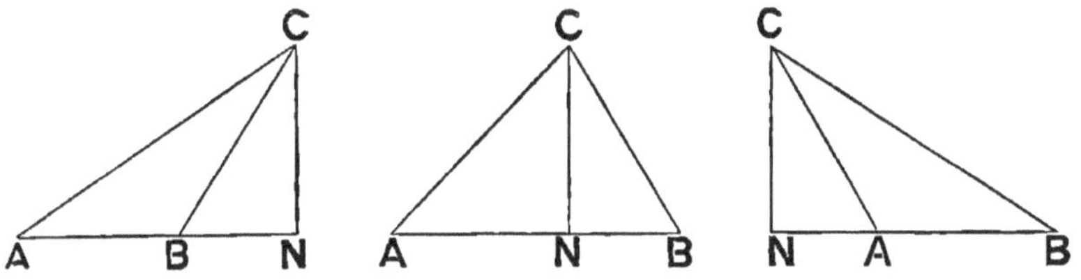

Drop CN ⊥ from C on AB, produced if necessary.

$$\left.\begin{aligned} \text{Now}\quad AN^2 \sim BN^2 &= b^2 \sim a^2 \\ \text{and}\quad AN \underset{\sim}{+} BN &= c \end{aligned}\right\}$$

$$\text{Whence}\quad AN \overset{\sim}{\mp} BN = \frac{b^2 \sim a^2}{c}$$

the upper signs going together, and the lower together.

Hence AN and BN are known.

Then $\cos A = \pm AN/b$, and $\cos B = \pm BN/a$

the neg′ sign being taken when the $\wedge$ is obtuse.

These give A and B, and then $C = 180° - (A + B)$

2°, *given two sides and the included* $\wedge$—*say* b, c, A.

If all the other parts are wanted, we have, by § 68,

$$\frac{B \sim C}{2} = \tan^{-1}\left\{\frac{b \sim c}{b + c}\cot\frac{A}{2}\right\}$$

and $$\frac{B + C}{2} = 90^\circ - \frac{A}{2}$$

whence B and C are known.

Then $a = b \sin A / \sin B$, or $= c \sin A / \sin C$

But if a only is wanted, we have, by § 67,

$$a^2 = b^2 + c^2 - 2\,bc\cos A$$

Note—The adaptations of the above formulæ to logarithmic use are given in § 62, Examples 1 and 2 : the last will do as it stands if the numbers are small.

Or the solution may be effected thus—

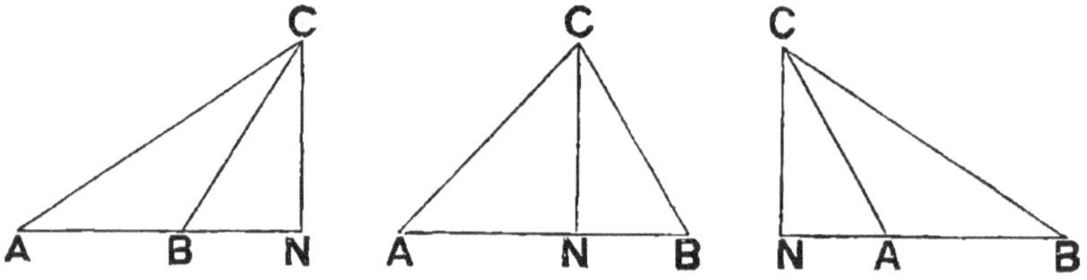

Drop CN $\perp$ on AB, produced if necessary.

Then $CN = b \sin A$

$$AN = \pm\, b \cos A$$

$$BN = c \mp AN = c \mp b \cos A$$

$$\tan B = \frac{b \sin A}{c \mp b \cos A}$$

$$a = (c \mp b \cos A) \sec B$$

In addition to the foregoing—which are the methods usually given in this case—the following compact solution is worth notice.

Given b, c, A; *and that* b > c.

$$\frac{b}{c} = \frac{\sin B}{\sin C} = \frac{\sin (A + C)}{\sin C} = (\cot C + \cot A) \sin A$$

$$\therefore \quad \cot C = \frac{b}{c \sin A} - \cot A$$

$$= \frac{b}{c \sin A}\left\{1 - \frac{c \cos A}{b}\right\}$$

$$= \frac{b}{c \sin A}(1 - \cos^2 \theta)$$

$$\text{i.e.} \quad \cot C = \frac{b \sin^2 \theta}{c \sin A}, \text{ where } \cos^2 \theta = \frac{c \cos A}{b}$$

Whence C is known, and then $B = 180^\circ - (A + C)$

and $a = b \sin A/\sin B$, or $= c \sin A/\sin C$

3°, *given two* ∧s *and a side—say* A, B, c.

$$C = 180^\circ - (A + B)$$

$$a = c \sin A/\sin C$$

$$b = c \sin B/\sin C$$

or $= c \sin B/\sin (A + B)$, if C is obtuse.

4°, *given two sides and an* ∧ *opposite one of them—say* a, b *and* A*—usually called 'the ambiguous case.'*

If $a \sin B > b$, then $\sin A > 1$, and ∴ there is *no* solution: i. e. the parts will not form a △.

If $a \sin B = b$, then $\sin A = 1$, and $A = 90^\circ$; so that $c = a \cos B$, and $C = 90^\circ - A$, complete the solution.

If $a \sin B < b$, then $\sin A < 1$, and there are *two* solutions, one an acute and the other an obtuse-angled △.

Suppose B_1, C_1, c_1 the parts of the acute-angled △;

and B_2, C_2, c_2 " " obtuse "

Then $\left.\begin{aligned} \sin B_1 &= b \sin A / a \\ C_1 &= 180^\circ - (A + B_1) \\ c_1 &= a \sin C_1 / \sin A \end{aligned}\right]$ is a set of solutions

And $\left.\begin{aligned} \sin B_2 &= b \sin A / a \\ C_2 &= 180^\circ - (A + B_2) \\ c_2 &= a \sin C_2 / \sin A \end{aligned}\right]$ is another set of solutions

∴ the preceding solutions belong, each of them, to one or other of the Δ^s (a, b, B_1) (a, b, B_2) where B_1, B_2 are acute and obtuse supplements; but both B_1 and B_2 come from the eq'n

$$\sin B = b \sin A / a.$$

∴ the solution (as often occurs in Algebra) gives the solution of a cognate Problem—*viz.* for the Δ which has an ∧ the supp't of the given ∧.

Respecting the double solution, and the solution covering the cognate case, we notice that—

If $b < a$, so that $B < A$, there is no obtuse value of B, so that both solutions belong to acute values of B.

If $b = a$, so that $B = A$, and A, B are both acute, there is one solution—*viz.* an isosceles Δ—and the other degenerates into a straight line.

If $b > a$, so that $B > A$, there is no obtuse value of A, but either value of B will do, and one solution belongs to each value.

The cases when $b > a$, and $b < a$, are here illustrated.

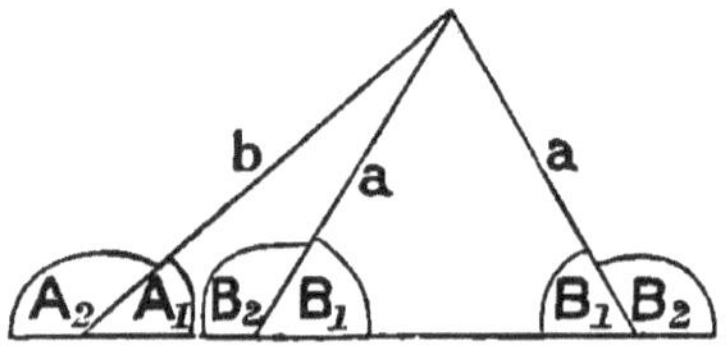

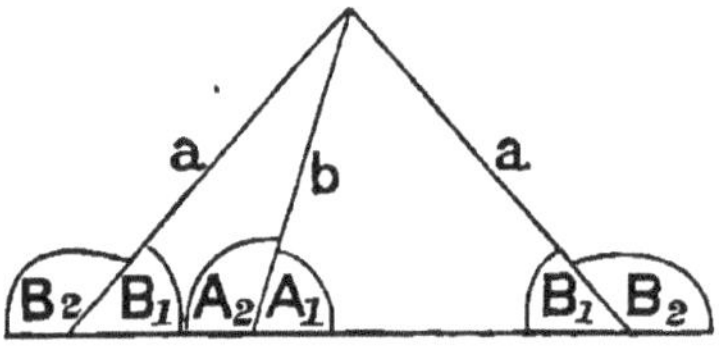

The double solution indicates a quadratic : thus from the eq'n

$$c^2 + b^2 - 2cb \cos A = a^2$$

we get $c_1 = b \cos A + \sqrt{a^2 - b^2 \sin^2 A}$

and $c_2 = b \cos A - \sqrt{a^2 - b^2 \sin^2 A}$

The 'ambiguous case' may also be treated geometrically as follows—

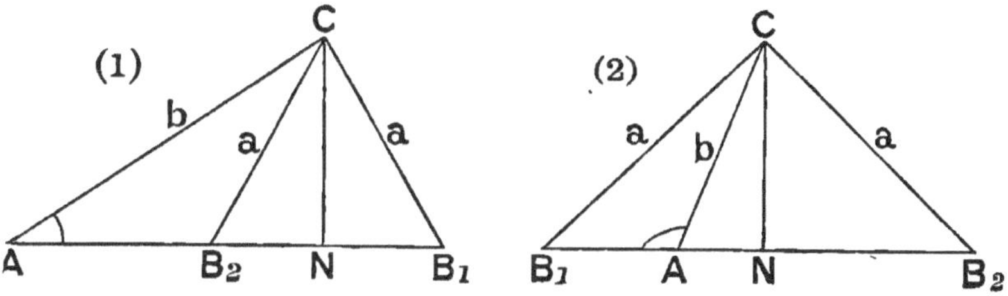

Drop CN ⊥ on AB.

Then in fig' (1)—the ordinary 'ambiguous case'—

$$\left.\begin{aligned} AN &= b \cos A \\ CN &= b \sin A \\ B_1 N &= \sqrt{a^2 - b^2 \sin^2 A} = B_2 N \end{aligned}\right]$$

$$\left.\begin{aligned} c_1 &= AN + B_1 N = b \cos A + \sqrt{a^2 - b^2 \sin^2 A} \\ c_2 &= AN - B_2 N = b \cos A - \sqrt{a^2 - b^2 \sin^2 A} \end{aligned}\right\}$$

And in fig' (2)—the 'cognate case'—

$$\left.\begin{aligned} c_1 &= B_1 N - NA = \sqrt{a^2 - b^2 \sin^2 A} + b \cos A \\ c_2 &= B_2 N + NA = \sqrt{a^2 - b^2 \sin^2 A} - b \cos A \end{aligned}\right\}$$

Note—Though all questions involving the solution of Δ^s are reducible to the preceding cases, the *data* may be in forms not exactly corresponding to any one of them : thus among the *data* may be *the sum of two sides* or *an altitude*, or *the area*. The necessary modifications in the solution will readily suggest themselves.

Examples

1. *If* $a/4 = b/5 = c/6$, *find* B : log 2 *is supposed known*; *and*

$$L\cos 27^\circ 53' = 9{\cdot}9464040, \quad L\cos 27^\circ 54' = 9{\cdot}9463371.$$

We have $\cos\frac{B}{2} = \sqrt{\frac{15 \times 5}{4 \times 6 \times 4}} = \sqrt{\frac{25}{32}}$

$$\therefore \quad L\cos\frac{B}{2} - 10 = 1 - \frac{7}{2}\log 2$$

$$\therefore \quad L\cos\frac{B}{2} = 9{\cdot}9463950$$

$$= L\cos 27^\circ 53' + \delta''$$

$\therefore$ diff′ for $\delta'' = {\cdot}0000090$

,, ,, $60'' = {\cdot}0000669$

$$\therefore \quad \delta'' = \frac{90}{669} \times 60'' = 8'', \text{ nearly}$$

$$\therefore \quad \frac{B}{2} = 27^\circ 53' 8''$$

$$\therefore \quad B = 55^\circ 46' 16''$$

NOTE—*Recollect that* $\log 2 = {\cdot}3010300$, *and* $\log 3 = {\cdot}4771213$.

2. *If* $A = 64^\circ 12'$, *and* $b : c = 9 : 7$, *find* B *and* C *to the nearest second*: *given* log 2, $L\tan 57^\circ 54' = 10{\cdot}2025255$

$$L\tan 11^\circ 17' = 9{\cdot}2999804, \quad L\tan 11^\circ 16' = 9{\cdot}2993216$$

$$\tan\frac{B-C}{2} = \tfrac{1}{8}\cot 32^\circ 6' = \tfrac{1}{8}\tan 57^\circ 54'$$

$$L\tan\frac{B-C}{2} = 10{\cdot}2025255 - {\cdot}9030900$$

$$= 9{\cdot}2994355$$

$$= L\tan 11^\circ 16' + \delta''$$

$\therefore$ diff′ for $\delta'' = {\cdot}0001139$

and ,, ,, $60'' = {\cdot}0006588$

$$\therefore \quad \delta'' = \tfrac{1139}{6588} \times 60'' = 10'', \text{ nearly}$$

$$\therefore \quad \frac{B - C}{2} = 11^\circ\, 16'\, 10''$$

$$\frac{B + C}{2} = 57^\circ\, 54'$$

$$\therefore \quad B = 69^\circ\, 10'\, 10''$$

$$C = 46^\circ\, 37'\, 50''$$

3. *Given* c, C, *and* h *the altitude of* C ; *find* a *and* b

We have $ab \sin C = 2\Delta = ch$

and $a^2 + b^2 - 2ab \cos C = c^2$

Put x for $a + b$, and y for $a - b$

then $x^2 - y^2 = 4ab = \dfrac{4ch}{\sin C}$

and $x^2 \sin^2 \dfrac{C}{2} + y^2 \cos^2 \dfrac{C}{2} = c^2$

$$\therefore \quad x^2 = c^2 + 2ch \cot \frac{C}{2}$$

and $y^2 = c^2 - 2ch \tan \dfrac{C}{2}$

i. e. $a + b = \sqrt{c}\, \sqrt{c + 2h \cot \dfrac{C}{2}}$

and $a - b = \sqrt{c}\, \sqrt{c - 2h \tan \dfrac{C}{2}}$

$\therefore$ a and b are included in

$$\frac{\sqrt{c}}{2} \left\{ \sqrt{c + 2h \cot \frac{C}{2}} \pm \sqrt{c - 2h \tan \frac{C}{2}} \right\}$$

Exercises

1. If $C = 90^\circ$, $a = 5$, $b = 20$, find B : given log 2, and that

$L \tan 75^\circ\, 58' = 10.6021537$, $L \tan 75^\circ\, 57' = 10.6016170$.

2. If $a = 18$, $b = 20$, $c = 22$; find $L \tan \dfrac{A}{2}$, the logs of 2 and 3 being supposed known.

3. Is the triangle in which $b = 14$, $c = 7$, and $C = 30°$, ambiguous?

4. Solve the triangle, by using a table-book, in which

$$a = 55, \quad A = 41° \, 13' \, 22'', \text{ and } B = 71° \, 19' \, 5''.$$

5. If $C = 120°$, find a, b, and the area, in terms of c and the median (m say) from C.

6. If the angles of a triangle are in A. P., and the greatest side is to the least as 5 to 4, find the angles: $\log 3$ supposed known, and it is given that

$$L \tan 10° \, 53' \, 40'' = 9{\cdot}2843610, \quad L \tan 10° \, 53' \, 30'' = 9{\cdot}2842475.$$

7. The sides of a triangle are 9, 10, 11; find the $L \tan$ (smallest angle): given $\log 7 = {\cdot}8450980$, and assuming $\log 2$ and $\log 3$.

8. If

$$\log a = {\cdot}9717937, \quad \log b = {\cdot}8729345,$$
$$L \sin A = L \sin 70° \, 12' \, 10'' = 9{\cdot}9735422,$$
$$L \sin 51° \, 41' \, 20'' = 9{\cdot}8746794, \text{ and } L \sin 58° \, 6' \, 30'' = 9{\cdot}9289325;$$

find B, C, and $\log c$.

9. If $c : a - b = 9 : 2$, and $C = 60°$; find A and B: $\log 3$ is supposed known, and

$$L \cos 78° \, 54' \, 10'' = 9{\cdot}2843730, \quad L \cos 78° \, 54' \, 20'' = 9.2842656.$$

10. If $a = 409$, $b = 317$, and $C = 41° \, 16'$; find A and B: use a table-book.

11. If $a = 19$, $b = 1$, and $A - B = 90°$; find C: $\log 3$ is known,

$$L \tan 41° \, 59' = 9{\cdot}9541834, \text{ diff}' \text{ for } 60'' = {\cdot}0002540.$$

12. One angle of a triangle is 60°, and the sides including it are as 5 to 3: show that the other angles are $\tan^{-1} \dfrac{3\sqrt{3}}{7}$, and $\tan^{-1} 5\sqrt{3}$.

13. If the vertical angle of a triangle is 36°, the base 4, and the altitude $\sqrt{5} - 1$, solve it.

14. Given $a + b$, c, and C, show that a and b are $m \sin^2 \dfrac{\phi}{2}$ and $m \cos^2 \dfrac{\phi}{2}$, where $m = a + b$, and ϕ is given by

$$\sin \phi = \pm \frac{1}{m} \sqrt{(m + c)(m - c)} \sec \frac{C}{2}$$

15. In the 'ambiguous case,' where a, b, A are the given parts (b being greater than a, and the double values of the other parts being denoted by suffixes) prove each of the following—

(1) $\cot\frac{C_1 + C_2}{2} = \tan A$

(2) $\frac{\sin C_1}{\sin B_1} + \frac{\sin C_2}{\sin B_2} = 2 \cos A$

(3) Distance between centres of circum-circles of the two triangles

$$= (c_1 \sim c_2)/(2 \sin A)$$

(4) $\frac{2b + c_1 + c_2}{1 + \cos A} + \frac{2b - c_1 - c_2}{1 - \cos A} = 4b$

(5) $\frac{\Delta_1}{\Delta_2} + \frac{\Delta_2}{\Delta_1} = \frac{2(a^2 + b^2 \cos 2A)}{b^2 - a^2}$

(6) $\Delta_1^2 + \Delta_2^2 - 2\Delta_1\Delta_2 \cos 2A = \frac{a^2}{b^2}(\Delta_1 + \Delta_2)^2$

(7) $c_1^2 - 2c_1c_2 \cos 2A + c_2^2 = 4a^2 \cos^2 A$

(8) If $\Delta_1 = k\Delta_2$, b/a lies between 1 and $\overline{k+1}/\overline{k-1}$

16. In a triangle ABC, if $AB = AC + \frac{1}{2}BC$, and P is the point in BC for which $PC = 3PB$, prove that angle APC is half angle ACP.

Math' Tri': '78.

17. Show that the three sides, area, and an altitude of a triangle, can always be represented respectively by the rational formulæ

$$(ux)^2 + (vy)^2, \quad (uy)^2 + (vx)^2, \quad (u^2 + v^2)(x^2 + y^2),$$
$$uvxy(u^2 + v^2)(x^2 - y^2), \quad 2uvxy$$

Dr. Hart: *E. T.* xxiii.

18. Given the base c, the altitude h, and the difference of the base angles α; find the other sides, and the vertical angle.

If θ_1, θ_2 are two values for this vertical angle,

prove that $\cot\theta_1 + \cot\theta_2 = 4h/(c \sin^2\alpha)$

Prove that only one of these is a proper solution; and that if it is θ_1, then

$$\tan\frac{\theta_1}{2} = (\sqrt{4h^2 + c^2\sin^2\alpha} - 2h)/\overline{c(1 - \cos\alpha)}$$

Account for the appearance of θ_2.

Wolstenholme: 546.

§ **71.** If in the use of any one of the preceding methods for a particular case, it should happen—as it sometimes does in practical applications—that the 'differences' are 'irregular' or 'insensible,' then the method will not give accurate results. Such cases may sometimes be treated by considering that—

if A is *very small*, i. e. $< 1'$,

then its radian measure (θ say) $< \cdot 0003$

Now as sin A lies between θ and $\theta - \theta^3/6$

and cos A " " 1 " $1 - \theta^2/2$

∴ the error, by putting θ for sin A, and 1 for cos A, will not affect the 7th dec′l place, and ∴ will give results as true as can be got by an ordinary table-book.

Examples

1. *If* $C = 90°$, *and* b, A *are given, where* A *is small, find* c *approximately.*

Here the formula $c = b/\cos A$ is not good, ∵ (see § 59) the diff′s for cosines of small ∧s are 'irregular.'

But then $c - b = b \dfrac{1 - \cos A}{\cos A} = 2\,b \sin^2 \dfrac{A}{2}$, nearly

$$\therefore \quad c = b\left(1 + 2 \sin^2 \frac{A}{2}\right), \text{ nearly}$$

2. *If* a, b, C *are given; and also that* $C = 180° - k°$, *where* k *is* **very small,** *find an approximation for* c.

By Example 2, p. 222, $c = (a + b)\left\{1 - \dfrac{4ab}{(a+b)^2} \sin^2 \dfrac{k}{2}\right\}^{\frac{1}{2}}$

$$= (a + b)\left\{1 - \frac{2ab}{(a+b)^2} \sin^2 \frac{k}{2}\right\}, \text{ nearly}$$

$$= a + b - \frac{2ab}{a+b} \sin^2 \frac{k}{2}$$

$$= a + b - \frac{ab}{2(a+b)} \sin^2 k, \text{ since } \cos \frac{k}{2} = 1$$

$$= a + b - \frac{ab}{2\,a + b)} \left(\frac{\pi k}{180}\right)^2$$

if $k \not> 10''$, see p. 179, Example 2, Cor′

The same substitutions may be used to find connections between small variations in sides and $\wedge^s$ of a Δ, consequent on an error in the estimation of the size of one of them.

Note—In what follows the letter δ will stand for the words '*a small increment of*': thus δa means *a small increment of the side* a. *All increments of angle are supposed to be in radian measure.*

Recollect that if A, B, C are $\wedge^s$ of a Δ, then

$$\delta A + \delta B + \delta C = 0, \text{ always.}$$

Examples

1. *In a triangle right-angled at* C, *suppose that while* b *is true, either* a *or* A *has a small error, producing a corresponding error in the other: required the connection between the errors.*

$$a = b \tan A$$

$$\therefore \quad a + \delta a = b \tan (A + \delta A)$$

$$\therefore \quad \delta a = b \frac{\sin \delta A}{\cos A \cos (A + \delta A)}$$

$$\therefore \quad \delta a = b \sec^2 A \,.\, \delta A, \text{ approximately.}$$

2. *Given* b, c, *and* A, *and that while* b *and* c *are true there is a small error in* A: *required to find the corresponding error in* B.

$$\sin B = \frac{b}{c} \sin C = \frac{b}{c} \sin (A + B)$$

$$\therefore \quad \sin (B + \delta B) = \frac{b}{c} \sin (A + \delta A + B + \delta B)$$

$$\sin B + \cos B \,.\, \delta B = \frac{b}{c} \left\{ \sin (A + B) + (\delta A + \delta B) \cos (A + B) \right\}$$

$$\therefore \quad \cos B \,.\, \delta B = -\frac{b}{c} (\delta A + \delta B) \cos C$$

$$\therefore \quad \delta B \left(\cos B + \frac{b}{c} \cos C \right) = -\frac{b}{c} \delta A \cos C$$

$$\therefore \quad \delta B \left(\cos B + \frac{\sin B}{\sin C} \cos C \right) = -\frac{\sin B \cos C}{\sin C} \delta A$$

$$\therefore \quad \delta B = -\frac{\sin B \cos C}{\sin A} \,.\, \delta A$$

Exercises

1. Given A, C, c; and that $B = A + n''$, where n is small; show that

$$b = \frac{c}{\sin C}\left(\sin A + \frac{n}{206265}\cos A\right)$$

2. If C is very obtuse, show that the number of seconds in A + B is nearly

$$206265\sqrt{2(a+b)(a+b-c)/(ab)}$$

3. If the sides and angles of a triangle ABC receive small increments, show that—

(1) when a and b are constant,

$$c \,.\, \delta B + b\cos A \,.\, \delta C = 0$$

(2) when a and A are constant,

$$\cos C \,.\, \delta b + \cos B \,.\, \delta c = 0$$

(3) when a and B are constant,

$$\tan A \,.\, \delta b - b \,.\, \delta C = 0$$

Joh' Camb': '43.

4. The angles A, B of a triangle have small unknown errors, show that sin (A + B) will have its least error when C is nearly a right angle.

5. The side b of a triangle is rightly known, but the angle B has an error whose cube may be neglected, show that the circum-diameter is

$$b \operatorname{cosec} B\left\{1 - \delta B \cot B + \frac{(\delta B)^2}{2}(\cot^2 B + \operatorname{cosec}^2 B)\right\}$$

Pemb' Camb': '85.

6. Solve a triangle when given the lengths of its three altitudes p_1, p_2, p_3; and show that a small error in the measurement of p_1 will cause errors

$$\frac{\sin A \,.\, \delta p_1}{p_1 \sin B \sin C} \text{ in the value of A}$$

$$\text{and } -\frac{a\,\delta p_1}{p_1}\cot B \cot C \text{ in the value of a}$$

Math' Tri': '66.

7. If the sides of a triangle are slightly changed, prove that

$$\delta D = \Pi \cot A \,.\, \Sigma(\sec A \,.\, \delta a)$$

CHAPTER XIV

Lines and Circles remarkably associated with the Triangle

§ **72.** *Def's*—With respect to a triangle ABC, and in addition to the notation fixed in § 62, the following lettering will be invariably used, unless the contrary is expressly stated—

I	for the position	of the	*in-centre*:
E_1	,,	,,	*ex-centre* within the angle A:
E_2	,,	,,	,, ,, ,, B:
E_3	,,	,,	,, ,, ,, C:
S	,,	,,	*circum-centre*:
O	,,	,,	*ortho-centre*:
G	,,	,,	*centroid*:
K	,,	,,	**Lemoine* (or *symmedian*) *point*:
N	,,	,,	*nine-point-centre*:
Ω	,,	,,	*positive* **Brocard point*:
Ω'	,,	,,	*negative* **Brocard point*:
r	for the length	of the	*in-radius*:
r_1	,,	,,	*ex-radius* corresponding to E_1:
r_2	,,	,,	,, ,, ,, E_2:
r_3	,,	,,	,, ,, ,, E_3:
R	,,	,,	*circum-radius*:
p_1, p_2, p_3	,,	,,	*altitudes* of A, B, C, respectively:
ω	for the magnitude	of the	**Brocard angle.*

---

* See Section ix of the General Addenda for the 3rd edition of *Euclid Revised*. This Section is published separately by the Clarendon Press (price 6*d.*) under the following title—*An Introduction* (*within the limits of Euclidian Geometry*) *to the Lemoine and Brocard Points, Lines, and Circles.*

The results now to be given—though many of them are mathematically of great importance—do not involve any fresh principles, but are rather amplifications of the principles already given in Chapter XI. They will therefore be considered as Examples and Exercises, and printed accordingly.

§ 73. *Properties of some* Lines *of importance in connection with a known triangle.*

Examples

1. *Lengths of the* internal and external bisectors *of an angle of a triangle.*

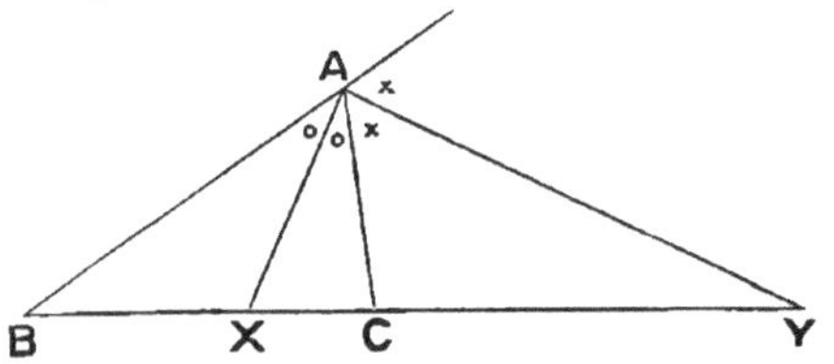

Let AX, AY be the respective intern′ and extern′ bisectors of $B\hat{A}C$; where X is in BC, and Y in BC produced.

Then, since area AXB + area AXC = Δ

and area AYB ∼ area AYC = Δ

∴, if $x = AX$, and $y = AY$, we have

$$\tfrac{1}{2} cx \sin\frac{A}{2} + \tfrac{1}{2} bx \sin\frac{A}{2} = \tfrac{1}{2} bc \sin A$$

and $$\tfrac{1}{2} cy \sin\left(90^\circ + \frac{A}{2}\right) \sim \tfrac{1}{2} by \sin\left(90^\circ - \frac{A}{2}\right) = \tfrac{1}{2} bc \sin A$$

whence $$x = \frac{2\,bc \cos\frac{A}{2}}{b + c} = \frac{2\sqrt{bc\,ss_1}}{b + c}$$

and $$y = \frac{2\,bc \sin\frac{A}{2}}{b \sim c} = \frac{2\sqrt{bc\,s_2 s_3}}{b \sim c}$$

Cor′ (1)—$XY^2 = x^2 + y^2$

$$= \frac{4\,b^2 c^2}{(b^2 \sim c^2)^2}\left\{(b \sim c)^2 \cos^2\frac{A}{2} + (b + c)^2 \sin^2\frac{A}{2}\right\}$$

$$\therefore\ XY^2 = \frac{4\,b^2\,c^2}{(b^2 \sim c^2)^2}\,(b^2 + c^2 - 2\,bc\cos A)$$

$$\therefore\ XY = \frac{2\,abc}{b^2 \sim c^2}$$

Cor' (2)— $BX = x\,\dfrac{\sin\frac{A}{2}}{\sin B} = \dfrac{ac}{b+c}$, and $BY = y\,\dfrac{\sin\left(90^\circ + \frac{A}{2}\right)}{\sin B} = \dfrac{ac}{b \sim c}$

2. *Lengths of the* median, *and* symmedian, *from the same corner of a triangle.*

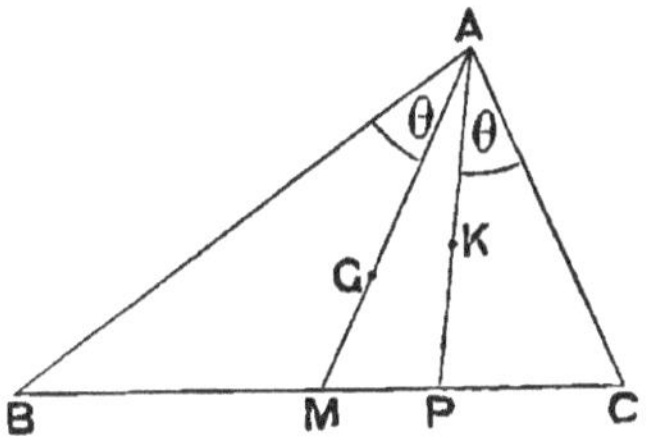

Let AGM be a median, AKP a symmedian of Δ ABC.

Then $B\hat{A}M = C\hat{A}P = \theta$ say.

Put m for AM.

Then $2\,m^2 + \dfrac{a^2}{2} = b^2 + c^2$

and $b^2 + c^2 - a^2 = 2\,bc\cos A$

$$\therefore\ 4\,m^2 = b^2 + c^2 + 2\,bc\cos A$$

i.e. median $= \frac{1}{2}\sqrt{b^2 + c^2 + 2\,bc\cos A}$

or $= \frac{1}{2}\sqrt{2\,(b^2 + c^2) - a^2}$

Again $\cos\theta = \dfrac{m^2 + c^2 - a^2/4}{2\,cm}$

$$\therefore\ 4\,cm\cos\theta = b^2 + 3\,c^2 - a^2$$

Also $4\,cm\sin\theta = 4\,\Delta = 2\,ac\sin B$, or $2\,ab\sin C$

Now $\dfrac{AP}{AM} = \dfrac{\sin(\theta + B)}{\sin(\theta + C)} = \dfrac{2\,ac\sin B\cos B + (b^2 + 3\,c^2 - a^2)\sin B}{2\,ab\sin C\cos C + (b^2 + 3\,c^2 - a^2)\sin C}$

$$= \frac{b}{c}\left\{\frac{a^2 + c^2 - b^2 + b^2 + 3\,c^2 - a^2}{a^2 + b^2 - c^2 + b^2 + 3\,c^2 - a^2}\right\}$$

$$\therefore\ \text{symmedian} = \frac{2\,bc}{b^2 + c^2} \times \text{median}$$

Note—The last result may be got still more briefly, by using the result (see E. R. p. 384)

$$PB : PC = c^2 : b^2$$

and applying Euler's formula given on p. 109 of E. R.

Cor'—If q_1, q_2, q_3 are respective $\perp^s$ from K on BC, CA, AB,

$$q_1/a = q_2/b = q_3/c = \lambda \text{ say}$$

and $aq_1 + bq_2 + cq_3 = 2\Delta$

$$\therefore \quad \lambda = 2\Delta/\Sigma a^2$$

$$\therefore \quad q_1 = 2a\Delta/\Sigma a^2; \quad \text{and sim'ly for } q_2, q_3$$

Now $\sin\theta = \Delta/(cm)$

$$KA = q_2 \operatorname{cosec}\theta = \frac{2b\Delta}{\Sigma a^2}\cdot\frac{cm}{\Delta} = \frac{2bc}{\Sigma a^2} \times \text{median}$$

And $$KP = 2bc\left\{\frac{1}{b^2+c^2} - \frac{1}{\Sigma a^2}\right\}m = \frac{2a^2bc}{(b^2+c^2)\Sigma a^2} \times \text{median}$$

$$\therefore \sin KPC = \frac{q_1}{KP} = \frac{2a\Delta}{\Sigma a^2} \times \frac{(b^2+c^2)\Sigma a^2}{2a^2bcm} = \frac{\Delta(b^2+c^2)}{abcm} = \frac{b^2+c^2}{ab}\sin\theta$$

Whence $CP = \dfrac{ab^2}{b^2+c^2}$, and $BP = \dfrac{ac^2}{b^2+c^2}$

3. *The* **pedal triangle.** (*See Euclid Revised, p.* 174.)

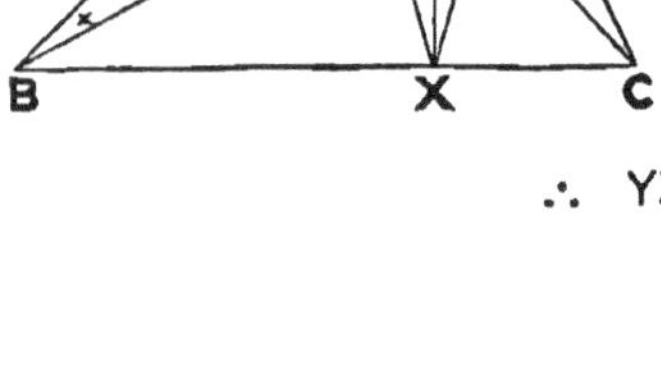

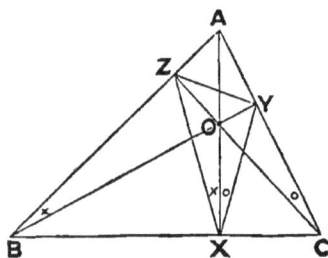

Let X, Y, Z be the respective feet of $\perp^s$ from A, B, C on opposite sides; so that XYZ is the pedal Δ of ABC.

Then AO is clearly diam' of a ⊙ round AYOZ.

$$\therefore \quad YZ = AO \sin A$$
$$= AY \operatorname{cosec} AOY \sin A$$
$$= c \cos A \operatorname{cosec} C \sin A$$
$$= a \cos A = R \sin 2A$$

$$\text{Sim'ly} \quad ZX = b \cos B = R \sin 2B$$
$$\text{and} \quad XY = c \cos C = R \sin 2C$$
$$\therefore \text{ perim' } XYZ = \Sigma(a \cos A)$$
$$= R\Sigma \sin 2A$$

$$\text{Again} \quad Y\hat{X}Z = A\hat{B}Y + A\hat{C}Z = 180^\circ - 2A$$
$$\therefore \; 2 \text{ area } XYZ = bc \cos B \cos C \sin 2A$$
$$\therefore \text{ area } XYZ = 2\Delta \Pi \cos A$$
$$= \frac{\Pi(a \cos A)}{2R}$$
$$= \frac{R^2}{2} \Pi \sin 2A$$

From above, if x denote YZ, and X the $Y\hat{X}Z$,

$$\text{then} \quad \left.\begin{aligned} x &= a \cos A \\ X &= 180^\circ - 2A \end{aligned}\right\}$$

$$\text{and} \quad \left.\begin{aligned} a &= x \operatorname{cosec} \frac{X}{2} \\ A &= 90^\circ - \frac{X}{2} \end{aligned}\right\}$$

With sim'r results for the other sides and $\wedge^s$.

Now every Δ may be considered the pedal of some other Δ.

$\therefore$ from all formulæ connecting the sides and $\wedge^s$ of a Δ are deducible other formulæ, by putting

either $a \cos A$ for a, and $180^\circ - 2A$ for A, &c,

or $a \operatorname{cosec} \frac{A}{2}$,, ,, $90^\circ - \frac{A}{2}$,, ,,

Also, since half $\wedge^s$ of pedal Δ are comp'ts of $\wedge^s$ of original Δ,

$\therefore$ cosines of ,, ,, ,, ,, sines ,, ,, ,,

i. e. all homogeneous relations between sines, or sides, of a Δ have analogues between cosines of half $\wedge^s$ of its pedal Δ.

E. g. from the formula $a^2 = b^2 + c^2 - 2bc \cos A$

we deduce $\cos^2 \frac{A}{2} = \cos^2 \frac{B}{2} + \cos^2 \frac{C}{2} - 2 \cos \frac{B}{2} \cos \frac{C}{2} \sin \frac{A}{2}$

So also such formulæ as

$$\Sigma \sin A = 4 \Pi \cos \frac{A}{2}$$

$$\text{and} \quad \Sigma \sin 2A = 4 \Pi \sin A$$

are seen to be mutually deducible.

4. Cross ratios. (*See Euclid Revised, p.* 332.)

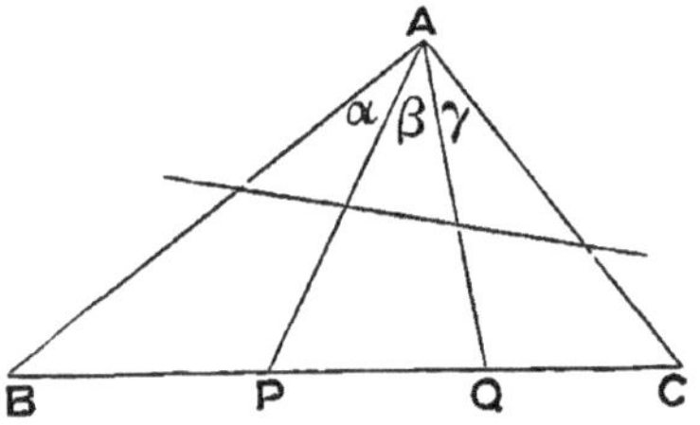

(1) If P, Q are any p'ts in side BC of Δ ABC; and if α, β, γ stand respectively for $\wedge^s$ BAP, PAQ, QAC; then any cross ratio of the pencil A (BPQC) is expressible in terms of α, β, γ.

E. g. $$\frac{BQ \,.\, CP}{BP \,.\, CQ} = \frac{BQ}{AB} \cdot \frac{AB}{BP} \cdot \frac{CP}{AC} \cdot \frac{AC}{CQ}$$

$$= \frac{\sin(\alpha+\beta)}{\sin Q} \cdot \frac{\sin P}{\sin \alpha} \cdot \frac{\sin(\beta+\gamma)}{\sin P} \cdot \frac{\sin Q}{\sin \gamma}$$

$$= \frac{\sin(\alpha+\beta)\,\sin(\beta+\gamma)}{\sin\alpha\,\sin\gamma}$$

Cor' (1)—If $\quad \sin\alpha\,\sin\gamma = \sin\beta\,\sin(\alpha+\beta+\gamma)$

any transversal of the pencil is cut harmonically.

Cor' (2)—If P, Q are p'ts of trisection of BC, then

$$4 = \frac{BQ \,.\, CP}{BP \,.\, CQ} = \frac{\sin(\alpha+\beta)}{\sin\alpha\,\sin\beta} \cdot \frac{\sin(\beta+\gamma)}{\sin\beta\,\sin\gamma} \cdot \sin^2\beta$$

$$\therefore \quad 4(1 + \cot^2\beta) = (\cot\alpha + \cot\beta)(\cot\beta + \cot\gamma)$$

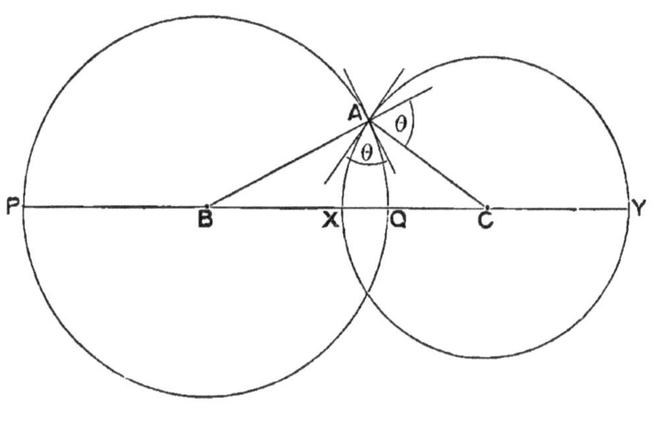

(2) Taking a $\triangle$ ABC; if with centre B, and radius BA, a $\odot$ is described, cutting BC in Q, and CB produced in P; and if with centre C, and radius CA, a $\odot$ is described cutting BC in X, and BC produced in Y; then

$$\sin^2\frac{A}{2} = \frac{(c+a-b)(a+b-c)}{4bc}$$

$$= \frac{PX \, . \, QY}{XY \, . \, PQ}$$

Hence, if P, X, Q, Y are collinear p'ts in order; and if $\odot$s on PQ, XY as diam's cut at an $\wedge$ θ; then, since $\theta = 180^\circ - A$ in above, we have

$$\text{the cross ratio } \frac{PX \, . \, QY}{XY \, . \, PQ} = \cos^2\frac{\theta}{2}$$

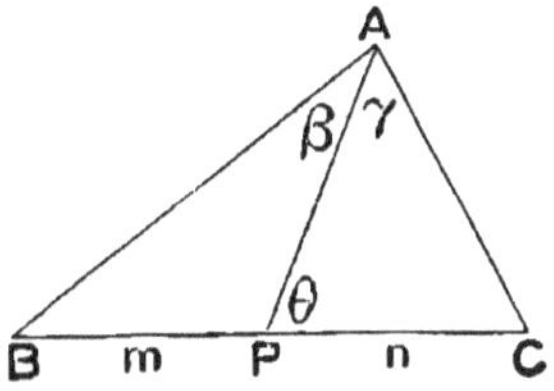

5. *Any line from* A *meets* BC *in* P: *if* PB, PC, *and the angles* APC, PAB, PAC, *are denoted respectively by* m, n, θ, β, γ; *then*

$$m \cot\beta - n \cot\gamma = (m+n)\cot\theta = n \cot B - m \cot C$$

For $\dfrac{AP}{m} = \dfrac{\sin(\theta-\beta)}{\sin\beta} = (\cot\beta - \cot\theta)\sin\theta$

and $\dfrac{AP}{n} = \dfrac{\sin(\theta+\gamma)}{\sin\gamma} = (\cot\gamma + \cot\theta)\sin\theta$

Also $\dfrac{m}{AP} = \dfrac{\sin(\theta-B)}{\sin B} = (\cot B - \cot\theta)\sin\theta$

and $\dfrac{n}{AP} = \dfrac{\sin(\theta+C)}{\sin C} = (\cot C + \cot\theta)\sin\theta$

whence above results follow at once.

Cor'—If AP is a median,

$$2\cot\theta = \cot B - \cot C$$

$$\therefore \; \Sigma\cot\theta = 0, \text{ for the } \textit{three} \text{ medians}$$

6. *Parts of a triangle in terms of its* altitudes.

Since $2\Delta = p_1 a = p_2 b = p_3 c$

all formulæ, involving a, b, c *homogeneously*, will be true when for a, b, c are respectively put the reciprocals of p_1, p_2, p_3. Hence come such results as

$$\tan^2 \frac{A}{2} = \frac{\left(\frac{1}{p_3} + \frac{1}{p_1} - \frac{1}{p_2}\right)\left(\frac{1}{p_1} + \frac{1}{p_2} - \frac{1}{p_3}\right)}{\left(\frac{1}{p_1} + \frac{1}{p_2} + \frac{1}{p_3}\right)\left(\frac{1}{p_2} + \frac{1}{p_3} - \frac{1}{p_1}\right)}$$

As an Exercise the Student should express Δ, s, s_1, $\sin^2 \frac{A}{2}$, &c. in terms of these altitudes.

Exercises

1. AD is the bisector of A; and P, Q points in BC, on opposite sides of D, such that the angles PAD, QAD are each equal to θ; prove that

$$\frac{1}{AP} + \frac{1}{AQ} : \frac{1}{b} + \frac{1}{c} = \cos\theta : \cos\frac{A}{2}$$

2. If Δ_1 is the area of a triangle of which two sides and the included angle are respectively the bisectors of A, and A; and if Δ_2, Δ_3 are similarly constructed with respect to B, C; prove that

$$\frac{1}{\Delta_1} + \frac{1}{\Delta_3} = \frac{1}{\Delta_2}$$

where a, b, c are in order of magnitude.

3. If the bisector x, of the angle A, divides BC into parts m, n (m lying next B) prove that—

(1) $2\,Ra\,x \sin\frac{A}{2} = m^2 b + n^2 c$

(2) $p_1 = mn(m + n)\sin A/(m^2 + n^2 - 2mn\cos A)$

Also, if x makes an acute angle ϕ with BC, and c is greater than b, prove that

$$\tan\phi = \frac{c + b}{c - b}\tan\frac{A}{2}$$

4. If x_1, x_2, x_3 are the internal bisectors of A, B, C respectively; prove that—

(1) $\Pi\left\{x_1(b + c)/a\right\} = 8s\Delta$

(2) $\Sigma\left\{x_1^2(b + c)^2/(bc)\right\} = 4s^2$

5. Prove that the area of the triangle formed by joining the points where the internal bisectors of the angles of a triangle meet the opposite sides is

$$2\Delta \Pi \left\{a/(a+b)\right\}$$

6. If two internal bisectors of the angles of a triangle are equal show (by trigonometry) that the triangle is isosceles.

NOTE—*A simple geometrical proof of this will be found in E. R.* p. 413.

7. If l, m, n are the medians of a triangle, prove that

$$3\Delta = \sqrt{2\Sigma m^2 n^2 - \Sigma l^4}$$

8. If the median from A divides the angle A into parts θ, ϕ, prove that—

(1) $\tan\dfrac{\theta \sim \phi}{2} = \tan\dfrac{B \sim C}{2}\tan^2\dfrac{A}{2}$

(2) $\cot\theta \sim \cot\phi = \cot B \sim \cot C$

9. If C is a right angle, and ϕ the angle between the bisector of A and the median from A, prove that

$$\tan\phi = \tan^3\frac{A}{2}$$

10. If the median from $A = b = c/2$, prove that

$$\sin^2 C = \tfrac{5}{8} \quad \text{and} \quad \sin^2 B = \tfrac{5}{32}$$

11. BY, CZ are perpendiculars from B, C on the opposite sides, and AY, BZ are equal; show that the cross of the median from B with CZ, and the cross of the median from C with BY, are on the bisector of A.

12. If m is the median from A, prove that

$$\left\{m^2 - (a/2)^2\right\}\tan A = 2\Delta$$

Hence prove that, if P is any point on a circle inscribed in a square; and α, β the angles subtended at P by the diagonals—

$$\tan^2\alpha + \tan^2\beta = 8$$

13. If a_1, b_1, c_1 are the sides of the pedal triangle respectively next A, B, C, prove that $a_1 b^2 + b_1 a^2 = b_1 c^2 + c_1 b^2 = c_1 a^2 + a_1 c^2$

14. A new triangle PQR is formed by drawing tangents to the circumcircle at A, B, C: if XYZ is the pedal triangle of ABC, prove that

$$\text{area } XYZ : \text{area } PQR = 4\Pi\cos^2 A : 1$$

15. In Example **4** (2) show that the other five cross ratios of the range PXQY are expressible as the other five T. F.s of $\theta/2$.

Phil' Trans': 1871.

16. If x, y, z are the lengths of perpendiculars from A, B, C on any line, prove that

$$\Sigma(a^2 x^2) - 2\Sigma(bc\,yz\cos A) = 4\Delta^2$$

Ox' 1st Public Exam': '77.

Show that this theorem is equivalent to the following—

With A, B, C as centres; and ρ_1, ρ_2, ρ_3 as radii; circles are described: the condition that these circles may have a common tangent is

$$\Sigma\left(\frac{\rho_1}{p_1}\right)^2 - 2\Sigma\left(\frac{\rho_2\,\rho_3}{p_2\,p_3}\cos A\right) = 1$$

Editor's Question: E. T. xx.

17. If l, m, n are the medians, prove that—

(1) $\Sigma\left\{a\sin B\sqrt{4l^2 - a^2\sin^2 B} + a\sin C\sqrt{4l^2 - a^2\sin^2 C}\right\} = 18\Delta$

(2) $\Sigma\left\{2a\sin B\sqrt{4m^2 - b^2\sin^2 A}\right\} + \Sigma(a^2\sin 2C) = 24\Delta$

McKenzie: E. T. xxvii.

18. If the bisectors of A, B, C meet the opposite sides in D, E, F respectively; and if x, y, z are the respective perpendiculars from A, B, C on EF, FD, DE; prove that

$$\Sigma\left(\frac{p_1}{x}\right)^2 = 11 + 8\Pi\sin\frac{A}{2}$$

Nash: E. T. xlix.

Prove also that $\tan EDF = \tan\theta\cos\frac{B - C}{2}$

where $2\tan\frac{\theta}{2} = \sec\frac{A}{2}$

I. I. Walker: *E. T.* xxiv.

19. Taking the figure of Example 2, if ϕ denote the angle APC, and χ the angle AMC, prove that—

(1) $\cos\phi/\cos\chi = \cos A$

(2) $\tan\frac{\phi - \chi}{2} = \frac{c - b}{c + b}\tan\frac{A}{2}$

(3) $\tan\phi = \frac{c^2 + b^2}{c^2 - b^2}\tan A$

I. I. Walker: E. T. xiii.

20. Taking the definition of, and the theorems about, the *mean centre* of any number of points, given on pages 110 to 112 of *Euclid Revised*, show that the *mean centre* of points A, B, C—

(1) for multiples sin A, sin B, sin C is I:

(2) ,, −sin A, sin B, sin C is E_1:

(3) ,, sin 2 A, sin 2 B, sin 2 C is S:

(4) ,, tan A, tan B, tan C is O:

(5) ,, $\sin^2$ A, $\sin^2$ B, $\sin^2$ C is K

(6) ,, sin 2 B + sin 2 C, &c. is N.

21. If M is the mean centre of A, B, C for multiples l, m, n; and if MA = x, MB = y, MC = z, and $2\sigma = lx + my + nz$; prove that

$$\text{area ABC} = (l + m + n)\sqrt{\sigma(\sigma - lx)(\sigma - my)(\sigma - nz)}\Big/(l\,m\,n)$$

Brill: E. T. xlviii.

§ 74. *Properties of the* circum-, in-, *and* ex-circles.

Examples

1. *Length of the* circum-radius.

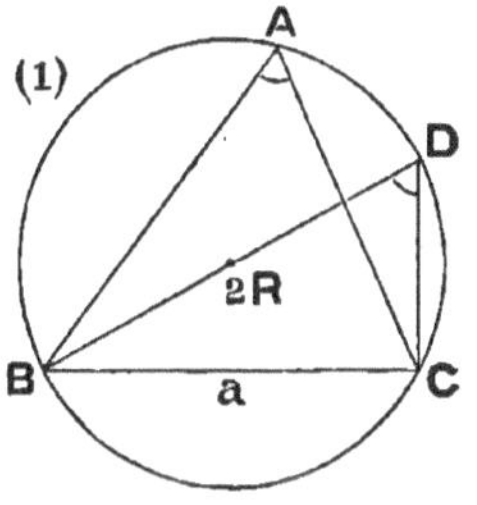

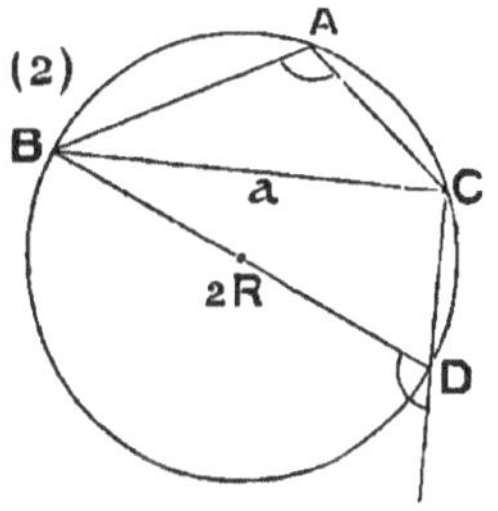

Let $\hat{A}$ of △ ABC be acute in fig′ (1), and obtuse in fig′ (2).

Draw the diam′ BD of the ⊙ round ABC; then $B\hat{C}D$ is right.

Now $B\hat{D}C = \hat{A}$ in fig′ (1); but $= 180° - \hat{A}$ in fig′ (2)

∴, always, $a = 2R \sin BDC = 2R \sin A$

∴ $abc = 2R\, bc \sin A = 4R\Delta$

$$\text{or } R = \frac{abc}{4\Delta} = \frac{a}{2\sin A} = \frac{b}{2\sin B} = \frac{c}{2\sin C}$$

2. *Length of the* in-radius.

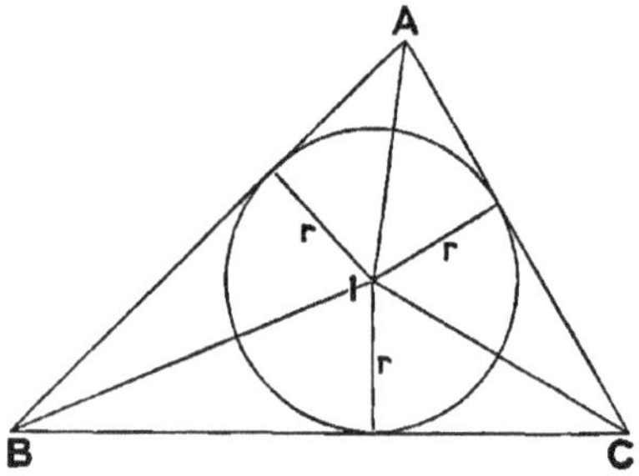

By *Euc'* iv. 4, I is known to be the p't in which the bisectors of A, B, C concur.

Now area BIC + area CIA + area AIB = Δ

$$\therefore \quad \frac{r}{2}(a + b + c) = \Delta$$

$$\text{i.e.} \quad r = \frac{\Delta}{s} = \sqrt{\frac{s_1 s_2 s_3}{s}} \quad . \; . \; . \; . \; . \; . \; . \quad (1)$$

Also that tangents from A, B, C to the in-$\odot$ are respectively s_1, s_2, s_3, are well-known geometrical results. (See E. R. p. 211)

$$\therefore \quad r = s_1 \tan\frac{A}{2} = s_2 \tan\frac{B}{2} = s_3 \tan\frac{C}{2} \quad . \; . \; . \; . \; . \; . \quad (2)$$

$$\text{Again} \quad r = \frac{\Delta}{s} = \frac{ab \sin C}{a + b + c} = \frac{a \sin B \sin C}{\sin A + \sin B + \sin C}$$

$$= \frac{4 a \sin\frac{B}{2} \cos\frac{B}{2} \sin\frac{C}{2} \cos\frac{C}{2}}{4 \cos\frac{A}{2} \cos\frac{B}{2} \cos\frac{C}{2}}$$

$$= \frac{a \sin\frac{B}{2} \sin\frac{C}{2}}{\cos\frac{A}{2}} \quad . \; . \; . \; . \; . \; . \; . \; . \quad (3)$$

Note—The last result may also be seen from the arrangement

$$r = \sqrt{\frac{s_1 s_2 s_3}{s}} = a \sqrt{\frac{s_1 s_3}{ac} \cdot \frac{s_1 s_2}{ab} \cdot \frac{bc}{s s_1}}$$

3. *Lengths of the* ex-radii.

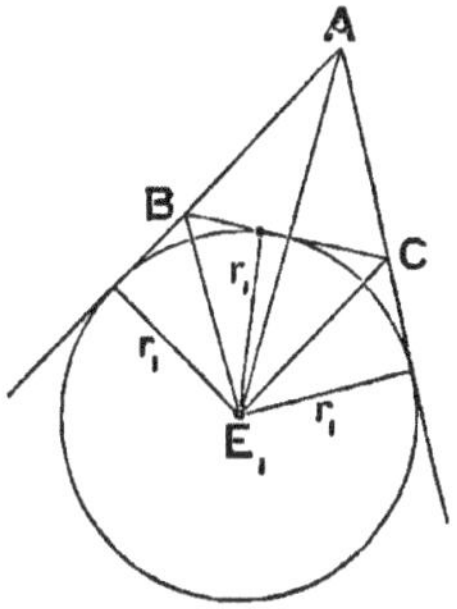

It is a well-known geometrical result (See E. R. pp. 75, 76) that E_1 is the p′t where the intern′ bisector of A, and the extern′ bisectors of B, C, concur.

Now area AE_1C + area AE_1B – area $BE_1C = \Delta$

$$\therefore \quad \frac{r_1}{2}(b + c - a) = \Delta$$

$$\text{i.e.} \quad r_1 = \frac{\Delta}{s_1} = \sqrt{\frac{s\,s_2\,s_3}{s_1}} \quad \ldots\ldots\ldots \quad (1)$$

Also that tangents from A, B, C to the ex-⊙ are respectively s, s_3, s_2 are well-known geometrical results. (See E. R. p. 211)

$$\therefore \quad r_1 = s\tan\frac{A}{2} = s_3\cot\frac{B}{2} = s_2\cot\frac{C}{2} \quad \ldots\ldots \quad (2)$$

$$\text{Again} \quad r_1 = \frac{\Delta}{s_1} = \frac{ab\sin C}{b + c - a} = \frac{a\sin B\sin C}{\sin B + \sin C - \sin A}$$

$$= \frac{4\,a\sin\frac{B}{2}\cos\frac{B}{2}\sin\frac{C}{2}\cos\frac{C}{2}}{4\sin\frac{B}{2}\sin\frac{C}{2}\cos\frac{A}{2}}$$

$$= \frac{a\cos\frac{B}{2}\cos\frac{C}{2}}{\cos\frac{A}{2}} \quad \ldots\ldots \quad (3)$$

Note (1)—The last result also appears from the arrangement

$$r_1 = \sqrt{\frac{s\,s_2\,s_3}{s_1}} = a\sqrt{\frac{s\,s_2}{ac}\cdot\frac{s\,s_3}{ab}\cdot\frac{bc}{s\,s_1}}$$

Note (2)—Of course sim′r results hold for r_2 and r_3.

4. *Connections between the circum-, in-, and ex-radii.*

$$\frac{r}{4R} = \frac{\Delta}{s} \cdot \frac{\Delta}{abc} = \frac{s_1 s_2 s_3}{abc} = \sqrt{\frac{s_1 s_2}{ab} \cdot \frac{s_2 s_3}{bc} \cdot \frac{s_3 s_1}{ca}}$$

$$\therefore \quad r = 4R \sin\frac{A}{2} \sin\frac{B}{2} \sin\frac{C}{2} \quad \ldots \ldots \quad (1)$$

$$\frac{r_1}{4R} = \frac{\Delta}{s_1} \cdot \frac{\Delta}{abc} = \frac{s s_2 s_3}{abc} = \sqrt{\frac{s_2 s_3}{bc} \cdot \frac{s s_2}{ac} \cdot \frac{s s_3}{ab}}$$

$$\therefore \quad r_1 = 4R \sin\frac{A}{2} \cos\frac{B}{2} \cos\frac{C}{2} \quad \ldots \ldots \quad (2)$$

Whence $r_1 - r = 4R \sin^2\frac{A}{2}$ . . . . (3)

and $r_2 + r_3 = 4R \cos^2\frac{A}{2}$ . . . . (4)

$\therefore$ also $r_1 + r_2 + r_3 - r = 4R$ . . . . (5)

$$\Delta\left(\frac{1}{r_1} + \frac{1}{r_2} + \frac{1}{r_3}\right) = s_1 + s_2 + s_3 = s = \frac{\Delta}{r}$$

$$\therefore \quad \frac{1}{r_1} + \frac{1}{r_2} + \frac{1}{r_3} = \frac{1}{r} \quad \ldots \ldots \quad (6)$$

$$\sqrt{r r_1 r_2 r_3} = \frac{\Delta^2}{\sqrt{s s_1 s_2 s_3}} = \Delta \quad \ldots \ldots \quad (7)$$

From (6) and (7) $\Delta = \dfrac{r_1 r_2 r_3}{\sqrt{\Sigma r_1 r_2}}$ . . . . (8)

$$4R = r_1 + r_2 + r_3 - r = \frac{(r_1 + r_2 + r_3)\,\Sigma(r_1 r_2) - r\,\Sigma(r_1 r_2)}{\Sigma(r_1 r_2)}$$

$$= \frac{\Sigma r_1^2 (r_2 + r_3) + 3 r_1 r_2 r_3 - r_1 r_2 r_3}{\Sigma(r_1 r_2)}$$

$$\therefore \quad R = \frac{\Pi(r_1 + r_2)}{4\Sigma(r_1 r_2)} \quad \ldots \ldots \quad (9)$$

$$\left.\begin{aligned} a + c - b &= \frac{2\Delta}{r_2} \\ a + b - c &= \frac{2\Delta}{r_3} \end{aligned}\right\}$$

$$\therefore \quad a = \Delta\left(\frac{1}{r_2} + \frac{1}{r_3}\right) = \frac{r_1(r_2 + r_3)}{\sqrt{\Sigma r_1 r_2}} \quad \ldots \ldots \quad (10)$$

$$\Sigma r_1 r_2 = \Delta^2 \Sigma \frac{1}{s_1 s_2} = s(s_1 + s_2 + s_3) = s^2 \quad . \quad . \quad . \quad . \quad (11)$$

$$\frac{1}{r^2} + \frac{1}{r_1^2} + \frac{1}{r_2^2} + \frac{1}{r_3^2} = \frac{s^2 + s_1^2 + s_2^2 + s_3^2}{\Delta^2} = \frac{\Sigma a^2}{\Delta^2} \quad . \quad . \quad (12)$$

$$\tan^2 \frac{A}{2} = \frac{r_1^2}{s^2} = \frac{r_1^2}{\Sigma r_1 r_2} = \frac{r r_1}{r_2 r_3} \quad . \quad . \quad . \quad . \quad . \quad . \quad (13)$$

$$\Sigma \frac{\Delta}{p_1} = \Sigma \frac{a}{2} = s = \frac{\Delta}{r}$$

$$\therefore \quad \Sigma \frac{1}{p_1} = \frac{1}{r} \quad . \quad . \quad . \quad . \quad . \quad . \quad . \quad . \quad . \quad . \quad . \quad . \quad (14)$$

$$\frac{1}{r_2} + \frac{1}{r_3} = \frac{s_2 + s_3}{\Delta} = \frac{a}{\Delta} = \frac{2a}{p_1 a} = \frac{2}{p_1} \quad . \quad . \quad . \quad . \quad . \quad (15)$$

$$\sin^2 \frac{A}{2} = \frac{s_2 s_3}{bc} = \frac{\frac{\Delta}{r_2} \cdot \frac{\Delta}{r_3}}{\frac{2\Delta}{p_2} \cdot \frac{2\Delta}{p_3}} = \frac{p_2 p_3}{4 r_2 r_3} \quad . \quad . \quad . \quad . \quad (16)$$

From the diagrams of Examples 2 and 3, pp. 252, 253, it is clear that

$$r\left(\cot \frac{B}{2} + \cot \frac{C}{2}\right) = a = r_1 \left(\tan \frac{B}{2} + \tan \frac{C}{2}\right) \quad . \quad . \quad . \quad (17)$$

The following well-known geometrical theorems give important connections which should be recollected. The simplest respective proofs of them will be found in *Euclid Revised*, pp. 295, 350.

To prove them trigonometrically will afford good Exercises to the Student.

$$\left.\begin{aligned} SI^2 &= R^2 - 2Rr \\ SE_1^2 &= R^2 + 2Rr_1 \end{aligned}\right\} \quad (\textit{Chapple's Theorem}) \quad . \quad (18)$$

$$\left.\begin{aligned} NI &= \frac{R}{2} - r \\ NE_1 &= \frac{R}{2} + r_1 \end{aligned}\right\} \quad (\textit{Feuerbach's Theorem}) \quad . \quad (19)$$

The following may be usefully added: they can be very easily proved by pure geometry.

$$\left.\begin{aligned} OS^2 &= R^2 + 2\rho^2 \\ OI^2 &= 2r^2 + \rho^2 \end{aligned}\right\} \quad . \quad . \quad . \quad . \quad . \quad . \quad . \quad . \quad . \quad . \quad (20)$$

where ρ is the radius of the polar ⊙, and ∴ is real only when the △ is obtuse-angled.

5. *Lengths of the in-radius* ρ, *and the ex-radius* ρ_1, *of the* pedal triangle.

Let σ be the semi-perim′ of the pedal Δ

$$\text{Then} \quad \sigma = 2\,R\,\Pi \sin A \quad \text{(See Example 3 of § 73)}$$

$$= 2\,R\,\frac{abc}{8\,R^3}$$

$$\therefore \quad R\sigma = \Delta \quad \ldots \ldots \quad (1)$$

$$\text{Now} \quad \rho = \frac{2\,\Delta\,\Pi \cos A}{\sigma}$$

$$\therefore \quad \rho = 2\,R\,\Pi \cos A \quad \ldots \ldots \quad (2)$$

$$\text{Again} \quad \rho_1 = \frac{\frac{R^2}{2}\,\Pi \sin 2A}{\frac{R}{2}(\sin 2B + \sin 2C - \sin 2A)}$$

$$= \frac{R\,\Pi \sin 2A}{4 \sin A \cos B \cos C}$$

$$\therefore \quad \rho_1 = 2\,R \cos A \sin B \sin C \quad \ldots \ldots \quad (3)$$

Cor′—Since a Δ is the pedal Δ of its principal *ex-central $\Delta\,E_1\,E_2\,E_3$; and the circum-⊙ of the pedal Δ is the N. P. ⊙ of the original Δ*;

$\therefore$ the circum-rad′ of $E_1\,E_2\,E_3 = 2\,R$*

$\therefore$ area principal ex-central $\Delta = 2\,R\,s$

Exercises

1. Prove that—

(1) $\Sigma \frac{2}{ab} = \frac{1}{Rr}$

(2) $\Sigma a : \Sigma (a \cos A) = R : r$

(3) $2\,\pi s / \Pi \cot \frac{A}{2}$ = perimeter in-⊙

(4) $\pi \Delta / \Pi \cot \frac{A}{2}$ = area in-⊙

(5) $\Sigma \cot \frac{A}{2} = s/r$

---

* See *Euclid Revised*, pp. 174, 175, 212, 338, for the geometry of all this.

(6) $\Sigma \tan \frac{A}{2} + \frac{s}{r} = 4R\Sigma\left(\frac{1}{a}\right)$

(7) $\Sigma \tan^2 \frac{A}{2} = r(r_1^2 + r_2^2 + r_3^2)/(r_1 r_2 r_3)$

(8) $\Sigma(a \cot A) = 2(R + r)$

(9) $\Sigma(IA \sin BIC) = s$

(10) $\Sigma(OA \sin BOC) = 2a \sin B \sin C$

(11) $(ab - r_1 r_2)/r_3 = r$

(12) $\Sigma \frac{r_1}{bc} = \frac{1}{r} - \frac{1}{2R}$

(13) $\Sigma(p_2 + p_3)/r_1 = 6$

(14) $r\Sigma r_1 + \Sigma r_2 r_3 = \Sigma ab$

(15) $r = 2R\left(\cos^2 \frac{A}{2} - \sin^2 \frac{B}{2} - \sin^2 \frac{C}{2}\right)$

(16) $\frac{a \tan \frac{A}{2}}{r - r_1} = \frac{b \tan \frac{B}{2}}{r - r_2} = \frac{c \tan \frac{C}{2}}{r - r_3}$

(17) $\Sigma a^2 = 2s^2 - 2r^2 - 8Rr$

(18) $\Sigma ab = s^2 + r^2 + 4Rr$

NOTE—*The last two results will come by elim'g* Σab *and* Σa^2 *successively between* $\Sigma a = 2s$, and $sr^2 = s_1 s_2 s_3$.

(19) $IE_1/\sin \frac{A}{2} = IE_2/\sin \frac{B}{2} = IE_3/\sin \frac{C}{2}$

(20) $IE_1 = 2\sqrt{R(r_1 - r)}$

(21) $r . IE_1 . IE_2 . IE_3 = (abc/s)^2$

(22) $AI . AE_1 = bc$

(23) $\Sigma\left\{a^2 . IE_1^2 (b^2 - c^2) \operatorname{cosec}^2 \frac{A}{2}\right\} = 0$

(24) $\Pi E_1 E_2 = 8\Pi(r_1 \operatorname{cosec} A)$

(25) $(4R + r)(4R + r + s\sqrt{3})(4R + r - s\sqrt{3})$
$= \Sigma r_1^3 - 3r_1 r_2 r_3$

Ox' 1st Public Exam': '90.

(26) If $(r_2 + r_3 - r_1)(r_3 + r_1 - r_2)(r_1 + r_2 - r_3) + 8 r_1 r_2 r_3 = 0$

then $r + 4R = 2s$

Ox' Jun' Schol': '79.

(27) $\sqrt{abc}\,\Sigma \sqrt{\frac{a}{r_1}} = 16 R \sqrt{r}\,\Pi \cos\frac{\pi - A}{4}$

Ox' Jun' Schol': '74.

(28) $\tan ISE_1 = \frac{2(\sin B - \sin C)}{2\cos A - 1}$

Leudesdorf: E. T. xxxii.

(29) $\Sigma\left(a^2 \cot\frac{A}{2}\right) : \Sigma\left(a^2 \tan\frac{A}{2}\right) = R + r : R - r$

(30) $3\sqrt{\frac{r_1 r_2 r_3}{r}} - \sqrt{\frac{r r_2 r_3}{r_1}} - \sqrt{\frac{r r_1 r_3}{r_2}} - \sqrt{\frac{r r_1 r_2}{r_3}} = 2s$

(31) $a(s^2 - AP^2) = 4 s s_2 s_3$

where P is point in which ex-circle E_1 touches BC.

Math' Tri': '66.

2. If ρ_1, ρ_2, ρ_3 are radii of three circles touching each other externally; and ρ is radius of circle through their centres; prove that

$$\rho = \frac{\Pi(\rho_2 + \rho_3)}{4\sqrt{\Sigma \rho_1 . \Pi \rho_1}}$$

3. If x, y, z are the distances of any point from the sides of a triangle, prove that

$$\Sigma(x \sin A) = 2R\,\Pi \sin A$$

4. A circle circumscribes one triangle, and is inscribed in another: if the triangles are similar, prove that

any side of inner : homologous side of outer $= 4\,\Pi \sin\frac{A}{2} : 1$

5. If P, Q are any points in BC; and ρ_1, ρ_2, ρ_3 the respective circum-radii of ABP, APQ, AQC; prove that $R\rho_2 = \rho_1 \rho_3$.

6. In the figure of *Euclid* iv. 10, show that, approximately,

area large circle : area small circle = 1809 : 500

7. Prove that—

(1) $\Sigma(\text{area } BIC / \text{area } BE_1C) = 1$

(2) $\Sigma(a / \text{area } BE_1C) = 2/r$

8. If $A = 60^\circ$, show that the bisector of A goes through N.

9. If AS meets BC in D, prove that

$$DS = R \cos A / \cos(B - C)$$

10. If AS meets BC in D, and the circum-circle in D′, prove that

$$\Sigma (DD'/AD) = 1$$

W. J. C. Sharpe: E. T. xl.

11. If ρ_1, ρ_2, ρ_3 are the radii of the in-circles of BSC, CSA, ASB respectively; prove that

$$4R^2 \Sigma \rho_1 + 2R \Sigma a\rho_1 = abc$$

12. Show that r_1, r_2, r_3 are the roots of

$$x^3 - (4R + r)x^2 + s^2 x - rs^2 = 0$$

13. Prove that the area of the triangle formed by joining the points of contact of the in-circle is $\Delta r/(2R)$.

14. Find the sides and angles of the principal ex-central triangle; and prove that its area may be expressed as

$$2R\Delta/r, \text{ or } s^2\bigg/\left(2\Pi\cos\frac{A}{2}\right), \text{ or } 8sR^2\Delta/(abc), \text{ or } \Delta\bigg/\left(2\Pi\sin\frac{A}{2}\right),$$

$$\text{or } 8R^2\Pi\cos\frac{A}{2}, \text{ or } \tfrac{1}{2}abc\,\Sigma\left\{\left(\frac{1}{a}+\frac{1}{b}\right)\tan\frac{C}{2}\right\}$$

Show also that the radius of its in-circle is

$$4R\Pi\cos\frac{A}{2}\bigg/\Sigma\cos\frac{A}{2}$$

15. Prove that the area of the ex-central triangle IE_2E_3 is $2Rs_1$.

16. Prove that $\Sigma NA^2 = 3R^2 - NO^2$.

NOTE—*Recollect that* N, O, A, B, C *are the circum-, in-, and ex-centres of the pedal* Δ; *and that the circum-rad′ of that* Δ *is* $R/2$: *then use results* (5) *and* (18) *of Example* 4.

17. If x, y, z are the perpendiculars from A, B, C respectively, on the direction of SI; prove that $ax + by + cz = 0$, where the one of x, y, z, which is on the opposite side of SI from that of the other two, is considered negative.

Pet′ Camb′: ′49.

S 2

18. P is a point within a triangle; prove that—

if, 1°, PA cos BPC = PB cos CPA = PC cos APB,

then P is the in-centre;

and if, 2°, PA sec BPC = PB sec CPA = PC sec APB,

then P is the orthocentre.

19. If P is a point within a triangle, at which BC subtends the angle θ, and if PS produced meets the circum-circle in Q, prove that

$$SQ^2 - SP^2 = 2(\cot A - \cot\theta)\,\Delta\,PBC$$

Genese: E. T. lv.

Deduce *Chapple's Theorem* that $SI^2 = R^2 - 2Rr$

20. In fig′ (2) (page 233) of the 'ambiguous case,' if the angles ACB_2, B_2CN, NCB_1 are equal (α being their common value) and if r, r_2 have their usual meanings for triangle AB_1C, and r', r'_2 are corresponding quantities for AB_2C, prove that

$$\frac{r r_2' + r' r_2}{r r_2' - r' r_2} = 2\cos\alpha$$

NOTE—*Use forms* (3) *for* r, r_1, *in Examples* 2 *and* 3.

21. In the 'ambiguous case' prove that—

(1) $r r' : r_1 r_1' = b - a : b + a$

(2) I, I', E_3, E_3' are concyclic

(3) $r + r_1' = r' + r_1 = CN$ (fig′ p. 233)

(4) $r r' r_2 r_2' = \Delta\Delta'$

Tucker: E. T. xxvii, xxx, xlviii.

22. If AN meets BC in α, prove that—

(1) $AN : N\alpha = \cos(B - C) + 2\cos A : \cos(B - C)$

(2) $B\alpha : \alpha C = \sin 2A + \sin 2B : \sin 2A + \sin 2C$

23. Prove each of the following—

(1) $OS^2 = R^2(1 - 8\,\Pi\cos A)$

(2) $OI^2 = 4R^2\left(8\,\Pi\sin^2\frac{A}{2} - \Pi\cos A\right)$

(3) $OE_1^2 = 4R^2\left(8\sin^2\frac{A}{2}\cos^2\frac{B}{2}\cos^2\frac{C}{2} - \Pi\cos A\right)$

(4) $IG^2 = \frac{4}{9}R^2(1 + \Pi\cos A) - \frac{4}{3}Rr + \frac{2}{3}r^2$

24. Prove that the distances between the in- and ex-centres of the principal ex-central triangle are respectively

$$8\,R \sin\frac{B+C}{4},\quad 8\,R \sin\frac{C+A}{4},\quad 8\,R \sin\frac{A+B}{4}$$

25. From any point P perpendiculars PX, PY, PZ are dropped on the sides of a triangle; prove that

$$2 \text{ area } XYZ = (PS^2 \sim R^2)\,\Pi \sin A.$$

Interpret the geometrical meaning when P is on the circum-circle.
When will XYZ be maximum?

26. If ρ is the radius of the polar circle of a triangle, show that

$$\rho^2 = -\,4\,R^2 \cos A \cos B \cos C$$

Account for the negative sign.

27. Show that the polar circle cuts the circum-circle and N. P. circle each at the same angle θ, where

$$\cos^2\theta = -\cos A \cos B \cos C.$$

28. If r_1, r_2, r_3 are the roots of $x^3 - px^2 + qx - t = 0$, show that the ex-radii of the pedal triangle are the roots of

$$(pq - t)^3 x^3 - 2\,(pq - t)^2 (q^2 - pt)\,x^2 \\ + 16\,qt^2 (pq - t)\,x - 8\,t^2 \{4\,q^3 - (pq + t)^2\} = 0$$

Edwardes: E. T. xl.

NOTE—*Use Example* 5, *and its Cor'*; *and also results* (11) (10) (7) (6) (4) *of Example* 4.

29. If SIO is a right angle, prove that

$$4\,(2\,R - r)\,(R + r) = \Sigma a^2$$

30. If ρ is the in-radius of the pedal triangle, prove that—

(1) $OI^2 = 2\,(r^2 - R\rho)$

(2) $OE_1^2 = 2\,(r_1^2 - R\rho)$

(3) $OS^2 = R^2 - 4\,R\rho$

31. Prove that the sum of the squares of the distances of the centroids of triangles BIC, CIA, AIB, BE_1C, CE_2A, AE_3B from G is

$$\tfrac{2}{9}\Sigma ab + \tfrac{8}{9} R\,(2\,R - r)$$

S. Watson: E. T. xxxiii.

32. If r_1, r_2, r_3 are in H. P.; and r, r_2, R in G. P.; show that

$$\cos B = \frac{8}{9}, \quad \text{that} \quad R = 9r, \quad \text{and that} \quad \tan\frac{A}{2}\tan\frac{C}{2} = \frac{1}{3}.$$

33. P is any point within a triangle, and ρ_1, ρ_2, ρ_3 the circum-radii of BPC, CPA, APB; show that

$$\left(\frac{abc}{\rho_1\rho_2\rho_3}\right)^2 = \left(\frac{a}{\rho_1} + \frac{b}{\rho_2} + \frac{c}{\rho_3}\right)\left(-\frac{a}{\rho_1} + \frac{b}{\rho_2} + \frac{c}{\rho_3}\right)\left(\frac{a}{\rho_1} - \frac{b}{\rho_2} + \frac{c}{\rho_3}\right)\left(\frac{a}{\rho_1} + \frac{b}{\rho_2} - \frac{c}{\rho_3}\right)$$

NOTE—*Use the 2nd result of* § 27.

34. S_1, S_2, S_3 are the images of S with respect to BC, CA, AB respectively: show that if the circum-circles of $S_1 S_2 S_3$ and ABC cut at an angle θ,

$$2\cos\theta = 1 - 8\prod\cos A$$

35. X, Y, Z are the centres of three circles touching each other externally in pairs, and with respective radii x, y, z: P is the centre of the fourth circle, touching and including the three: if α, β, γ denote PX, PY, PZ respectively, prove that

$$\Sigma\left(\frac{\alpha}{x}\right)^2 + 4 = 2\Sigma\left(\frac{\beta\gamma}{yz}\right)$$

Math' Tri': '64.

NOTE—*Use Exercise* 1 of p. 79.

36. If AI, BI, CI cut SO in Q_1, Q_2, Q_3 respectively; and if cos A, cos B, cos C are in A. P.; prove that SQ_1, SQ_2, SQ_3 are in H. P.

Pet' Camb': '66.

37. L, M, N are the points where the altitudes of A, B, C cut the sides of the pedal triangle: prove that

$$OL\cos(B - C) = OM\cos(C - A) = ON\cos(A - B)$$

and that, if ρ is the in-radius of the pedal triangle,

$$\Sigma\frac{\rho^2}{OL^2} = 1 + 2\prod\frac{\rho}{OL}$$

Pet' Camb': '66.

38. If ρ_1, ρ_2 are the radii of the circles touching b, c, and the circum-circle, internally and externally respectively, prove that

$$\rho_2 - \rho_1 = 4R\tan^2\frac{A}{2}$$

39. If x, y, z are the respective distances of I from E_1, E_2, E_3, prove that

$$32\,R^3 - 2\,(x^2 + y^2 + z^2)\,R - xyz = 0$$

40. If x, y, z are the respective distances of O from A, B, C, prove that

$$xyz\,(ax + by + cz)^3 + abc\,(x^2 + y^2 + z^2)\,(ax + by + cz)^2 - 4\,a^3\,b^3\,c^3 = 0$$

41. If l, m, n are the segments of the pedal line of P, included within the angles A, B, C respectively; and if $\Sigma\, a^2 = 2\,\sigma^2$; prove that—

(1) $\sqrt{a^2 - l^2} + \sqrt{b^2 - m^2} + \sqrt{c^2 - n^2} = 0$

the negative sign being taken for one of the radicals

(2) $\Sigma\left\{(\sigma^2 - a^2)\,l^2\right\} = 4\,s^2$

(3) $\Sigma\left\{(\sigma^2 - a^2)\,a^2\,PA^2\right\} = a^2\,b^2\,c^2$

Editor: E. T. xxix.

42. With a, b, c as diameters circles are drawn; and δ is the diameter of the circle touching, and including, all three: prove the relation

$$\sqrt{\delta/s_1 - 1} + \sqrt{\delta/s_2 - 1} + \sqrt{\delta/s_3 - 1} = \sqrt{s/(\delta - s)}$$

Math' Tri': '73.

43. If a triangle is acute-angled can OA, OB, OC be taken as sides of a new triangle? If this is possible; and α, β, γ are the angles of this new triangle, respectively opposite the lengths OA, OB, OC; prove that

$$2\left(1 + \Sigma\,\frac{\cos\alpha}{\cos A}\right) = \Pi\,\sec A$$

Math' Tri': '81.

44. ρ_1 is the in-radius of the triangle formed by joining the points of contact of the ex-circle E_1; and ρ_2, ρ_3 are similar quantities: prove that

$$\rho_1\left(1 - \tan\frac{A}{4}\right) = \rho_2\left(1 - \tan\frac{B}{4}\right) = \rho_3\left(1 - \tan\frac{C}{4}\right)$$

Tucker: E. T. xviii.

45. If ρ_1 is the in-radius of triangle $E_1\,BC$, and ρ_2, ρ_3 are the analogous radii, prove that

$$\sqrt{\frac{a}{\rho_1} + \frac{b}{\rho_2} + \frac{c}{\rho_3}} = \sqrt{\frac{a}{r_1}} + \sqrt{\frac{b}{r_2}} + \sqrt{\frac{c}{r_3}}$$

S. Watson: E. T. xix.

46. If P is any point on the circum-circle, prove that

$$\Sigma\,(PA^2 \sin 2A) = 4\,\Delta$$

and deduce that, if Q is any point on the N. P. circle,

$$\Sigma\,\{QA^2 \sin A \cos (B - C)\} = R^2\,(4\,\Pi \sin A + \Pi \sin 2A)$$

Wolstenholme: 579.

47. Three circles are so placed within a triangle that each touches two sides and the two remaining circles: if x, y, z are the respective radii of the circles touching the sides of the angles A, B, C, prove that

$$x\left(1 + \tan\frac{A}{4}\right)^2 = y\left(1 + \tan\frac{B}{4}\right)^2 = z\left(1 + \tan\frac{C}{4}\right)^2$$
$$= r\left(1 + \tan\frac{A}{4}\right)\left(1 + \tan\frac{B}{4}\right)\left(1 + \tan\frac{C}{4}\right)$$

NOTE—*This is the trigonometrical statement of what is known as* MALFATTI'S PROBLEM—*a geometrical proof of which by Dr. Hart is given in Vol' I of the Q. J. Trigonometrical solutions are given in 'Gaskin's solutions of the Geometrical Problems proposed at St. John's, Cambridge,'* p. 82; *and in Hymer's Trigonometry*, p. 153.

Putting 2α, 2β, 2γ *for* A, B, C, *the fig' gives*

$$y \cot\beta + z \cot\gamma + 2\sqrt{yz} = r(\cot\beta + \cot\gamma)$$

and two sim'r eq'ns: from these

$$\sqrt{x}\cos\alpha + \sqrt{z}\sin\alpha = \sqrt{y}\cos\beta + \sqrt{z}\sin\beta$$

and two sim'r eq'ns: the rest follows easily.

48. Prove that $IO \,.\, IS \cos SIO = R\rho - Rr + r^2$ where ρ is the in-radius of the pedal triangle.

Also, by finding a similar result for $IO \,.\, IS \sin SIO$, show that

$$4\,(\text{area } SIO)^2 = R^2 r^2 + 6R^2 r\rho - R^2\rho^2 - 2R^3\rho - 2Rr^3 - 2Rr^2\rho - \rho^4$$

Editor: E. T. xxx.

49. Apply the result of Example 5 on p. 210 to show that the area of the triangle SIO is

$$-2R^2 \sin\frac{A-B}{2}\sin\frac{B-C}{2}\sin\frac{C-A}{2}$$

§ 75. *Some fundamental trigonometrical properties of the* Lemoine and Brocard Geometry.

Examples

1. *If through* A, B, C *lines* AXY (*or* XAY), BYZ (*or* YBZ), CZX (*or* ZCX) *are drawn, so as to make the same angle* θ (*measured one way round*) *with* AB, BC, CA, *respectively, and forming a triangle* XYZ; *then triangles* XYZ, ABC are similar, and

any side of one : *homologous side of other* $= \sin(\omega \pm \theta) : \sin\omega$

where ω *is the* Brocard angle *of triangle* ABC.

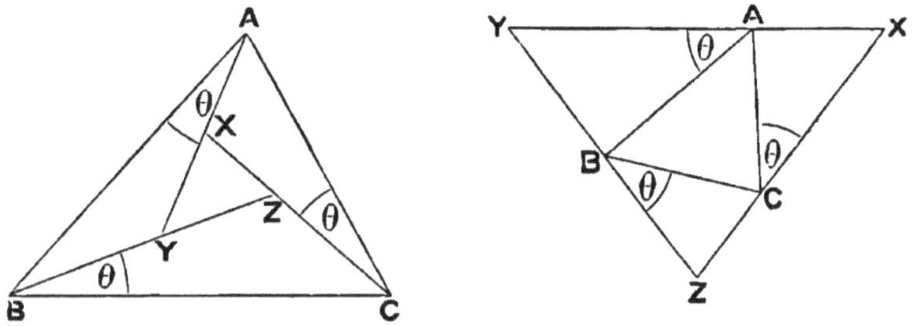

It is at once obvious that $\hat{X} = \hat{A},\ \hat{Y} = \hat{B},\ \hat{Z} = \hat{C}$.

$$\text{Also}\quad \frac{YZ}{BC} = \frac{BZ \pm BY}{BC} = \frac{BZ}{BC} \pm \frac{BY}{BA}\,\frac{BA}{BC}$$

$$= \frac{\sin(C \pm \theta)}{\sin C} \pm \frac{\sin\theta}{\sin B}\cdot\frac{\sin C}{\sin A}$$

$$= \cos\theta \pm \sin\theta\,.\,\Sigma\cot A$$

$$= \cos\theta \pm \sin\theta\cot\omega$$

$$\therefore\quad YZ : BC = \sin(\omega \pm \theta) : \sin\omega$$

Cor' (1)—When $\theta = \omega$, and AX &c are drawn *within* ABC, the Δ XYZ shrinks up into a Brocard point.

Cor' (2)—When $\theta = \pi/2$, the ratio becomes $\cot\omega$.

Cor' (3)—When $\theta = \omega$, and AX &c are drawn *without* ABC, the ratio becomes $2\cos\omega$.

Note—The particular case of this theorem, when the Δ XYZ shrinks into a point, probably gave rise to the discovery of the Brocard points; and thence to the rest of this remarkable branch of modern Geometry.

2. *Magnitude of the* Brocard Angle.

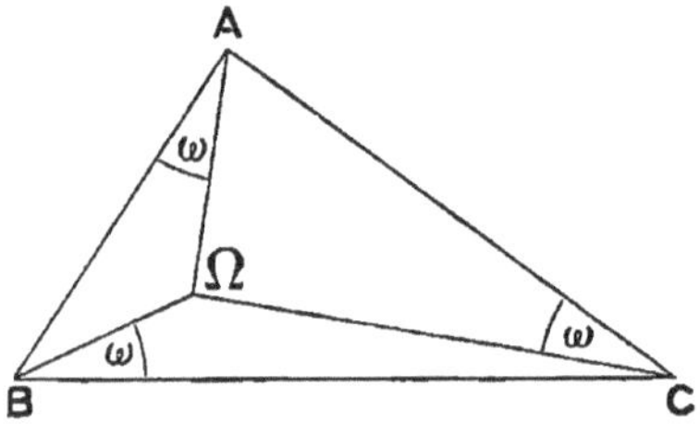

If Ω is the pos' Brocard p't of the △ ABC, then

$$\omega = \Omega \widehat{AB} = \Omega \widehat{BC} = \Omega \widehat{CA}$$

$$\therefore \quad \frac{\sin(A-\omega)}{\sin\omega} = \frac{\Omega C}{\Omega A} = \frac{\Omega C}{\Omega B} \cdot \frac{\Omega B}{\Omega A} = \frac{\sin\omega}{\sin(C-\omega)} \cdot \frac{\sin\omega}{\sin(B-\omega)}$$

$$\therefore \quad \sin^3\omega = \sin(A-\omega)\sin(B-\omega)\sin(C-\omega)$$

whence, as in Example 7, page 83,

$$\cot\omega = \Sigma\cot A \quad \ldots \quad (1)$$

or

$$\operatorname{cosec}^2\omega = \Sigma\operatorname{cosec}^2 A \quad \ldots \quad (2)$$

other useful forms of this are

$$\cot\omega = \frac{1 + \Pi\cos A}{\Pi\sin A} \quad \ldots \quad (3)$$

$$\cot\omega = \Sigma\frac{2R\cos A}{a} = \frac{R}{abc}\Sigma(2bc\cos A) = \frac{\Sigma a^2}{4\Delta} \quad \ldots \quad (4)$$

Otherwise—The following is an elegant proof* of this important property

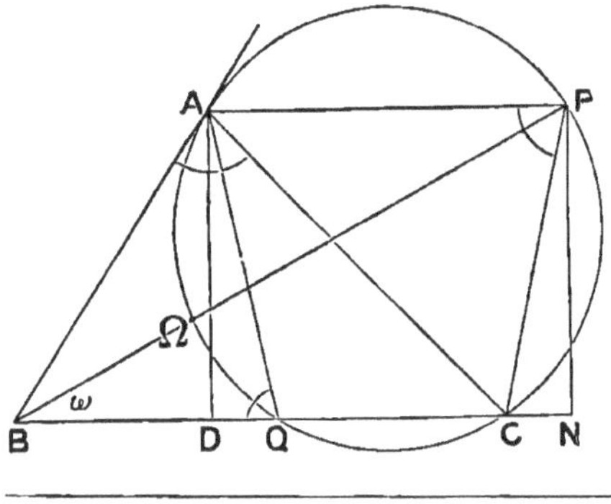

Let ⊙ which touches AB at A, and goes thro' C, cut the ∥ to BC thro' A in P.

Then PB cuts this ⊙ in Ω,†

so that $P\widehat{B}C = \omega$.

Drop AD, PN ⊥s on BC.

---

* Due to Mr. R. F. Davis, who kindly sent it to me immediately on its discovery in August, 1890. R. C. J. N.

† See *Euclid Revised*, p. 394.

Then, if Q is the p't in which BC cuts ⊙, APCQ is a symmetrical trapezium

$$\therefore \quad AD = PN, \text{ and } QD = CN.$$

$$\therefore \quad BN = QD + BD + CD$$

Dividing each side by AD (or PN) and noticing that $B\hat{A}C = A\hat{P}C = A\hat{Q}B$

we get $\cot \omega = \cot A + \cot B + \cot C$

3. *Length of the radius* R_1 *of the* first Lemoine (*or* T. R., i.e. triplicate ratio) circle.

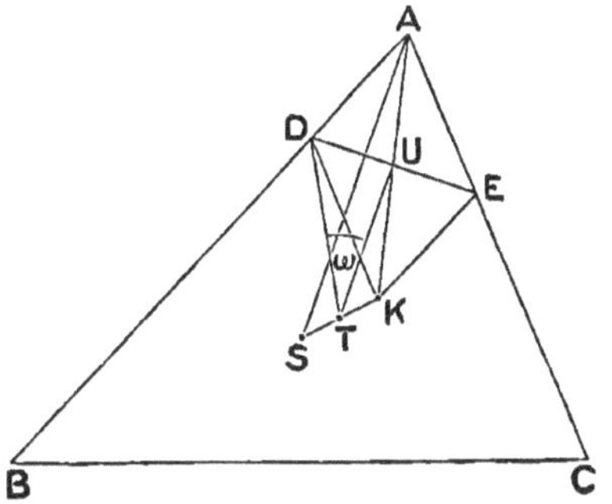

If KD, KE, ∥[s] to the sides AC, AB, meet AB, AC respectively in D, E; and if U is the cross of KA, DE; then D, E are p'ts on the T. R. ⊙; and T, the mid p't of KS, is its centre.

Also, from the geometry of that ⊙, $D\hat{T}U = \omega$.

$$\therefore \quad R_1 = TD = TU \sec \omega = \frac{R}{2} \sec \omega$$

4. *Length of the radius* R_2 *of the* second Lemoine (*or* cosine) circle.

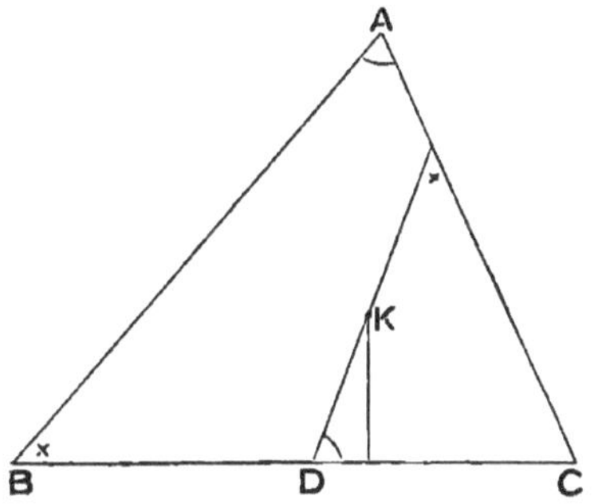

If an anti-∥ thro' K to AB meets BC in D, then D is a p't on the ⊙, K being its centre.

Now (See § 73, Ex' 2, Cor')

⊥ from K on BC $= 2a\Delta/\Sigma a^2$

$$\therefore \quad R_2 = KD = \frac{a}{\sin A} \cdot \frac{2\Delta}{\Sigma a^2} = R \tan \omega \quad \text{(See § 75, Ex' 2)}$$

Cor'— $4R_1^2 - R_2^2 = R^2(\sec^2 \omega - \tan^2 \omega) = R^2$

5. *Length of the radius* R_3 *of the* Taylor circle.

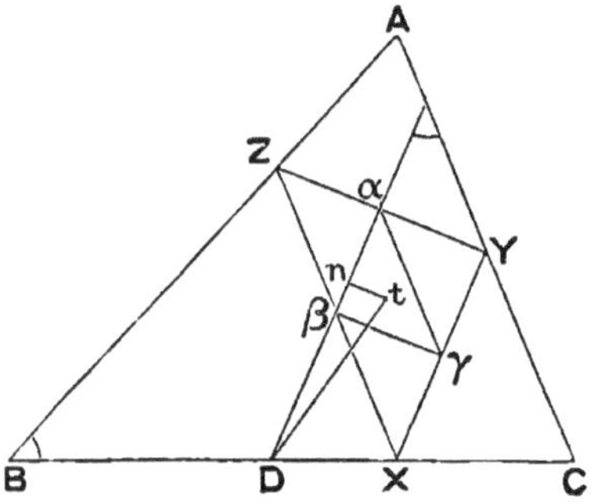

If XYZ is the pedal Δ; α, β, γ the mid p'ts of its sides YZ, ZX, XY, respectively; and if αβ meets BC in D; then D is a p't on the ⊙; and t, the in-centre of Δ αβγ, is its centre.

Drop tn ⊥ on αβ; and put ϕ for $\widehat{Dtn}$

$$\text{Then } R_3 \sin\phi = Dn = \tfrac{1}{4} \text{ perim' of pedal } \Delta$$
$$= R\,\Pi \sin A \quad \text{(See § 73, Ex' 3)}$$
$$\text{And } R_3 \cos\phi = tn = \tfrac{1}{2} \text{ in-rad' of pedal } \Delta$$
$$= R\,\Pi \cos A \quad \text{(See § 74, Ex' 5, Cor' 2)}$$
$$\therefore\ R_3 = R\sqrt{\Pi \sin^2 A + \Pi \cos^2 A}$$
$$\text{and } \tan\phi = \Pi \tan A$$

Note—ϕ is the ∧ subtended by any one of the 3 anti-||s to the sides of ABC, formed by producing the sides of αβγ, at any p't of the ⊙ lying on the *same* side of that anti-|| as t. If ϕ' is the ∧ subtended on the side *opposite* to t, we have

$$\tan\phi' = \tan(\pi - \phi) = -\Pi \tan A$$

6. *Length of the radius* R_4 *of a* Tucker circle.

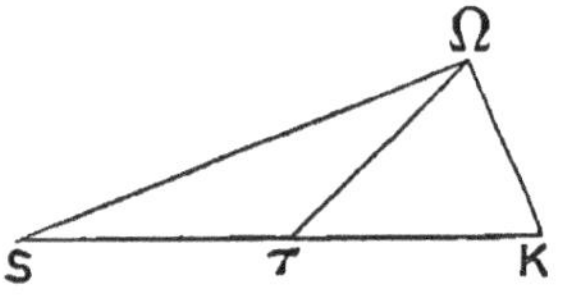

The centre T of a Tucker ⊙ is in SK (or SK produced) the particular ⊙ being defined by the ratio in which SK is divided by T.

Also, if Ω is a Brocard p't then Ω is a permanent centre of similarity of the circum ⊙, cosine ⊙, and *any* Tucker ⊙.

Now $S\hat{\Omega}K$ is right

$\therefore$ if $K\tau : S\tau = p : q$

$(p+q)^2\,\Omega\tau^2 = p^2.\,\Omega S^2 + q^2.\,\Omega K^2$ (See E. R. p. 109)

And $\dfrac{\Omega\tau}{R_4} = \dfrac{\Omega S}{R} = \dfrac{\Omega K}{R_2}$

$$\therefore\ (p+q)^2 R_4{}^2 = p^2 R^2 + q^2 R_2{}^2$$

$$= R^2(p^2 + q^2\tan^2\omega)$$

$$\therefore\ R_4 = \frac{R}{p+q}\sqrt{p^2 + q^2\tan^2\omega}$$

Note (1)—Particular cases of Tucker ⊙s are—

(1) the circum-⊙, when $q = 0$

(2) „ cosine „ „ $p = 0$

(3) „ T. R. „ „ $p = q$

(4) „ Taylor „ „ $p : -q = \Pi\cos A : 1 + \Pi\cos A$

the neg′ sign of q in the last showing that τ is in SK *produced*.

These will appear by substituting in the expression just found for R_4—the three first at once, and the fourth after a little reduction.

Note (2)—It will easily be found, in the usual algebraic way, that the *minimum* Tucker ⊙ has its radius $R\sin\omega$, and that for it

$$p : q = \sin^2\omega : \cos^2\omega$$

There is no *maximum* Tucker ⊙.

Note (3)—R_4 can also be expressed in terms of θ, the common value of the twelve ∧s of Exercise 19, p. 402 in *Euclid Revised.*

For Ω being a centre of similarity of the Δs FDE, ABC, in the fig′ of that Ex′, we have that the ratio of similarity of these Δs

$$= \Omega F : \Omega A = \sin\Omega AF : \sin\Omega FB = \sin\omega : \sin(\theta + \omega)$$

$\therefore$ R_4 and R are in the same ratio

i. e. $R_4 = R\sin\omega/\sin(\theta+\omega)$

By putting 0, ω, $\pi/2$, successively, for θ in this expression, we get the radii of the circum-, T. R., and cosine-⊙s.

The *minimum* ⊙ is got by putting $\pi/2$ for $\theta + \omega$.

7. *Length of* $\Omega\Omega'$.

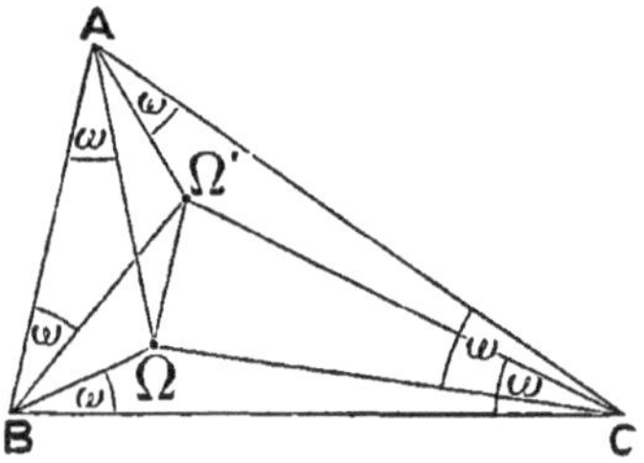

$$\Omega A = b\frac{\sin\omega}{\sin A}$$

$$= 2R\sin\omega\cdot\frac{b}{a}$$

$$\Omega' A = c\frac{\sin\omega}{\sin A}$$

$$= 2R\sin\omega\cdot\frac{c}{a}$$

$$\therefore\quad \Omega\Omega'^2 = 4R^2\sin^2\omega\left\{\frac{b^2+c^2-2bc\cos(A-2\omega)}{a^2}\right\}$$

$$= 4R^2\sin^2\omega\left[1+\frac{2bc}{a^2}\{\cos A-\cos(A-2\omega)\}\right]$$

$$= 4R^2\sin^2\omega\left\{1-\frac{4bc}{a^2}\sin(A-\omega)\sin\omega\right\}$$

But $\cot\omega - \cot A = \cot B + \cot C$

$$\therefore\quad \sin(A-\omega) = \frac{\sin^2 A}{\sin B\sin C}\sin\omega$$

$$\therefore\quad \frac{bc}{a^2}\sin(A-\omega) = \sin\omega$$

$$\therefore\quad \Omega\Omega' = 2R\sin\omega\sqrt{1-4\sin^2\omega}$$

8. *Length of the radius* ρ *of the* Brocard circle.

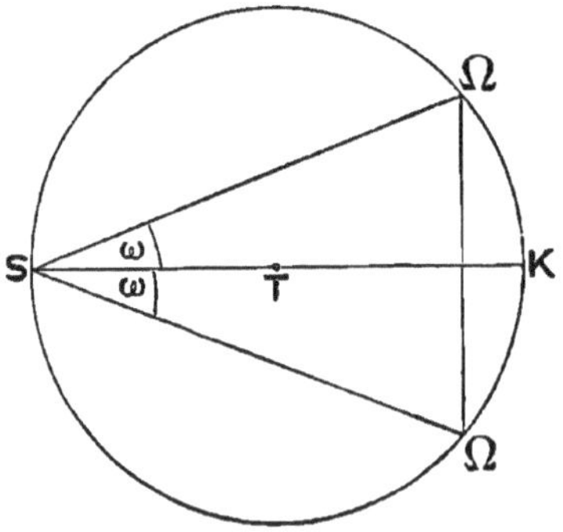

This ⊙ has SK as diam′, and goes thro′ Ω and Ω'.

Also $\Omega\hat{S}\Omega' = 2\omega$.

$$\therefore \quad \rho = \frac{\Omega\Omega'}{2 \sin 2\omega}$$

$$= \frac{R}{2} \sec \omega \sqrt{1 - 4 \sin^2 \omega}, \text{ by last Example}$$

$$= \frac{R}{2} \sqrt{\sec^2 \omega - 4 \tan^2 \omega}$$

$$= \frac{R}{2} \sqrt{1 - 3 \tan^2 \omega}$$

Cor'—

$$\rho^2 = \frac{R^2}{4} \sec^2 \omega - R^2 \tan^2 \omega$$

$$= R_1^2 - R_2^2$$

$$\therefore \quad \pi R_1^2 = \pi R_2^2 + \pi \rho^2$$

$\therefore$ area T. R. ⊙ = area cosine ⊙ + area Brocard ⊙.

Note—The centre of the Brocard ⊙ is T, the centre of the T. R. ⊙.

Exercises

1. If x, y, z are the perpendiculars from K on the sides of the medial triangle, prove that

$$a^2 x / \cos A = b^2 y / \cos B = c^2 z / \cos C$$

Deduce *Schlömilch's Theorem*—If D is the mid′ point of BC, and d the mid′ point of the altitude of A; and if E, e and F, f are similar pairs of points; then D d, E e, F f cointersect in K.

2. On the sides of triangle ABC, triangles PAB, QBC, RCA are described external and similar to itself; so that the angles adjacent to A, B, C are respectively C, A, B; prove that the circles round the three new triangles cointersect in the positive Brocard point of ABC.

3. If KP, KQ, KR are perpendiculars from K on BC, CA, AB respectively; prove that—

(1) $PB/a + QC/b + RA/c = \frac{3}{2} = PC/a + QA/b + RB/c$

(2) $PB \cdot QC \cdot RA + PC \cdot QA \cdot RB = (3\,abc \tan^2 \omega)/4.$

4. Prove that the tangent from A to the T. R. circle is

$$R \tan \omega \sqrt{b^2 + c^2} / a.$$

5. Prove that the perpendiculars from the centre of the T. R. circle on the sides are proportional to $\cos(A - \omega)$, $\cos(B - \omega)$, $\cos(C - \omega)$.

6. If ρ_1, ρ_2, ρ_3 are the circum-radii of $A\Omega B$, $B\Omega C$, $C\Omega A$, show that $\rho_1 \rho_2 \rho_3 = R^3$.

Tucker.

7. Show that the Taylor circle intercepts on the side AB a length

$$a \cos A \cos (B - C)$$

8. If L, M, N are the feet of the perpendiculars from Ω (or Ω') on the sides of the triangle, show that area $LMN = \Delta \sin^2 \omega$.

9. If t is the centre of the Taylor circle, prove that—

(1) the perpendicular from t on $BC = \frac{R}{2} \{\cos A - \cos 2A \cos (B - C)\}$

(2) $St : Kt = 1 + \Pi \cos A : - \Pi \cos A$

R. F. Davis.

10. If θ, ϕ, χ are the angles made by KA, KB, KC with KS, prove that—

(1) $$\frac{\sqrt{2(b^2 + c^2) - a^2}}{\sin(B - C)} \sin \theta = \frac{\sqrt{2(c^2 + a^2) - b^2}}{\sin(C - A)} \sin \phi = \frac{\sqrt{2(a^2 + b^2) - c^2}}{\sin(A - B)} \sin \chi$$

(2) $$\frac{b^2 + c^2 - 2a^2}{a \sin(B - C)} \tan \theta = \frac{c^2 + a^2 - 2b^2}{b \sin(C - A)} \tan \phi = \frac{a^2 + b^2 - 2c^2}{c \sin(A - B)} \tan \chi$$

Simmons: E. T. 1.

11. OA, OB, OC are produced to α, β, γ, so that $A\alpha$, $B\beta$, $C\gamma$ are respectively equal to BC, CA, AB: prove that—

(1) triangles ABC, $\alpha\beta\gamma$ have the same centroid

(2) area $\alpha\beta\gamma = 2\Delta(2 + \cot \omega)$

(3) $\alpha\beta^2 + \beta\gamma^2 + \gamma\alpha^2 = 8\Delta(3 + 2 \cot \omega)$

(4) the Brocard angle of $\alpha\beta\gamma = \cot^{-1}(2 \cot \omega + 3)/(\cot \omega + 2)$

Neuberg.

CHAPTER XV

Quadrilaterals and other Polygons

§ **76**. The only new principle involved in this chapter is the idea of a *negative area*. Otherwise it will consist of Examples and Exercises on preceding results applied to Polygons.

Def'—If a point, in going continuously and completely round the boundary of a closed figure, crosses its own path anywhere, that path is called an **autotomic** (*self-cutting*) **circuit**: the sign of any area enclosed by an autotomic circuit is considered **positive** if that area lies on the *left-hand* of the travelling point, but **negative** if it lies on the *right-hand*; and the total area of the whole circuit is the algebraic sum of the several areas which it encloses.

Illustrations

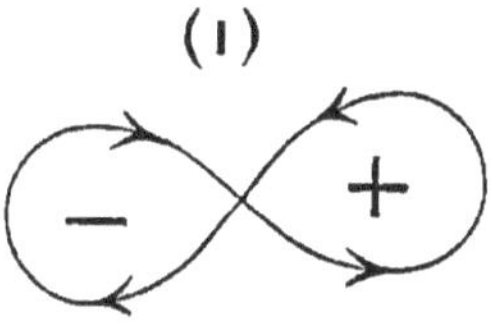

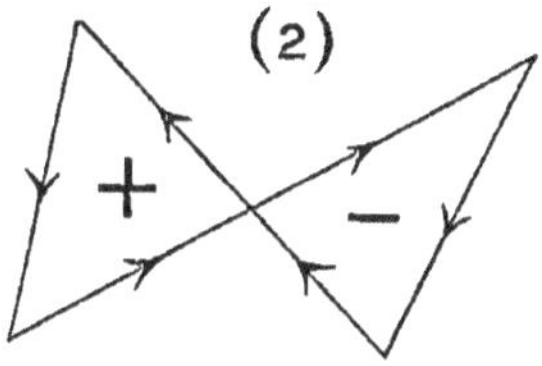

Let fig' (1) represent the trace made by a skater, in going round a 'figure of eight' on one foot, as indicated by the arrow heads: then the right-hand loop is a pos' area, and the left-hand loop a neg' area.

Let fig' (2) represent a cross-quadrilateral, whose sides are taken in the order indicated by the arrow heads; then the left hand △ is a pos' area, and the right-hand △ a neg' area.

Examples

1. Area *of any quadrilateral (convex, or crossed) in terms of its sides, and a pair of opposite angles.*

Let Ω be the area of a quad′ ABCD; d, e, f, g its successive sides AB, BC, CD, DA; x its diag′ AC; σ its semi-perim′; θ, ϕ the respective $\wedge^s$ at B, D.

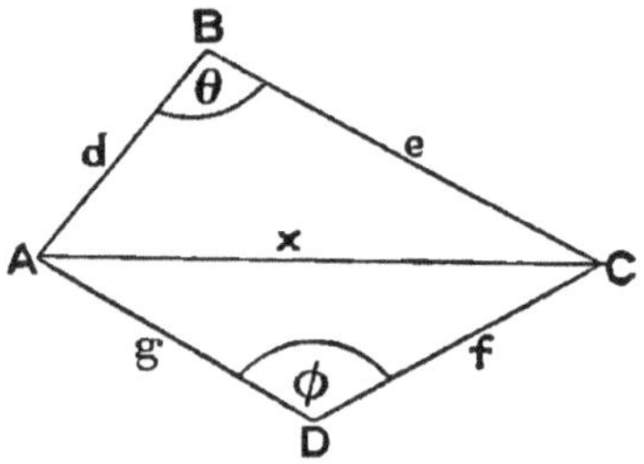

1°, let the quad′ be *convex*.

Then $\Omega = \triangle ABC + \triangle CDA$

$$\text{Now} \quad d^2 + e^2 - 2\,de\cos\theta = x^2 = f^2 + g^2 - 2\,fg\cos\phi$$

$$\therefore \quad d^2 + e^2 - f^2 - g^2 = 2\,(de\cos\theta - fg\cos\phi)$$

$$\text{Also} \quad 4\,\Omega = 2\,(de\sin\theta + fg\sin\phi)$$

Whence, by squaring and adding, we get

$$16\,\Omega^2 + (d^2 + e^2 - f^2 - g^2)^2 = 4\left\{(de)^2 + (fg)^2 - 2\,defg\cos(\theta + \phi)\right\}$$

$$= 4\left\{(de + fg)^2 - 2\,defg\,(1 + \cos\overline{\theta + \phi})\right\}$$

$$\therefore \ 16\,\Omega^2 = 4\,(de + fg)^2 - (d^2 + e^2 - f^2 - g^2)^2 - 16\,defg\cos^2\frac{\theta + \phi}{2}$$

$$= \left\{(d+e)^2 - (f-g)^2\right\}\left\{(f+g)^2 - (d-e)^2\right\} - 16\,defg\cos^2\frac{\theta+\phi}{2}$$

$$\therefore \ \Omega = \left\{(\sigma - d)(\sigma - e)(\sigma - f)(\sigma - g) - defg\cos^2\frac{\theta + \phi}{2}\right\}^{\frac{1}{2}}$$

Particular cases (which the Student should work out independently) are—

(1) when quad′ is cyclic, and circumscribes a $\odot$:

$$\text{then} \quad \theta + \phi = \pi$$

$$\text{and} \quad d + f = g + e$$

$$\text{so that} \quad f + g + e - d = 2\,f, \text{ \&c.}$$

and Ω becomes $\sqrt{defg}$

(2) when quad′ is cyclic, but does not circumscribe a ⊙ :

$$\Omega \text{ becomes } \sqrt{(\sigma - d)(\sigma - e)(\sigma - f)(\sigma - g)}$$

(3) when quad′ circumscribes a ⊙, but is not cyclic:

$$\Omega \text{ becomes } \sqrt{defg - defg\cos^2\frac{\theta+\phi}{2}}$$

$$\text{i. e. } \sqrt{defg}\cdot\sin\frac{\theta+\phi}{2}$$

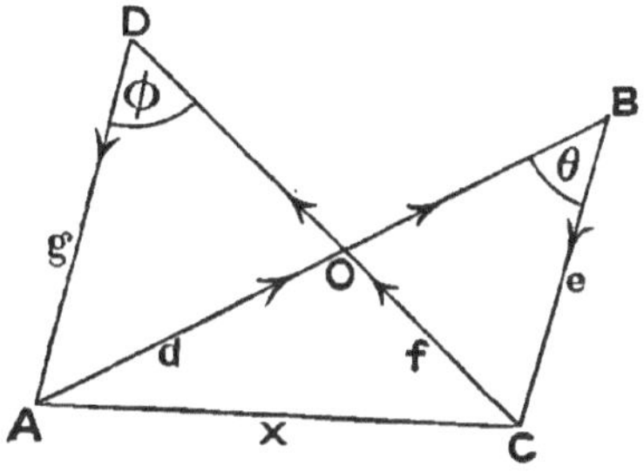

2°, let the quad′ be *crossed*, so that ABCD is an autotomic circuit—the sides AB, CD crossing in O.

Then $\Omega = \Delta\,AOD - \Delta\,COB$
$\quad = \Delta\,ADC - \Delta\,ABC$

Now $f^2 + g^2 - 2fg\cos\phi = x^2 = d^2 + e^2 - 2de\cos\theta$

$\therefore\; f^2 + g^2 - d^2 - e^2 = 2(fg\cos\phi - de\cos\theta)$

Also $4\,\Omega = 2(fg\sin\phi - de\sin\theta)$

Whence, by squaring and adding, we get

$$16\,\Omega^2 + (f^2 + g^2 - d^2 - e^2)^2 = 4\left\{(fg)^2 + (de)^2 - 2\,defg\cos(\theta - \phi)\right\}$$

$$= 4\left\{(fg - de)^2 + 2\,defg\,(1 - \cos\overline{\theta - \phi})\right\}$$

$$\therefore\; 16\,\Omega^2 = 4(fg - de)^2 - (f^2 + g^2 - d^2 - e^2)^2 + 16\,defg\sin^2\frac{\theta-\phi}{2}$$

$$= \left\{(f+g)^2 - (d+e)^2\right\}\left\{(e-d)^2 - (f-g)^2\right\} + 16\,defg\sin^2\frac{\theta-\phi}{2}$$

$$\therefore\; \Omega = \left\{\sigma(\sigma - d - e)(\sigma - d - f)(\sigma - d - g) + defg\sin^2\frac{\theta-\phi}{2}\right\}^{\frac{1}{2}}$$

A particular case (which the Student should work out independently) is when quad′ is cyclic: then $\theta = \phi$, and

$$\Omega \text{ becomes } \sqrt{\sigma(\sigma - d - e)(\sigma - d - f)(\sigma - d - g)}$$

2. *If* x, y *are the diagonals of any quadrilateral (convex or crossed) and* ω *the angle between them, then its* area *is* $\frac{1}{2}$ xy sin ω.

If also d, e, f, g *are the successive sides, then, for the acute value af* ω,

$$\cos\omega = \frac{(d^2 \sim e^2) + (f^2 \sim g^2)}{2\,xy}$$

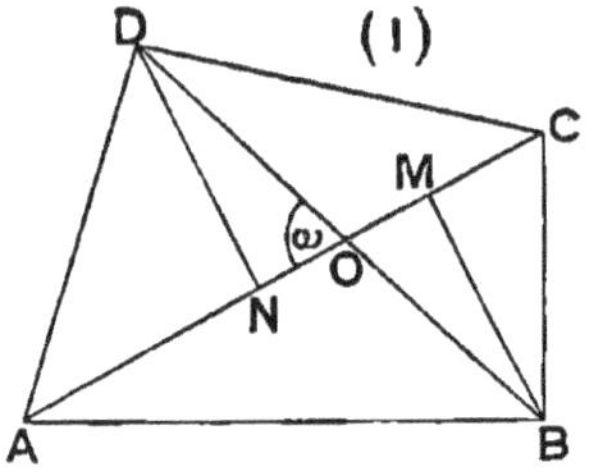

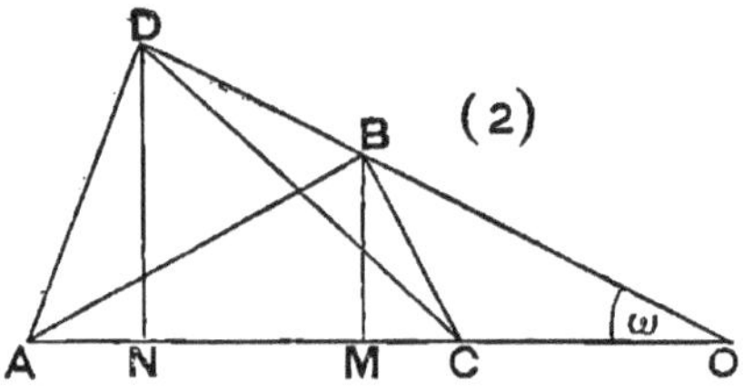

Let AB, BC, CD, DA, AC, BD be respectively d, e, f, g, x, y; and let AC, DB cut in O.

Drop BM, DN $\perp^s$ on AC.

In fig′ (1) the quad′ is convex, in fig′ (2) crossed.

Then, the upper signs referring to fig′ (1) and the lower to fig′ (2)

$$\text{area ABCD} = \tfrac{1}{2}(DN \pm BM)\,x$$

$$= \tfrac{1}{2}(DO \sin\omega \pm BO \sin\omega)\,x$$

$$= \tfrac{1}{2}\,xy \sin\omega$$

$$\text{Again}\quad d^2 \sim e^2 = AM^2 \sim CM^2 = x\,(AM \sim CM)$$

$$\text{and}\quad f^2 \sim g^2 = CN^2 \sim AN^2 = x\,(CN \sim AN)$$

$$\therefore\quad (d^2 \sim e^2) + (f^2 \sim g^2) = 2\,x\,.\,MN = 2\,xy\cos\omega$$

$$\therefore\quad \cos\omega = \frac{(d^2 \sim e^2) + (f^2 \sim g^2)}{2\,xy}$$

Cor′ (1)—If quad′ circumscribes a ⊙,

$$d + f = e + g$$

$$\text{so that}\quad d^2 + f^2 - e^2 - g^2 = 2\,(eg - df)$$

$$\text{and then}\quad \cos\omega = \frac{df \sim eg}{xy}$$

Note—If a quad′ is crossed it cannot circumscribe a ⊙.

Cor' (2)—If quad′ is cyclic, $xy = df \pm eg$ (*Ptolemy's Theorem*)

and then $\cos\omega = \dfrac{(d^2 \sim e^2) + (f^2 \sim g^2)}{2\,(df \pm eg)}$

and area $= \frac{1}{2}(df \pm eg)\sin\omega$

Note—In the case of a cyclic quad′ it is often useful to recollect the geometrical theorem. (E. R. p. 290)

$$x/y = \overline{dg \pm ef} \big/ \overline{fg \pm de}$$

Cor' (3)—If a quad′ is both cyclic and touches a ⊙,

$$\cos\omega = \overline{df \sim eg} \big/ \overline{df + eg}$$

3. *Length of the* circum-radius R *of a cyclic quadrilateral in terms of its sides.*

Using the notation and diagrams of the preceding Example, we have

$$\left.\begin{aligned} fg &= 2\,R\,.\,DN = 2\,R\,.\,DO\sin\omega \\ de &= 2\,R\,.\,BM = 2\,R\,.\,BO\sin\omega \end{aligned}\right] \text{for all cyclic quad's}$$

Whence, upper signs referring to a convex, and lower to a cross quad′, we get

$$fg \pm de = 2\,R\,y\sin\omega$$

Sim′ly $dg \pm ef = 2\,R\,x\sin\omega$

Also $df \pm eg = xy$ (*Ptolemy's Theorem*)

$\therefore$ $(fg \pm de)(dg \pm ef)(df \pm eg) = 4\,R^2(xy\sin\omega)^2 = 16\,R^2\,\Omega^2$

$\therefore$ for a convex quad′

$$R = \frac{1}{4}\sqrt{\frac{(de + fg)(df + eg)(dg + ef)}{(\sigma - d)(\sigma - e)(\sigma - f)(\sigma - g)}}$$

And for a cross quad′

$$R = \frac{1}{4}\sqrt{\frac{(fg - de)(df - eg)(dg - ef)}{\sigma(\sigma - d - e)(\sigma - d - f)(\sigma - d - g)}}$$

Note—The above results may also be found (as the Student can easily prove for himself) from the consideration that $R = x/(2\sin\theta)$, where x and $\sin\theta$ are given by the eq′ns

$$d^2 + e^2 - 2\,de\cos\theta = x^2 = f^2 + g^2 \mp 2\,fg\cos\theta$$

But the method used above is better.

Cor'—The expression for the circum-rad' of a convex cyclic quad' will give a relation among the sines of 4 $\wedge^s$ whose sum is 180°.

For, if $\alpha + \beta + \gamma + \delta = 180°$, then 2α, 2β, 2γ, 2δ can be arranged round the centre of the ⊙, so that they are respectively opposite d, e, f, g.

$$\therefore \quad R = \frac{d}{2s_1} = \frac{e}{2s_2} = \frac{f}{2s_3} = \frac{g}{2s_4}$$ [See fig' on p. 279]

where s_1, s_2, s_3, s_4 stand for $\sin\alpha$, $\sin\beta$, $\sin\gamma$, $\sin\delta$.

∴, substituting for d, e, f, g in the expression for R, we get that $16R^2$

$$= \frac{64R^6(s_1s_2+s_2s_3)(s_1s_3+s_2s_4)(s_1s_4+s_2s_3)}{R^4(-s_1+s_2+s_3+s_4)(s_1-s_2+s_3+s_4)(s_1+s_2-s_3+s_4)(s_1+s_2+s_3-s_4)}$$

So that the relation is

$$(-s_1+s_2+s_3+s_4)(s_1-s_2+s_3+s_4)(s_1+s_2-s_3+s_4)(s_1+s_2+s_3-s_4)$$
$$= 4(s_1s_2+s_2s_3)(s_1s_3+s_2s_4)(s_1s_4+s_2s_3).$$

4. *If a quadrilateral is inscribed in a circle, radius* R, *and circumscribes a circle, radius* r; *and if* d *is the distance between the centres of the circles, then*

$$\frac{1}{(R+d)^2} + \frac{1}{(R-d)^2} = \frac{1}{r^2}$$

By *Poncelet's Theorem*, if two ⊙s can have any one quad' so placed that it is inscribed in the one, and circumscribes the other, then for *every* position of a quad' ABCD, inscribed in the outer ⊙, so that AB, BC, CD touch the inner ⊙, DA will also touch it.

We may ∴ take the position of symmetry when AC is the line of centres.

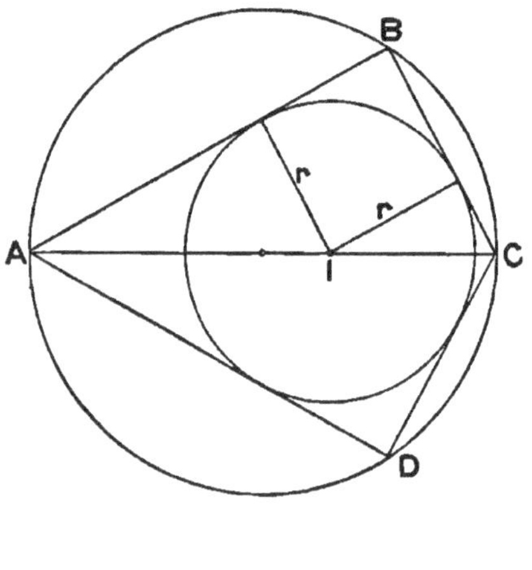

Let I be the centre of in-⊙; and suppose I nearer C than A.

Then $AI = R + d$

and $CI = R - d$

$$\therefore \quad \frac{r}{R+d} = \sin BAC$$

and $$\frac{r}{R-d} = \sin BCA$$

$$= \cos BAC$$

$$\therefore \quad \frac{1}{(R+d)^2} + \frac{1}{(R-d)^2} = \frac{1}{r^2}$$

Note—The analogy between this result and *Chapple's Theorem* for a Δ, viz.

$$d^2 = R^2 - 2Rr$$

will be seen by noticing that the latter can be put into the form

$$\frac{1}{R+d} + \frac{1}{R-d} = \frac{1}{r}$$

Sim'r results can be got for the pentagon, hexagon, &c; but they are rather complex.

In the case of the hexagon a connection of a different kind, between R, r, d, is given in Exercise 41 following.

5. Ptolemy's Theorem *by trigonometry.*

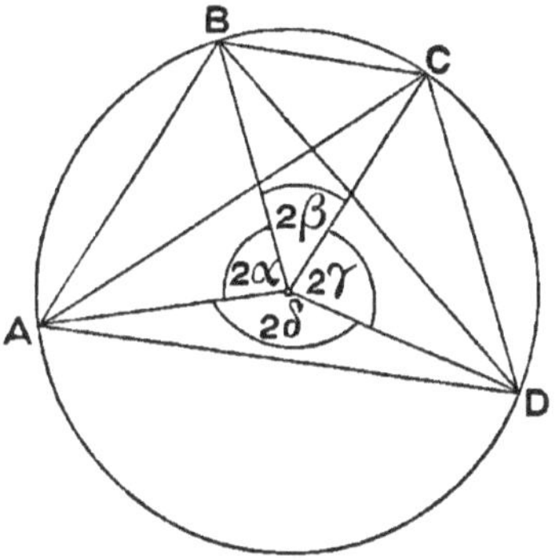

Let ABCD be a cyclic quad'.

Take the rad' of its circum-⊙ as unity; and let 2α, 2β, 2γ, 2δ be the $\angle$s which AB, BC, CD, DA respectively subtend at the circum-centre.

Then the halves of AB, BC, CD, DA, AC, BD are respectively

$$\sin\alpha,\ \sin\beta,\ \sin\gamma,\ \sin\delta,\ \sin(\gamma+\delta),\ \sin(\beta+\gamma)$$

$$\therefore\ \tfrac{1}{2}(AB\,.\,CD + BC\,.\,DA)$$

$$= 2\sin\alpha\sin\gamma + 2\sin\beta\sin\delta$$

$$= \cos(\alpha-\gamma) - \cos(\alpha+\gamma) + \cos(\beta-\delta) - \cos(\beta+\delta)$$

$$= \cos(\alpha-\gamma) + \cos(\beta-\delta), \quad \text{since } \alpha+\gamma = \pi - (\beta+\delta)$$

Also $\tfrac{1}{2}AC\,.\,BD = 2\sin(\beta+\gamma)\sin(\gamma+\delta)$

$$= \cos(\beta-\delta) - \cos(\beta+2\gamma+\delta)$$

$$= \cos(\beta-\delta) + \cos(\alpha-\gamma)$$

$$\therefore\ AB\,.\,CD + BC\,.\,DA = AC\,.\,BD$$

6. *Given a side* a, *and the number of sides* n *of a regular polygon, to find its* circum-radius R, *its* in-radius r, *and its* area.

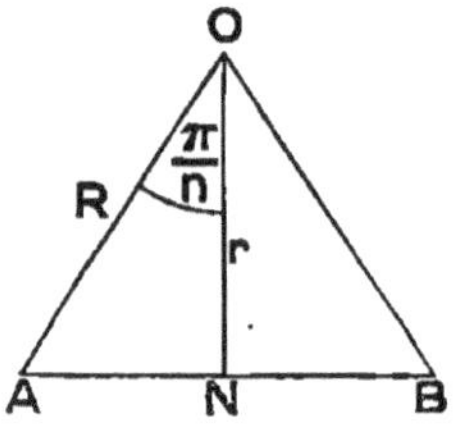

Let AB be a side of the pol′ ; O the common centre of its in- and circum-⊙s; ON ⊥ to AB.

Then OA = R, ON = r

$$A\hat{O}B = \frac{2\pi}{n},\quad A\hat{O}N = \frac{\pi}{n}$$

$$\therefore \quad \frac{a}{2} = AN = R \sin\frac{\pi}{n}, \text{ or } = r\tan\frac{\pi}{n}$$

$$\therefore \quad R = \frac{a}{2}\operatorname{cosec}\frac{\pi}{n}$$

$$r = \frac{a}{2}\cot\frac{\pi}{n}$$

Also area pol′ $= n \,.\, \Delta\, AOB$

$$= \tfrac{1}{2}\, n a r$$

$$= \frac{na^2}{4}\cot\frac{\pi}{n}$$

The area may also be expressed either

$$\text{as } \frac{n}{2} R^2 \sin\frac{2\pi}{n}, \text{ or as } nr^2 \tan\frac{\pi}{n}$$

Cor′ (1)—Area ⊙ (rad′ r) $= \text{Lim}'_{n=\infty}\left(nr^2\tan\frac{\pi}{n}\right)$

$$= \text{Lim}'_{n=\infty}\, \pi r^2 \left\{\frac{\tan\frac{\pi}{n}}{\frac{\pi}{n}}\right\}$$

$$= \pi r^2$$ (see pp. 169, 7)

Cor′ (2)—Perim′ pol′ $= na = 2nr\tan\frac{\pi}{n} = 2nR\sin\frac{\pi}{n}$

Exercises

1. If x, y are the diagonals, and d, e, f, g the sides of a quadrilateral circumscribing a circle, show that its area is

$$\tfrac{1}{2}\sqrt{(xy)^2 - (df - eg)^2}$$

2. If θ is the angle between two sides a, b of a cyclic quadrilateral; and if the diagonal conterminous with a, b is a diameter; show that the area of the quadrilateral is

$$\{2\,ab - (a^2 + b^2)\cos\theta\}/(2\sin\theta)$$

3. The sides of a cyclic quadrilateral are in a G. P. whose ratio is r : if θ is the angle between the first pair of sides, and ϕ that between the second pair, prove that

$$\frac{\tan\theta}{\tan\phi} = \frac{r^2 - 1}{r^2 + 1}$$

4. A quadrilateral is inscribed in a circle, radius R, and circumscribes a circle, radius r: if d is the distance between the centres, prove that the rectangle under the diagonals is

$$8\,R^2 r^2/(R^2 - d^2)$$

Prove also that, if Ω is the area of the quadrilateral, σ its semi-perimeter, and α, β, γ, δ its angles, then

$$\sigma/\Sigma \operatorname{cosec}\alpha = r = \Omega/\sigma$$

5. If R is the radius of a circle circumscribing a regular pentagon, each of whose sides is a, prove that

$$R : a = 17 : 20 \quad \textit{nearly}$$

6. Show that the area of any polygon ABCD &c, circumscribing a circle is

$$(\tfrac{1}{2}\text{ its perimeter})^2/\Sigma \cot\frac{A}{2}$$

7. Show that the area of a regular sixteen sided polygon, inscribed in a circle of radius unity, is

$$\sqrt{2 - \sqrt{2 + \sqrt{2}}}$$

8. If a, b, c, d are the successive sides of a convex quadrilateral; and if a, c, produced, cross at an angle θ, and b, d, produced, cross at an angle ϕ; prove that

$$a^2 + c^2 - 2\,ac\cos\theta = b^2 + d^2 - 2\,bd\cos\phi$$

9. Show that the area of a regular pentagon, of which a side is a, is

$$\frac{a^2}{2}(\sin 36^\circ + 3\cos 18^\circ)$$

10. A hexagon, which has five sides each of which is a, is inscribed in a circle of radius r; prove that the sixth side is

$$2r\sin\left(5\sin^{-1}\frac{a}{2r}\right) \quad \text{or} \quad \frac{a^5}{r^4} - \frac{5a^3}{r^2} + 5a$$

11. A regular polygon, each side being a, is inscribed in a circle of radius r; if x is a side of a regular polygon of double the number of sides, show that x is given by the equation

$$x^4 - 4r^2x^2 + r^2a^2 = 0$$

R. U. Schol: '82.

12. A circle has a regular polygon of n sides inscribed in it, and another regular polygon of n sides circumscribed about it; find n if

area in-polygon : area circum-polygon = 3 : 4

13. In a regular hexagon (each side of which is a) a circle is inscribed, then in that circle another regular hexagon is inscribed, then in that a circle, and so on until there are n hexagons: show that the sum of the areas of *all* the hexagons is

$$6\sqrt{3}\{1 - (3/4)^n\}a^2$$

14. Show that a diagonal of a regular pentagon, inscribed in a circle of radius R, is

$$\frac{R}{2}\sqrt{10 + 2\sqrt{5}}$$

15. If p, P are the perimeters of regular polygons of n sides, respectively inscribed in and circumscribed about the same circle; and if p′, P′ are similar quantities for polygons of 2 n sides; prove that

$$P' = \frac{2pP}{p + P}, \quad \text{and} \quad p' = p\sqrt{\frac{2P}{p + P}}$$

16. Prove by trigonometry the results at the end of page 10.

17. It is a known geometrical result (see *Euclid Revised*, page 289) that the sum of the rectangles under the pairs of opposite sides of a *non*-cyclic quadrilateral is greater than the rectangle under its diagonals: show that the excess is $4\,abcd\cos^2\frac{\theta + \phi}{2}$; where a, b, c, d are the sides, and θ, ϕ a pair of opposite angles.

18. If a, b are a pair of opposite sides of a quadrilateral; x a third side; α the angle between a and x; β the angle between b and x; and γ the angle between the diagonals; prove that x is given by the equation

$$x^2 - cx + ab \sin(\alpha + \beta + \gamma)/\sin\gamma = 0$$

where $c = \{a \sin(\alpha + \gamma) + b \sin(\beta + \gamma)\}/\sin\gamma$

Give a geometrical meaning to c.

19. From any point P, within a quadrilateral ABCD, perpendiculars are dropped on the sides; if $\alpha, \beta, \gamma, \delta$ are their feet, prove that

$$\Sigma(PA^2 \sin 2A) = 4 \text{ area } ABCD - 8 \text{ area } \alpha\beta\gamma\delta$$

Sid' Camb'. '48.

20. A square (area S) circumscribes a quadrilateral (area Ω) and the latter has diagonals x, y; prove that

$$(x^2 + y^2 - 4\Omega) S = x^2 y^2 - 4\Omega^2$$

21. If a, b, c are three sides of a quadrilateral; α the angle between a, b; β the angle between b, c; and γ the angle between a, c produced; prove that the fourth side is

$$(a^2 + b^2 + c^2 - 2ab\cos\alpha - 2bc\cos\beta - 2ca\cos\gamma)^{\frac{1}{2}}$$

Extend the theorem to any polygon.

T. C. D. '31.

22. Two points A, B are joined by two arcs ACB, ADB of circles, both of radius unity: any two points on them C, D, are joined by an arc CD, also of radius unity: prove that (if $AC = \alpha$, $CB = \alpha'$, $AD = \beta$, $DB = \beta'$, $CD = \gamma$) then

$$\cos\alpha + \cos\alpha' + \cos\beta + \cos\beta'$$
$$= 1 + \cos\gamma + \cos(\alpha + \alpha') + \cos(\alpha - \beta')$$

R. U. Schol': '89.

NOTE—*If* X *is mid p't of* AB, *and* Y *of* CD, *then this result is the trigonometrical equivalent of the well-known geometrical theorem* (E. R. p. 108)

$$AC^2 + CB^2 + BD^2 + DA^2 = AB^2 + CD^2 + 4XY^2.$$

23. In the same circle are inscribed a regular hexagon and a regular octagon: prove that

$$\text{area hexagon} : \text{area octagon} = \sqrt{27} : \sqrt{32}.$$

24. The sides of a regular pentagon and decagon, inscribed in a circle of radius R, are respectively a, a′; and the circles inscribed in them have radii r, r′: prove that

$$(1)\quad a^2 - a'^2 = R^2$$

$$(2)\quad a/r + a'/r' = 2R/r'$$

25. A cyclic quadrilateral circumscribes a circle; if R is its circum-radius, Ω its area, and a any side, prove that

$$16R^2 = \Omega^2 . \Sigma \frac{1}{a^2} + \Sigma a^2$$

26. If a, b, c, d are the successive sides of a cyclic quadrilateral; and if b, d produced cross at an angle θ; and a, c produced cross at an angle ϕ; prove that

$$\frac{\sin\theta}{\sin\phi} = \frac{a^2 \sim c^2}{b^2 \sim d^2}$$

27. A convex quadrilateral has two adjacent sides equal, and the other two equal (*Professor Sylvester's* **Kite**) and e is the ratio of two unequal sides; if θ is the acute angle between a pair of adjacent unequal sides, and ϕ the angle between a pair of opposite unequal sides; prove that

$$\tan\frac{\phi}{2} = \frac{1 - e}{1 + e}\tan\frac{\theta}{2}$$

Sylvester: *E. T.* xxiii.

28. If a, b, c, d are the successive sides of any quadrilateral; x, y its diagonals; and ϕ the sum of a pair of opposite angles; prove that

$$x^2 y^2 = a^2 c^2 + b^2 d^2 - 2abcd\cos\phi$$

R. U. Schol': '86.

29. A series of n circles, each of radius a, is ranged touching each other, so that their centres are concyclic: find the area enclosed within the *external* curvilinear polygon formed by their circumferences.

30. ABCDE is a regular pentagon: X, Y, P, Q are the respective mid points of AB, AE, BC, DE: I is the in-centre of the pentagon: if AI crosses XY, PQ in M, N, prove that

$$2(IM - IN) = \text{the in-radius.}$$

31. The diagonals of a quadrilateral are equal, and at right angles: if the sides are a, b, c, d, show that its area is

$$\tfrac{1}{4}\left\{a^2 + c^2 + \sqrt{4b^2d^2 - (a^2 - c^2)^2}\right\}$$

and that this is expressible in terms of any three of the sides.

What does the area become if the quadrilateral is cyclic?

32. If a, b, c, d are sides of a quadrilateral circumscribing a circle; α the angle between a, d; and x the length of the tangent to the circle from the vertex of α; prove that

$$(a + c) x^2 = ad (2x - a + b) \cos^2 \frac{\alpha}{2}$$

33. P is any point within a quadrilateral ABCD: PA, PB, PC, PD are respectively denoted by f_1, f_2, f_3, f_4: the respective perpendiculars from A, B, C, D, A, B on the bisectors of the angles subtended at P by AB, BC, CD, DA, BD, AC, are denoted by p_1, p_2, p_3, p_4, p_5, p_6, prove that

$$\frac{p_1 p_3}{f_1 f_3} + \frac{p_2 p_4}{f_2 f_4} = \frac{p_5 p_6}{f_1 f_2}$$

and that, when P is at the cross of the diagonals, the dexter member becomes unity.

Renshaw: *E. T.* xxv.

34. ABCDE is a cyclic pentagon: prove that

$$EA^2 . BC . CD . DB + EC^2 . AB . BD . DA$$
$$= EB^2 . AC . CD . DA + ED^2 . AB . BC . CA.$$

Brill: *E. T.* xliii.

NOTE—*Deduce from Salmon's Theorem* (*Conics*, 6th ed', p. 87) *that, if* A, B, C, D *are concyclic, and* E *any other point, then*

$$EA^2 . \triangle BCD + EC^2 . \triangle ABD = EB^2 . \triangle ACD + ED^2 . \triangle ABC$$

35. With the notation of Exercise 28, show that, if x is conterminous with a, d, then

$$\{x^2 (ab + cd) - (ac + bd)(bc + ad)\}^2 \sin^2 \frac{\phi}{2}$$
$$+ \{x^2 (ab - cd) - (ac - bd)(bc - ad)\}^2 \cos^2 \frac{\phi}{2}$$
$$= 4 a^2 b^2 c^2 d^2 \sin^2 \phi$$

Wolstenholme: 597.

36. Show how the construction of a regular heptagon may be made to depend on the trisection of the angle $\cos^{-1}\left(\frac{1}{2\sqrt{7}}\right)$.

Cayley: *E. T.* xl.

NOTE—*If* $x = 4\cos^2\theta$, *then* $\sin 7\theta$, *expressed in terms of* x, *will give* $x^3 - 5x^2 + 6x - 1 = 0$, *where* θ *is any* $\wedge$ *a side of the heptagon subtends at the in-centre.*

37. If c_1, c_2, c_3, c_4 are the cosines of the angles of a convex quadrilateral, prove that

$$\Sigma c_1^4 + 4\Sigma c_2^2 c_3^2 c_4^2 + 8\Pi c_1 = 2\Sigma c_1^2 c_2^2 + 4\Sigma c_1^2 . \Pi c_1$$

Jes', Christ's and Emm', Camb': '90.

38. A quadrilateral ABCD is inscribed in a circle centre S, and circumscribes a circle centre I: if SI produced meets the circum-circle in X, Y, prove that

$$\frac{1}{IA^2} + \frac{1}{IC^2} = \frac{1}{IX^2} + \frac{1}{IY^2}$$

Trin' Camb': '90.

NOTE—*Use result of Example* 4.

39. Prove that the distance between the mid points of the interior diagonals of a cyclic quadrilateral, whose sides are d, e, f, g, is

$$\sqrt{\{(d^2 - f^2)^2 eg + (e^2 - g^2) df\}/\{4(ef + dg)(de + fg)\}}$$

Sid' Camb': '91.

40. A cyclic pentagon ABCDE (sides a, b, c, d, e) circumscribes a circle; prove that

$$a\left(\frac{c - d}{b + e - a}\right) + b\left(\frac{d - e}{c + a - b}\right) + c\left(\frac{e - a}{d + b - c}\right) + d\left(\frac{a - b}{e + c - d}\right) + e\left(\frac{b - c}{a + d - e}\right) = 0$$

A. Russell: *E. T.* l.

NOTE—$\tan\frac{A}{2} = area/\sigma(\sigma - b - d)$ $\tan\frac{D}{2} = area/\sigma(\sigma - e - b)$

whence $\sin A - \sin D = \overline{d - e}/\overline{c + a - b} . \sin(A + D)$ *and sim'r results*: *then* $2R = -a/\sin(E + C) =$ &c.

41. A cyclic hexagon circumscribes a circle; if R is the circum-radius, r the in-radius, and d the distance between the centres; prove that

$$\sin\cos^{-1}\frac{r}{R - d} + \cos\sin^{-1}\frac{r}{R + d} = 1$$

Genese: *E. T.* xxvii.

NOTE—*Take a position of symmetry, as in Example* 4, *and use Exercise* 34, *on p.* 218, *as a* Lemma.

The elementary solution given in the E. T. Reprint is wrong.

CHAPTER XVI

Application of Trigonometry to Surveying

§ **77.** One of the most important applications of Trigonometry is to the practical work of Land-surveying. Before giving examples of such application, a few technicalities are here premised.

Def'—Any measured straight line is, in surveying, called a **base.**

Def'—If one point is visible from another, through a tube of small bore, then the **angle** round which that tube must be turned, in a vertical plane, to become horizontal, is called the **elevation**, or **depression**, of the first point with regard to the second—the former term being used when the first point is *above* the second, the latter when the first point is *below* the second.

Def'—The angle which the join of two points **subtends** at a third is the angle between the joins of the two points to the third.

Def'—The angle which any direction must be turned through to make it coincide with any Compass point is termed the **bearing** of that direction **from** that Compass point. Or, again, the word **bearing** is used to indicate simply the *direction* in which an object is seen, or in which it is inclined.

Note—Thus it might be said that—

(1) *the bearing of a tower from the North is* 15°; meaning that the direction in which the tower is seen is inclined to the N-point of the Compass at 15°:

(2) *the bearing of a ship is N. E.*; meaning that the direction in which the ship is seen is the N. E. point of the Compass.

(3) *a wall bears* 35° *East of South*; meaning that the line of the wall makes an angle of 35°, on the East side of the S. point of the Compass, with the direction of that point.

In the Problems of Surveying (often rather absurdly called *Heights and Distances*) we are supposed to be able to measure bases, that is distances between accessible points; and to observe (by instruments giving their magnitude) either the angular elevation of a point above, or its angular depression below, the horizontal plane through the eye; or, again, to measure the angle which the join of two visible points subtends at the eye—such measurement and observation being made by means of the practical instruments used by Surveyors.

These instruments are *Gunter's Chain*, *Newton's** *Sextant*, the *Theodolite*, the *Vernier*, the *Spirit Level*, and some others of less importance: descriptions of them will be found in any Treatise on Surveying. The Student will be here supposed to know the *Points* of the *Mariner's Compass*: as there are 32 such points, the angle between any two consecutive points is $360^\circ/32$, that is $11^\circ\frac{1}{4}$. These things assumed, the Problems of Land-Surveying are merely applications of the trigonometrical solution of triangles.

Examples

1. *To find the height* h *of an inaccessible point—say the summit of a mountain—above the horizontal plane in which the observer is situated.*

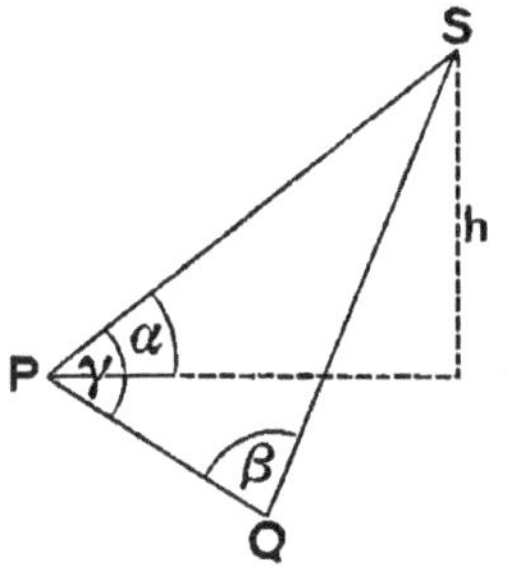

Let S be the summit, P the position of the observer.

Let the observer make the following measurements.

1°, a base (a say) between P and any well defined p't Q:

* Not *Hadley's*—as it is generally termed—see *Herschel's Astronomy*, 10th ed' p. 116.

2°, the elevation $\hat{\alpha}$ of S at P :

3°, „ $\hat{\beta}$ which SP subtends at Q :

4°, „ $\hat{\gamma}$ „ SQ „ „ P.

$$\text{Then} \quad h = SP \sin \alpha = \frac{a \sin \alpha \sin \beta}{\sin (\beta + \gamma)}$$

Note—If Q can be taken in the join of P to the foot of h, only two $\wedge$s need be observed.

2. *To find the distance apart of two inaccessible points—say the summits of two mountains.*

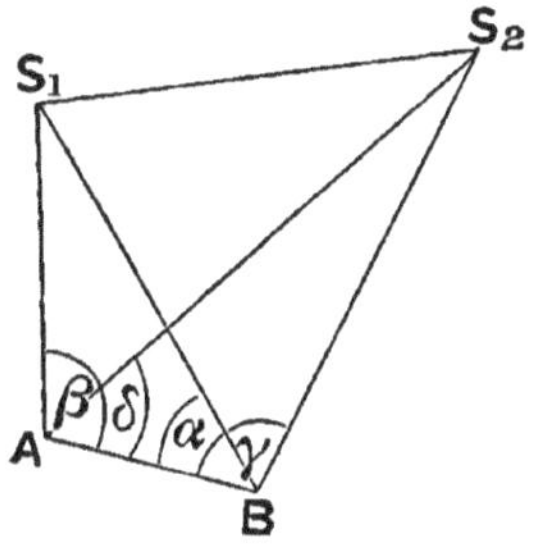

Let S_1, S_2 be the summits.
Let the observer measure—

1°, a base (a say) between two well-defined p'ts A, B ; not necessarily in the same plane with S_1, S_2 :

2°, the $\hat{\alpha}$ which S_1A subtends at B :

3°, „ $\hat{\beta}$ „ S_1B „ „ A :

4°, „ $\hat{\gamma}$ „ S_1A „ „ B :

5°, „ $\hat{\delta}$ „ S_2B „ „ A.

$$\text{Then} \quad S_1A = \frac{a \sin \alpha}{\sin (\alpha + \beta)}$$

$$S_2A = \frac{a \sin \gamma}{\sin (\gamma + \delta)}$$

whence, as the $\wedge$ which $S_1 S_2$ subtends at A can also be observed, the solution of the Δ S_1AS_2 is brought under the cond'ns of the case solved on p. 230.

3. A, B, C *are three Stations, whose distances from each other are known: to find how an Observer, at a fourth Station* P, *can calculate his distance from each of the three.**

(*Hipparchus' Problem*)

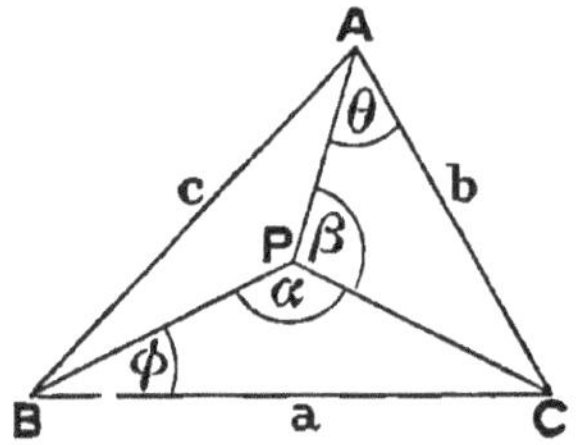

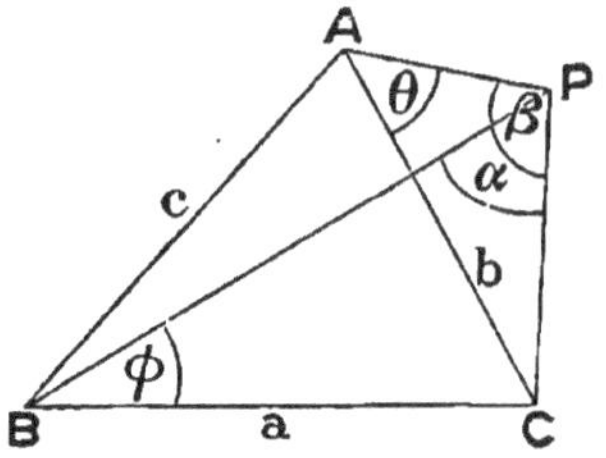

$$B\hat{P}C = \alpha, \text{ and } A\hat{P}C = \beta, \quad \text{can be observed.}$$

$$\text{Let} \quad C\hat{A}P = \theta, \text{ and } C\hat{B}P = \phi.$$

Then, denoting the sides and $\wedge^s$ of $\triangle$ ABC as usual, we have

$$\theta + \phi + \alpha + \beta + C = 2\pi$$

$$\therefore \quad \theta + \phi \text{ is known}$$

$$\text{Again,} \quad a\frac{\sin\phi}{\sin\alpha} = PC = b\frac{\sin\theta}{\sin\beta}$$

$$\therefore \quad \frac{\sin\theta}{\sin\phi} = \frac{a\sin\beta}{b\sin\alpha} = \tan\omega \text{ say}$$

$$\therefore \quad \frac{\tan\frac{\theta-\phi}{2}}{\tan\frac{\theta+\phi}{2}} = \tan\left(\omega - \frac{\pi}{4}\right)$$

$$\therefore \quad \theta - \phi \text{ is known}$$

$$\therefore \quad \theta \text{ and } \phi \text{ are known}$$

* Sometimes called by Surveyors the *Stasimetric Problem*—especially useful in Maritime Surveying: it was originally solved by the ancient astronomer *Hipparchus*: see *Philosophical Transactions*, vol' vi, p. 2093.

The Problem, and some cognate Problems, are very ingeniously treated in a volume by the late *Professor Wallace* of Edinburgh.

$$\therefore \quad PA = b\,\frac{\sin(\theta + \beta)}{\sin\beta}, \text{ is known}$$

$$PB = a\,\frac{\sin(\phi + \alpha)}{\sin\alpha}, \quad " \quad "$$

$$\text{and} \quad PC = a\,\frac{\sin\phi}{\sin\alpha}, \quad " \quad "$$

Note—When P is concyclic with A, B, C, we have $\theta = \phi$, or $\pi - \phi$, and the solution fails—in fact the Prob' is then indeterminate; and P may be any p't on the ⊙ ABC.

4. *To find the distance* δ *at which the summit of an object, of known height* h, *can be seen—the Earth being supposed a sphere of radius* R.

Let S be the summit: P the position of the observer on the Earth's surface when he can *just* see S, so that SP is a tang' to the Earth's surface; C the centre of the Earth; A the p't in which SC cuts that surface.

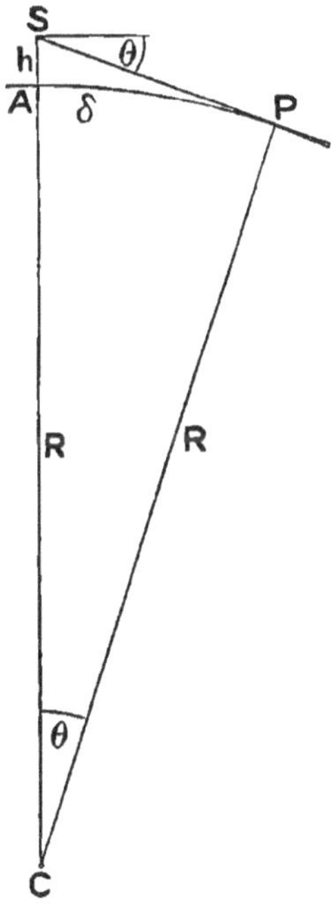

Then $SA = h$, arc $PA = \delta$

$\therefore$, if $PS = x$, and $PCS = {}^r\theta$,

we have

$$\left.\begin{aligned} x^2 &= h(h + 2R) \\ x &= (R + h)\sin\theta \\ \delta &= R\theta \end{aligned}\right]$$

$$\therefore \quad x = \sqrt{2Rh}\left(1 + \frac{h}{2R}\right)^{\frac{1}{2}}$$

$$= \sqrt{2Rh}\left(1 + \frac{h}{4R} - \frac{h^2}{32R^2}\ \&c\right)$$

Now, as R is nearly 4000 miles, and $h \not> 5$ miles at most, and is usually much less, the error by assuming that $x = \sqrt{2Rh}$, and $\sin\theta = \theta$ will not be great:

then $(R + h)\,\theta = \sqrt{2Rh}$

and $\therefore \quad \delta = R\sqrt{2Rh}/\overline{R + h}$

For small heights, $R/(R + h) = 1$, nearly

and then $\delta = \sqrt{2Rh}$, the formula generally used.

Cor'—If we assume that R = 3960 miles, which is very near the truth, then

$$\therefore \quad \delta^2 = 2\,h \times 3960$$

$$\therefore \quad 2\,\delta^2 = 3\,h \times (1760 \times 3)$$

whence we get the following fairly approximate rule—

Distance of horizon in miles = $\sqrt{\tfrac{3}{2} \times \textit{height of object in feet.}}$

For example: the top of an object 150 feet high can be seen about 15 miles off.

Def'—The angle which a tangent to the Earth's surface, from a point above that surface, must be turned through (in a vertical plane) to become horizontal, is called the **dip of the horizon** at that point: it is evidently equal to the angle which the tangent subtends at the Earth's centre.

5. *A right pyramid, on a square base, stands on a horizontal plane; and, when the sun has an elevation* α, *the distance of the extreme point of its shadow from the corners of the base nearest to it are* a, b, c: *show that the height of the pyramid is*

$$a \tan \alpha \sin \phi \operatorname{cosec} \theta$$

where θ *and* ϕ *are given by the equations*

$$\tan(45^\circ - \theta) = \frac{b^2 - a^2}{c^2 - a^2}$$

$$\tan^2\left(\phi + \frac{\theta}{2}\right) = \tan\frac{\theta}{2} \tan\left\{\frac{\theta}{2} + \tan^{-1}\frac{c^2 - b^2}{2\,a^2}\right\}$$

Pet' Camb': 50'

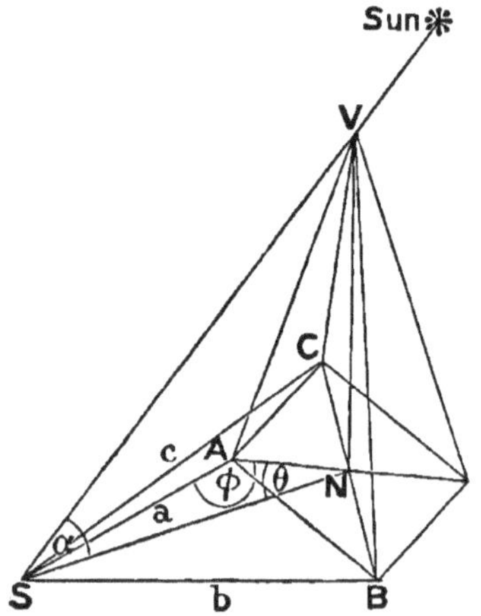

Let V be the vertex of the pyramid; S the extremity of its shadow.

Let A, B, C be the corners of the base whose respective dist's from S are a, b, c.

Then, if VN is the alt' of V, N is the cross of the diag's of the base.

Now if $V\hat{S}N = \alpha$, $S\hat{A}N = \phi$, $S\hat{N}A = \theta$,

then $VN = SN \tan \alpha = a \tan \alpha \sin \phi \operatorname{cosec} \theta$

Also $BC = 2\,AN = 2\,a \dfrac{\sin(\phi + \theta)}{\sin \theta}$

$$AB = a\sqrt{2}\,\frac{\sin(\phi + \theta)}{\sin \theta} = AC$$

$$b^2 = a^2 + AB^2 - 2\,a\,AB \cos(\phi - 45^\circ)$$

$$\therefore\quad b^2 - a^2 = 2\,a^2 \frac{\sin^2(\phi + \theta)}{\sin^2 \theta} - 2\,a^2 \frac{\sin(\phi + \theta)}{\sin \theta}(\cos \phi + \sin \phi)$$

$$\therefore\quad \frac{b^2 - a^2}{2\,a^2} = \frac{\sin(\phi + \theta)}{\sin \theta}\left(\frac{\sin \phi \cos \theta - \sin \phi \sin \theta}{\sin \theta}\right)$$

$$= \frac{\sin(\phi + \theta)\sin \phi}{\sin \theta}(\cot \theta - 1) \quad \dots \dots \dots \quad (1)$$

Again $c^2 = a^2 + AC^2 - 2\,a\,AC \cos(\phi + 45^\circ)$

$$\therefore\quad \frac{c^2 - a^2}{2\,a^2} = \frac{\sin(\phi + \theta)\sin \phi}{\sin \theta}(\cot \theta + 1) \quad \dots \dots \dots \quad (2)$$

(1) ÷ (2) gives $\dfrac{b^2 - a^2}{c^2 - a^2} = \dfrac{1 - \tan \theta}{1 + \tan \theta} = \tan(45^\circ - \theta)$

(2) − (1) gives $\dfrac{c^2 - b^2}{2\,a^2} = \dfrac{2 \sin(\phi + \theta)\sin \phi}{\sin \theta} = \tan \chi$, say

Now $\tan^2\left(\phi + \dfrac{\theta}{2}\right) - \tan^2 \dfrac{\theta}{2} \equiv \dfrac{\sin(\phi + \theta)\sin \phi}{\cos^2 \dfrac{\theta}{2} \cos^2\left(\phi + \dfrac{\theta}{2}\right)}$

$$= \tan \chi \tan \frac{\theta}{2} \sec^2\left(\phi + \frac{\theta}{2}\right)$$

$$\therefore\quad \tan^2\left(\phi + \frac{\theta}{2}\right)\left\{1 - \tan \chi \tan \frac{\theta}{2}\right\} = \tan \frac{\theta}{2}\left\{\tan \frac{\theta}{2} + \tan \chi\right\}$$

$$\therefore\quad \tan^2\left(\phi + \frac{\theta}{2}\right) = \tan \frac{\theta}{2} \tan\left(\frac{\theta}{2} + \chi\right)$$

$$= \tan \frac{\theta}{2} \tan\left\{\frac{\theta}{2} + \tan^{-1}\frac{c^2 - b^2}{2\,a^2}\right\}$$

Exercises

1. A surveyor has a Gunter's chain, and a logarithmic table-book, but no instrument for measuring angles: show how he can find the distance of an inaccessible object in his own horizontal plane.

2. At each of two points A, B, on the bank of a river, the join of two points P, Q, on the opposite bank (all four being in the same horizontal plane) subtends the same angle α; also the angles QAB, PBA are respectively β, γ; and QBA is greater than PBA: prove that

$$PQ/AB = \sin\alpha/\sin(\alpha+\beta+\gamma)$$

$$\text{and} \quad = \{\sin(\alpha+\beta)\sin(\alpha+\gamma) - \sin\beta\sin\gamma\}/\sin^2(\alpha+\beta+\gamma)$$

I. I. '89

3. The elevation of a column is observed; and, on approaching a feet nearer, is found to be doubled; also, on approaching b feet nearer, it is doubled again: find the distance of the middle point of observation from the column, in terms of a and b.

4. ABCD is the rectangular floor of a room; the altitude of the room at C subtends 18° at A, and 30° at B; also AB is 48 feet: show that the altitude is nearly 18 feet 10 inches.

5. A flagstaff, x feet high, stands on the top of a vertical tower, y feet high; if α is the greatest angle which the flagstaff subtends at a point on the horizontal plane through the base of the tower, show that

$$x : y = 2\sin\alpha : 1 - \sin\alpha$$

6. A wall 20 feet high bears 59° 5′ E. of S.; find the width of its shadow on a horizontal plane, when the Sun is due South, and at an elevation of 30°.

Assume that $L\sin 59°\,5' = 9{\cdot}9334445$ $\quad \log 29719 = 4{\cdot}4730342$

and that the logs of 2 and 3 are known.

Science and Art Exam': '86

7. A Captain of a ship, sailing due North, observes two lighthouses bearing N. E. and N. N. E. respectively: after sailing 20 miles he saw them in a line due East: find the distance between the lighthouses in miles, accurately to four decimal places.

Given $\log 8{\cdot}2843 = {\cdot}9182558$ $\quad \log 8{\cdot}2842 = {\cdot}9182506$

$L\tan 22°\,30' = 9{\cdot}6172243$ and assuming log 2.

Woolwich: '85

8. The elevation of a tower, at a place due South of it, is 45°; and at another place, due West of the former, and distance a from it, the elevation is 15°; show that the height of the tower is

$$a\,(3^{\frac{1}{4}}-3^{-\frac{1}{4}})/2$$

9. The elevation of the summit of a mountain is observed at three collinear points A, B, C (in any horizontal plane) to be α, β, γ respectively: prove that the square of the height of the mountain above the plane is

$$-\,AB\,.\,BC\,.\,CA/(BC\cot^2\alpha + CA\cot^2\beta + AB\cot^2\gamma)$$

R. U. Schol': '91

NOTE—*The sign of the lines is taken into account in this result, so that* BA = –AB. *Use Euler's theorem, given on p.* 109 *of E. R.*

10. A statue of height h, on a pillar of unknown height, subtends an angle α to an observer standing on the horizontal plane through the base of the pillar; the observer comes k nearer to the pillar, and finds that the statue subtends an angle β: find his distance from the pillar, in terms of h, k, α, β.

11. The dome of the capitol of Washington is 396 feet 4 inches high, and the statue of Freedom on it is 18 feet high: show that the best position for an observer to view the statue (his eyes being supposed 5 feet 4 inches from the ground) is about 400 feet from the Capitol.

12. ABCD is a horizontal line; from a point, vertically over D, the distances AB, BC are observed to subtend the same angle α: if AB = a, and BC = b, show that the height of the observer above D is

$$\{2ab\,(a+b)\tan\alpha\}/\{(a-b)^2 + (a+b)^2\tan^2\alpha\}$$

13. A man standing due South of a tower, of height h, observes its elevation to be α; he walks due West to a point A, where the elevation is β; he goes on, in the same direction, to B, where the elevation is γ: find the length of AB, in terms of h, α, β, γ.

14. The angles subtended by the base of a round tower at three collinear stations are observed, and the distances between the stations measured; hence find the diameter of the tower.

Ox' 1st Public Exam': '73

15. The elevation of a balloon is simultaneously observed from three collinear points A, O, B: if OA = a, OB = b; and the elevations at A, O, B are respectively $\cot^{-1}\alpha$, $\cot^{-1}\beta$, $\cot^{-1}\alpha$; show that the height of the balloon is $\sqrt{ab/(\alpha^2-\beta^2)}$.

Ox' 1st Public Exam': '74

16. A rock is observed from the deck of a ship to bear N. N. W.; and, after the ship has sailed 10 miles E. N. E., the rock bears due West; find its distance from the ship at each observation.

Christ's Camb': '46

17. While sailing S. W., a man observed two ships at anchor, one N. N. W., the other W. N.W., and, after sailing 5 miles, he saw the ships N. by W., and N. W. respectively; find their bearing, and distance from each other.

Christ's Camb': '45

18. AB (length a) and CD (length c) are two vertical posts at the edge of a river: P is the position of an observer on the edge exactly opposite A: if AB = AC, and the elevations of B and D at P are equal, find the breadth of the river; and prove that

$$\cos APD = a^2/c^2 = \cos CPB$$

19. A balloon being supposed to ascend from the Earth with a given uniform velocity, at a given inclination to the horizon, and towards a given point of the compass, two angles of elevation are observed from the same place, at a given interval of time: find an equation to give the height of the balloon at the time of either observation, its angular bearing at that time being given.

Joh' Camb': '33

20. The summit S of a mountain is just seen from a boat in a position A at sea; the boat is rowed directly towards S, and when at B the elevation of S is α: if the Earth is supposed a sphere centre O, and if SO cuts its surface in C, show that if arc AC = a, and arc BC = b, then approximately,

$$\text{Earth's radius} = \frac{a^2 - b^2}{2b} \cot \alpha$$

$$\text{and height of mountain} = \frac{a^2 b}{a^2 - b^2} \tan \alpha$$

Joh' Camb': '33

21. From the summit of two rocks A, B, at sea, at a given distance δ apart, the dips α, β, of the horizon are observed; and it is remarked that the summit of B is in a horizontal line through the summit of A: show that the radius of the Earth is

$$\delta/\cos^{-1}(\sec \alpha \cos \beta)$$

Pet' Camb': '49

22. A square tower stands on a horizontal plane: from a point of the plane three of its corners have elevations 45°, 60°, 45°, respectively: find the ratio of the height to the breadth of the tower.

Ox' 1st Public Exam': 89

23. A flagstaff, on a tower, is observed from two points in the same horizontal plane, one due South, and the other due East of it, and subtends the same angle at each point: the elevations of the top of the staff at the same places are respectively $\tan^{-1}\alpha$, $\tan^{-1}\beta$: if c is the distance between the points of observation, prove that the height of the flagstaff is

$$c\,(\alpha\beta - 1)\big/\sqrt{\alpha^2 + \beta^2}$$

Ox' 1st Public Exam': '78

24. Two vertical walls are h_1 and h_2 feet high, and are mutually perpendicular: when the Sun is due South their shadows are observed to be a_1 and a_2 feet broad, respectively: if then θ is the Sun's elevation, and ϕ the inclination of the first wall to the plane of the meridian, prove that

$$\cot\theta = \sqrt{\left(\frac{a_1}{h_1}\right)^2 + \left(\frac{a_2}{h_2}\right)^2}$$

$$\cot\phi = \frac{h_1 a_2}{h_2 a_1}$$

Joh' Camb': '31

25. A wheel (circumference a) makes n revolutions in going over an arc of a circular railway: if the chord of that arc is c, show that the radius of the railway is approximately $\frac{1}{2}\sqrt{(na)^3/6\,(na - c)}$

26. If there is a level stratum of cloud at the height of a mile above the Earth's surface; and if the elevation of an edge of the cloud is observed to be α; prove that the distance of that edge from the observer is

$$\frac{1}{\sin\alpha}\left(1 - \frac{\cot^2\alpha}{2R}\right) \text{ nearly}$$

where R is the radius of the Earth.

Math' Tri': '46. *R. U. Schol'*: '89

27. A cube stands on a horizontal plane; and the elevations of three of its corners, as observed from a point in its plane, are α, β, γ; prove that

$$\sin^2 2\beta = (1 - \sin^2\beta\cot^2\alpha)^2 + (1 - \sin^2\beta\cot^2\gamma)^2$$

Caius Camb': '78

28. A, B, C are three points in a horizontal plane: X, Y, Z are the summits of three mountains, whose respective heights are x, y, z: if AZY, BZX, CYX are respectively straight lines, prove that

$$x(y - z)/BC = y(z - x)/CA = z(x - y)/AB$$

Math' Tri': '70. *E. T.* xlvi

29. A man, standing on a plane, observes a row of equidistant pillars, the tenth and seventeenth of which subtend the same angles as they would if they stood in the position of the first, and were respectively a half and a third of their heights: show that (neglecting the height of the man's eye) the line of the pillars is inclined to the line drawn to the first at an angle $\cos^{-1}\frac{5}{13}$ nearly.

Math' Tri': '68

30. Two ships are sailing in parallel directions: the Captain of one of them makes three observations of the other, at intervals of an hour, and finds the bearings α, β, γ, from the North: find the bearing from the North of the direction of sailing.

Trin' Camb': '49

31. A man on a hill observes that three towers on a horizontal plane subtend equal angles at his eye, and that the angles of depression of their bases are α, α' α'': prove that, if c, c', c'' are the respective heights of the towers, then

$$\frac{\sin(\alpha' - \alpha'')}{c \sin \alpha} + \frac{\sin(\alpha'' - \alpha)}{c' \sin \alpha'} + \frac{\sin(\alpha - \alpha')}{c'' \sin \alpha''} = 0$$

Math' Tri': '64

32. Solve *Hipparchus' Problem* (p. 290) in the particular case when the angles which AB, BC, CA subtend at P, are equal.

33. The area of a quadrilateral field is determined by chaining the four sides, and measuring one angle; prove that, on the hypothesis that an error in a side is proportional to the length of that side, but that an error in an angle is the same for all angles, the angle A is preferable to the angle C, if the area of BAD is less than the area of BCD.

Math' Tri': '87

34. On the side of a hill are two places B, C, inaccessible to each other, but known to be at the same distance a from a station A: at the lower place C the angle α between the projections of CA, CB on the horizontal plane through C is measured, and the elevations λ, μ of A, B: prove that

$$BC = 2a\left\{\cos(\lambda - \mu)\cos^2\frac{\alpha}{2} - \cos(\lambda + \mu)\sin^2\frac{\alpha}{2}\right\}$$

Joh' Camb': '43

35. In the last Exercise, if, without the limitation that AB, AC are equal, the angle ACB is ω, and θ is the inclination of the hill (considered as a plane) to the horizon, prove that

$$\sin^2\theta \sin^2\omega = \sin^2\lambda + \sin^2\mu - 2\sin\lambda \sin\mu \sin\omega.$$

36. It is observed that the elevation of the summit of a mountain at each corner of a plane horizontal triangle ABC is α: show that the height of the mountain is

$$\frac{\alpha}{2} \tan\alpha \operatorname{cosec} A$$

If there is a small error n'' in the elevation at C, prove that the true height is

$$\frac{\alpha}{2}\frac{\tan\alpha}{\sin A}\left\{1 + \frac{\cos C}{\sin A \sin B}\frac{\sin n''}{\sin 2\alpha}\right\}$$

Math' Tri': '82

37. A tower AB stands on a horizontal plane and supports a spire BC: an observer at a place P on a mountain whose side may be treated as an inclined plane observes that AB, BC each subtend an angle α at his eye: he then moves to a place Q, measuring the distance PQ (2 a) and observes that AB, BC again subtend angles α at his eye: he then measures the angle AQP (β) and CQP (γ): show that, if x and y are the heights of AB, BC respectively,

$$x\cos\beta = y\cos\gamma = a\sqrt{\left(1 - \frac{\cos\beta\cos\gamma\cos^2\alpha}{\cos^2\frac{\beta+\gamma}{2}\cos^2\frac{\beta-\gamma}{2}}\right)}$$

Also, if M is the middle point of PQ, and D is the point on the line of greatest slope through M at which AB, BC each subtend an angle δ, and MD is measured (b), prove that the inclination θ of the mountain to the horizon is given by

$$\sqrt{\left\{\frac{x^2y^2}{(x-y)^2} - \left(\frac{a^2+b^2}{2b}\right)^2\right\}}\,.\sin\theta + \frac{a^2+b^2}{2b}\cos\theta$$
$$= \frac{xy(x+y)\sin 2\delta}{x^2+y^2-2xy\cos 2\delta}$$

Math' Tri': '84

NOTE—*The 1st part comes easily by noticing that* PQ *is horizontal, and* ⊥ *to plane* AMC. *For the 2nd, by an obvious extension of Apollonius' Locus (E. R. p. 294) to space, the Locus of* all *p'ts at which* AB, BC *subtend equal* ∧s *is a sphere, whose centre* S *is in* ABC *produced, and whose rad'* SB *is given by* $SB^2 = SA \,.\, SC = (SB + x)(SB - y)$; *this sph' cuts the pl' of the mountain in the* ⊙ PDQ. *Now take the diff' of the* ⊥s *from* D, *and the centre of* ⊙ PDQ, *on* ABC.

CHAPTER XVII

Maxima and Minima—Inequalities

§ **78.** The solutions of *maxima* and *minima* Problems by Trigonometry depend generally on artifices. To attempt to enumerate these or give rules for their discovery, would be in vain. But the subject may be illustrated by giving two which are frequently useful: these are—

1°, the algebraic arrangement of quantities as a sum of squares;

2°, Fermat's principle that—*a geometrical magnitude, varying according to an assigned law, will, in a maximum or minimum position, have consecutive values equal.*

Thus (as illustrations of 1°) from the results of Example 1, in § 76, we see that, if the sides of a quad′ are given, then its area is *max′* or *min′* if it is cyclic, according as it is convex or crossed. See also Example 1 following.

Again (as illustrations of 2°) a purely geometrical application of Fermat's principle is given in *Euclid Revised*, p. 426, to prove the property of *Philo's Minimum Line*; and it is easy to transform that proof into a trigonometrical form, by considering that,

$$\text{if } {}^r\theta = A\hat{P}\alpha, \text{ in the fig′ there given,}$$

$$\left.\begin{aligned} &\text{then } AX = \alpha X \cot A = P\alpha\,.\,\theta \cot A = PA\,.\,\theta \cot A \\ &\text{and } \beta Y = BY \cot \beta = PB\,.\,\theta \cot \beta = PB\,.\,\theta \cot B \end{aligned}\right\} \text{ ultimately}$$

$$\text{whence } PA\,.\,AN = PB\,.\,BN$$

$$\therefore\ PA\,.\,AB - PA\,.\,BN = BN\,.\,AB - BN\,.\,PA$$

$$\text{and } \therefore\ PA = BN$$

See also Example 2 following.

The treatment of inequalities is very similar to that of *maxima* and *minima*. The Student should keep in mind what is given in *Algebra under this head: especially he should recollect that

$$a_1 + a_2 + a_3 + \&c + a_n > n \sqrt{a_1 a_2 a_3 \&c\, a_n}$$

and that the Limits of value of the fraction

$$\overline{ax^2 + 2bx + c} \Big/ \overline{lx^2 + 2mx + n}$$

are given by the eq'n in y

$$(b - my)^2 = (a - ly)(c - ny)$$

Some examples of trigonometrical inequalities have already been incidentally given in §§ 44–49.

Examples

1. *If the distances of a point* P, *within a triangle* ABC, *from* A, B, C *are* x, y, z; *and the angles* BPC, CPA, APB *are* θ, ϕ, χ—*respectively in each case—find the* maximum *value of*

$$x \sin\theta + y \sin\phi + z \sin\chi$$

and the position of P *when this is the case.*

We have $c^2 = x^2 + y^2 - 2xy\cos\chi$

$$= x^2 + y^2 - 2xy\cos(\theta + \phi)$$

$$= (x\sin\theta + y\sin\phi)^2 + (x\cos\theta - y\cos\phi)^2$$

$\therefore$ $x\sin\theta + y\sin\phi$ has its *max'* value c, when $x\cos\theta = y\cos\phi$

Sim'ly $y\sin\phi + z\sin\chi$,, ,, ,, a, ,, $y\cos\phi = z\cos\chi$

and $z\sin\chi + x\sin\theta$,, ,, ,, b, ,, $z\cos\chi = x\cos\theta$

$\therefore$ the *max'* value of $\Sigma(x\sin\theta)$ is the semi-perim' of $\triangle ABC$; and the position of P for which this is the case is when

$$x\cos\theta = y\cos\phi = z\cos\chi$$

i. e. when P is the in-centre of ABC. (See Ex' 18, p. 260)

---

* See *C. Smith's Algebra*, Chapter XXVI; or *Chrystal's Algebra*, Chapter XXIV: in the latter the connection between *inequalities* and *maxima* and *minima* is brought out.

2. *Given the angles and perimeter of a convex quadrilateral; to find when its area is* maximum.

Catalan: E. T. l. *(Prof′ Genese's Solution)*

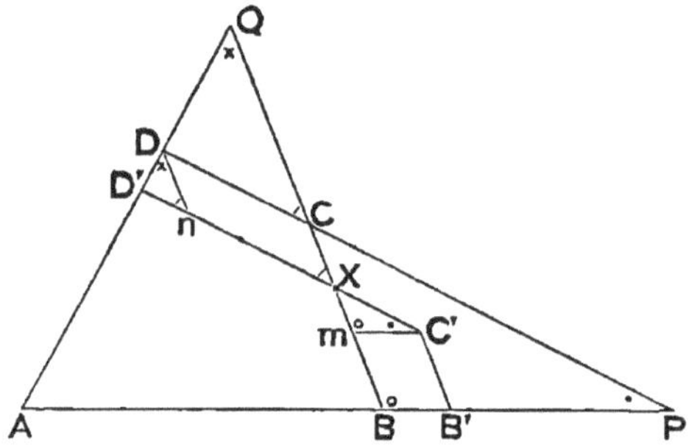

Let ABCD be the *max′* position of a quad′ whose ∧s and perim′ are given; and let AB′C′D′ be a consecutive position.

Draw C′m, Dn ∥ to AB, BC respectively. The rest of the fig′ is self-explanatory.

By Fermat's principle, area ABCD = area AB′C′D′

$$\therefore \text{ area } CDD'X = \text{area } BB'C'X'$$

$$\therefore \quad (CD + XD')\, CX \sin X = (C'B' + XB)\, C'X \sin X$$

Now ultimately $CD = XD'$, and $CB = C'B' = XB$

$$\therefore \quad XC/XC' = CB/CD, \text{ ultimately}$$

Also, since the perim′ is the same in *all* positions,

$$BB' + C'X + D'n = CX + Xm + DD'$$

$$XC' \frac{\sin C}{\sin B} + XC' + XC \frac{\sin Q}{\sin D} = XC + XC' \frac{\sin P}{\sin B} + XC \frac{\sin C}{\sin D}$$

$$\therefore \quad \frac{\sin B + \sin C - \sin P}{\sin C + \sin D - \sin Q} = \frac{XC \sin B}{XC' \sin D} = \frac{CB \sin B}{CD \sin D}$$

$$\therefore \quad \frac{CB \,.\, CP \sin C}{PC + PB - BC} = \frac{CD \,.\, CQ \sin C}{QD + QC - CD}$$

∴ ⊙ touching BC and PB, BC produced

= " " CD " QC, QD "

∴ these ⊙s are the same

i. e. quad′ ABCD circumscribes a ⊙

∴ its area is *max′* when $AB + CD = BC + DA$

3. *If the sum of any number of varying angles is given, then the product (and the sum) of the* sines *of those angles will be* **maximum**, *when the angles are all equal.*

For if $x + y = \alpha$, a fixed $\wedge$

$$2 \sin x \sin y = \cos (x - y) - \cos \alpha$$

and $\therefore$ is *max′* when $x = y$

Now let there be n $\wedge^s$, $\theta_1, \theta_2, \theta_3$, &c θ_n, whose sum is fixed: then, by the foregoing, if any two of them (say θ_1, θ_2) are unequal, we can, without altering the rest or the total sum, increase the product $\sin \theta_1 \sin \theta_2$, by making θ_1, θ_2 equal.

$\therefore$ $\Pi \sin \theta$ is *max′* when $\theta_1 = \theta_2 = \&c = \theta_n$

Sim′ly $\Sigma \sin \theta$,, ,, ,, ,, ,,

Note—Obviously $\Pi \cos \theta$ and $\Sigma \cos \theta$ can be treated in exactly the same way.

4. *If* A, B, C *are angles of a triangle, the* **minimum** *value of* $\Sigma \cot^2 A$ *is unity.*

This appears at once by considering that

$$\Sigma \cot^2 A \equiv 1 + \tfrac{1}{2} \Sigma (\cot A - \cot B)^2$$

and $\therefore$ is *min′* when $\cot A = \cot B = \cot C$

Cor′—The *min′* value of $\Sigma \tan^2 \frac{A}{2}$ is unity

5. *If* A, B, C *are angles of a triangle; and* x, y, z *any variables; then*

$$\Sigma x^2 \nless 2 \Sigma (yz \cos A)$$

For $\Sigma x^2 - 2 \Sigma (yz \cos A)$

$$\equiv (x - y \cos C - z \cos B)^2 + (y \sin C - z \sin B)^2$$

which is pos′, unless $\left. \begin{array}{l} y \sin C = z \sin B \\ \text{and } x = y \cos C + z \cos B \end{array} \right\}$

i. e. unless $x/a = y/b = z/c$

in which case it vanishes

$\therefore$ either $\Sigma x^2 > 2 \Sigma (yz \cos A)$

or $\Sigma x^2 = 2 \Sigma (yz \cos A)$

Exercises

1. Given the sum of any number of angles, show that the sum (and product) of their tangents is *minimum* when the angles are all equal.

2. If $a \tan\theta + b \tan\phi = c$, where a, b, c are fixed positive quantities; and θ, ϕ are each less than a right angle; prove that

$$a \sec\theta + b \sec\phi \text{ is } minimum \text{ when } \theta = \phi.$$

Hence show that, if $x + y = c$, the *minimum* value of

$$\sqrt{x^2 + a^2} + \sqrt{y^2 + b^2} \text{ is } \sqrt{(a+b)^2 + c^2}$$

3. If A, B, C are angles of a triangle, find the *maximum* value of—

(1) $\Sigma \sin A$ (2) $\Sigma \cos A$ (3) $\Pi \sin\frac{A}{2}$ (4) $\Pi \cos\frac{A}{2}$

4. If $\sin^2\theta + \sin^2\phi + \sin^2\chi = 1$; and θ, ϕ, χ are positive; prove that

$$\Sigma\theta > \pi/2.$$

5. Find what value of x gives the *maximum* value of—

(1) $\sin^2 x \sin(\alpha - 2x)$

(2) $\sin^m \dfrac{x}{m} \sin^n \dfrac{\alpha - x}{n}$

(3) $\dfrac{\sin x \cos x}{1 + \sin x + \cos x}$

6. In any triangle, except the equilateral, prove that

$$\Sigma(a \cos A) < s$$

7. Show that, if $A + B + C = 180°$, then

$$\Sigma \sin A < \tfrac{27}{4}, \text{ but } > 6\sqrt{3}\,\Pi \sin A$$

8. Show that $\Sigma \sin A \sim \Sigma \cos A >$, or < 1, according as the triangle is acute or obtuse angled.

9. If ρ is the radius of the N. P. circle, and r, s as usual, show that

$$s^2 \text{ lies between } 27\rho^2 \text{ and } 27r^2$$

10. Prove that the sum of the cosines of the halves of any two angles of a triangle is greater than the cosine of half the third angle.

11. If the equation $\tan\frac{\theta}{2} = (\alpha - 1 + \tan\theta)/(\alpha + 1 + \tan\theta)$ has real roots, show that $\alpha^2 > 1$.

12. Prove that, for real values of x, the Limits of

$$\sqrt{x^2 - 2x\cos\alpha + 1}\Big/\sqrt{x^2 - 2x\cos\beta + 1}$$

are $\sqrt{1 - \cos\alpha}\Big/\sqrt{1 - \cos\beta}$ and $\sqrt{1 + \cos\alpha}\Big/\sqrt{1 + \cos\beta}$

13. Prove that, in an acute-angled scalene triangle—

(1) $\Sigma \tan A$ (or $\Pi \tan A$) $> 3\sqrt{3}$

(2) $\Sigma \cot A > \sqrt{3}$

(3) $\Sigma (\tan A \tan B) > 9$

(4) $\Pi \sec A > 8$

14. Show that the *maximum* value of the Brocard angle is 30°.

15. A circle, radius r, is inscribed in a sector, radius x, and chord y: if x + y remains constant, find when r is *maximum* and when *minimum*.

16. If a, b, c are the sides of a triangle in descending order of magnitude; prove that, of the six squares which can be described so as to have their corners on the sides (or sides produced) of the triangle,

$$\frac{\text{area greatest square}}{\text{,, least ,,}} = \left(\frac{c + b\sin A}{c + b\sin C}\right)^2$$

17. If D is the diameter of the circum-circle of a triangle ABC; and $\delta_1, \delta_2, \delta_3$ the diameters of the greatest circles touching a side externally, and the circum-circle; prove that

$$\sqrt{D\,\delta_1\,\delta_2\,\delta_3} = (abc)/(4s)$$

18. AB, BE, CF are the bisectors of the angles of a triangle ABC, terminated by the opposite sides: prove that—

(1) $\dfrac{\Delta\,DEF}{\Delta\,ABC} \ngtr \frac{1}{4}$

(2) $AD\,.\,BE\,.\,CF > \dfrac{64\,\Delta^3}{(a + b)(b + c)(c + a)}$

19. CP, CQ are drawn to meet AB in P, Q, and to be equally inclined to the bisector of the angle BCA; find when the circum-circle of PCQ is *minimum*.

20. The diagonals of a quadrilateral are a, b, and they form an angle α; prove that the area of the *maximum* rectangle circumscribing the quadrilateral is $\frac{1}{2}ab(1 + \sin\alpha)$

21. AB is the diameter of a fixed semi-circle, centre C; P is a fixed point on the radius perpendicular to AB; XY is a varying chord through P: prove that the area AXYB is *maximum* when the angles XCY, PAC are equal.

Q. C. B. '73

22. If the altitude AN of a triangle is fixed, and its in-circle goes through its ortho-centre; prove that its *maximum* in-radius is $\frac{1}{2}$ AN.

Knowles: E. T. xli

NOTE—*Prove* $IO^2 = 2r^2 - AO \cdot ON$, *and deduce.*

23. A, B are fixed points, and P a point moving so that PA + PB remains constant: if 2θ denote the angle APB, and if C is the point of AB dividing it into parts m, n; prove that when CP is *minimum* its value is

$$2mn\cos\theta \Big/ \sqrt{(m+n)^2\sin^2\theta + (m-n)^2\cos^2\theta}$$

24. If α, β, γ are the lengths of the *minimum* lines bisecting a triangle, prove that

$$\sqrt{\Sigma(\alpha^2\beta^2)} = 2\Delta$$

Murphy: E. T. xxv

25. XY is a chord of a fixed semi-circle; XP, YQ are perpendiculars on its diameter: if XY varies in length, but is always parallel to a fixed direction, prove that the area PXYQ is *maximum* when XY subtends a right angle at the centre.

26. If $x = \overline{b + c\sin\theta} \big/ \overline{a + c\cos\theta}$, where a, b, c are all positive, and c least, show that there is one value of θ for which x is *maximum*, and one value for which it is *minimum*; and that both are given by

$$a\cos\theta + b\sin\theta = -c.$$

Q. C. B. '89

27. ABCDEF is a hexagon in which each of the sides AB, DE is 2c; and each of the sides BC, CD, EF, FA is 2a; prove that when the hexagon has its *maximum* area this area is

$$2(\rho + c)^{\frac{3}{2}}(\rho - c)^{\frac{1}{2}}$$

where ρ is the positive root of the equation

$$\rho^2 - c\rho = 2a^2$$

Math' Tri': '87

CHAPTER XVIII

Expansions—Series—Factors

§ **79**. A digression is here introduced, which would have had its natural place in the Preliminary Chapter, had the necessity for it been foreseen in time.

Algebraic* Note

On the Convergence of Infinite Expressions

Def′—An expression, the number of whose terms is infinite, is said to be **convergent**, if its value, starting from any given term, when only a finite number n of its terms is included, *converges*, that is tends continually to a definite *limit* (not zero) as n increases indefinitely: otherwise such an expression is said to be **non-convergent**.

In the term *non-convergent* are included the cases when, as more and more of its terms are taken, the value of an infinite expression—

(1) *diverges*, that is increases without limit:

(2) *oscillates*, that is has sometimes one value, and sometimes another:

(3) *tends to zero as its limit.*

In such expressions the infinite number of terms may be arranged as—

(A) a *sum*, called a *series*; or (B) a *product*: i.e. they may take either of the forms

$$u_1 + u_2 + u_3 + \ldots.$$

$$\text{or} \quad (1 + u_1)(1 + u_2)(1 + u_3)\ldots.$$

where u_1, u_2, u_3, &c may be pos′ or neg′.

---

* This note is intended to bring out some points which cannot be conveniently disposed of by a reference to text-books on Algebra. The only English text-book on that subject, wherein the question of Convergence will be found thoroughly discussed, from the point of view here put forward, is *Chrystal's Algebra*; but the discussion of it given there (in Chapter XXVI) is rather elaborate and difficult for the ordinary Student at this stage of his career: hereafter he should not fail to study it carefully.

We proceed to consider these forms separately.

(A) **Series.**

Def'—If there is a *limit* to which a series tends continually, and from which it may be made to differ by less than any assignable quantity, by taking a sufficient number of its terms, this *limit* is called its **sum to infinity**; and the quantity by which it differs from this limit, when only a finite number r of its terms is taken, is called its **residue after r terms.**

Clearly a series is convergent if its residue may be ultimately neglected.

For example, the geometrical series

$$1 + x + x^2 + x^3 + \ldots + x^{r-1} + \ldots, \text{ where } x < 1,$$

has for the sum of its 1st r terms

$$\frac{1 - x^r}{1 - x}, \text{ which} \equiv \frac{1}{1 - x} - \frac{x^r}{1 - x}$$

and ∴ differs from $1/(1 - x)$ by the residue $x^r/(1 - x)$.

But, as r increases indefinitely, this residue diminishes indefinitely, and ∴ may ultimately be neglected.

∴, as r increases, the sum tends more and more to become equal to $1/(1 - x)$, and finally differs from it by less than any assignable magnitude.

∴ the series is *convergent*, and its *sum to infinity* is $1/(1 - x)$

It will ∴ be a sufficient test of the convergence of a series if we can show that, term for term, it < a G. P. whose common ratio < 1.

Again the series $1 + \frac{1}{2} + \frac{1}{3} + \frac{1}{4} + \ldots$ may be arranged in groups thus—

$$\left(1 + \tfrac{1}{2}\right) + \left(\tfrac{1}{3} + \tfrac{1}{4}\right) + \left(\tfrac{1}{5} + \tfrac{1}{6} + \tfrac{1}{7} + \tfrac{1}{8}\right) + \left(\tfrac{1}{9} \ldots + \tfrac{1}{16}\right) + \ldots$$

Now each group $> \frac{1}{2}$

∴ the series $> \frac{1}{2}$ taken as often as we please

i. e. > a quantity which can be made as great as we please

and ∴ it is *non-convergent*.

Taking a series of the form

$$u_1 + u_2 + u_3 + \ldots + u_r + \ldots$$

where *all* the terms are pos', we see that obviously—

1°, the terms must constantly diminish if it is to converge; for otherwise its sum is constantly increasing:

2°, if it is conv't when all its terms are pos', *à fortiori* it is conv't when some are made neg'; for this change will diminish the numerical value of the residue.

By comparing the series Σu_n with the G. P. Σr^n, it is an immediate deduction* that—

if the *Limit* of $u_{r+1}/u_r < 1$, when $r = \infty$,

then Σu_n is convergent.

Hence the fraction u_{r+1}/u_r is suitably called the *test-ratio of convergence.*

In the preceding series the terms are all pos′; but if the type of a series is

$$u_1 - u_2 + u_3 - \ldots \pm u_r \ldots$$

where the terms are alternately pos′ and neg′, the residue after r terms is

$$\pm (u_{r+1} - u_{r+2} + \ldots)$$

and ∴ lies between $\pm u_{r+1}$ and $\pm (u_{r+1} - u_{r+2})$

provided the numerical value of each term < that of the preceding term:

∴ if, after a certain term, each term < the preceding term, the series converges to a definite limit, *provided* that u_r diminishes indefinitely as r increases.

The above arrangement shows the closeness with which the limit is approached when r terms are taken.

For example, r terms of the series

$$1 - \tfrac{1}{2} + \tfrac{1}{3} - \tfrac{1}{4} + \ldots \pm \frac{1}{r} \ldots$$

differs from $\log_e 2$ by less than $1/(r+2)$, and ∴ it is convergent.

But this is only true when the terms are arranged in order as above; for if the series is arranged in form

$$(1 + \tfrac{1}{3} + \tfrac{1}{5} + \tfrac{1}{7} + \ldots) - (\tfrac{1}{2} + \tfrac{1}{4} + \tfrac{1}{6} + \ldots)$$

it represents the difference of two non-conv′t series, and ∴ is itself non-conv′t.

Hence there are two types of convergent series—

(α) those which are conv′t when all the terms are made pos′:

(β) those whose convergence depends on the *accidental arrangement* of the pos′ and neg′ terms.

The former are said to be **absolutely convergent**: the latter (sometimes called **semi-convergent**) may appropriately be said to be **tactically convergent.**

---

* See *C. Smith's Algebra*, p. 317; or *Chrystal's Algebra*, vol′ II, p. 107.

(B) **Products.**

$$\text{If}\quad (1 + u_1)(1 + u_2)\ldots(1 + u_r) \equiv \Pi_r$$

$$\text{and}\quad (1 - u_1)(1 - u_2)\ldots(1 - u_r) \equiv \Pi'_r$$

where $u_1, u_2, \ldots u_r$ are all pos′,

we see at once that, if Π_r, Π'_r tend to finite limits—

1°, each of the quantities $u_1, u_2, \ldots, u_r < 1$;

and 2°, when this is the case,

Π_r may tend to a finite limit, or to ∞,

and Π'_r " " " " , or to zero.

To find conditions, change the forms thus:

$$\log \Pi_r = \log(1 + u_1) + \log(1 + u_2)\ldots + \log(1 + u_r)$$

$$= u_1 - \tfrac{1}{2}u_1^2 + \tfrac{1}{3}u_1^3 - \ldots$$

$$+ u_2 - \tfrac{1}{2}u_2^2 + \tfrac{1}{3}u_2^3 - \ldots$$

. . . .

$\therefore$ $\log \Pi_r$ lies between 0 and $u_1 + u_2 \ldots + u_r$

$\therefore$, *provided* $u_1 + u_2 + \ldots + u_r$ *is conv′t*,

$\log \Pi_r$ is pos′ and $\not>$ a definite quantity

and then also Π_r " " " " "

i. e. Π_r tends to some limit less than a definite quantity, as r increases.

$\therefore$ Π_r then converges.

But if $u_1 + u_2 \ldots$ is non-conv′t, so is Π_r;

for $(1 + u_1)(1 + u_2)\ldots = 1 + u_1 + u_2 \ldots +$ products of u_1, u_2 &c.

$$\text{Again}\quad -\log \Pi'_r = u_1 + \tfrac{1}{2}u_1^2 + \tfrac{1}{3}u_1^3\ldots$$

$$+ u_2 + \tfrac{1}{2}u_2^2 + \tfrac{1}{3}u_2^3\ldots$$

. . . .

and $\therefore$ $> u_1 + u_2 + \ldots$

$\therefore$ if $u_1 + u_2 \ldots >$ a finite magnitude

$\log(\Pi'_r)^{-1} >$ finite mag′

and $\therefore$ $\Pi'_r = 0$

So that then (in accordance with def′) Π'_r is non-conv′t.

But if $u_1 + u_2 \ldots$ is conv′t, then

since $\log \Pi'_r < -(u_1 + u_2 \ldots)$

$\therefore$ Π'_r is conv′t.

Hence the convergence of Π_r and Π'_r is *absolute* or *tactical* as is that of

$$u_1 + u_2 + \ldots$$

Cor′—If either of Π_r, Π'_r converges, so do the other.

As Examples

$(1 + \frac{1}{2})(1 + \frac{1}{3})(1 + \frac{1}{4}) \ldots$ is non-conv′t $(= \infty)$ for then $\Pi_r = (r+2)/2$

$(1 - \frac{1}{2})(1 - \frac{1}{3})(1 - \frac{1}{4}) \ldots$ is non-conv′t $(= 0)$ „ $\Pi'_r = 1/(r+2)$

Again $\left(1 + \frac{x^2}{1^2}\right)\left(1 + \frac{x^2}{2^2}\right)\left(1 + \frac{x^2}{3^2}\right)\ldots$

and $\left(1 - \frac{x^2}{1^2}\right)\left(1 - \frac{x^2}{2^2}\right)\left(1 - \frac{x^2}{3^2}\right)\ldots$

are each *absolutely* conv′t, for

$$\frac{1}{1^2} + \frac{1}{2^2} + \frac{1}{3^2} + \ldots$$

$$< 1 + \left(\frac{1}{2^2} + \frac{1}{2^2}\right) + \left(\frac{1}{4^2} + \frac{1}{4^2} + \frac{1}{4^2} + \frac{1}{4^2}\right) \text{ \&c in sim′r groups}$$

$$< 1 + \frac{2}{2^2} + \frac{4}{4^2} + \frac{8}{8^2} + \ldots$$

$$< 1 + \tfrac{1}{2} + \tfrac{1}{4} + \tfrac{1}{8} + \ldots$$

$$< 2$$

But $\left(1 - \frac{x}{2}\right)\left(1 + \frac{x}{3}\right)\left(1 - \frac{x}{4}\right)\left(1 + \frac{x}{5}\right)\ldots$ is *tactically* conv′t

---

Since no sine or cosine > 1, each of the infinite expressions

$$u_1 \cos^r \theta + u_2 \cos^s \phi + \ldots$$

and $(1 + u_1 \sin^r \theta)(1 + u_2 \sin \phi)\ldots$

is *absolutely* conv′t, provided that $u_1 + u_2 + u_3 + \ldots$ is *absolutely* conv′t.

§ **80. Theorem**—*If* n *is a positive integer, expressions can be found for* $\sin nx$ *and* $\cos nx$ *in terms of* $\sin x$ *and* $\cos x$, *where* x *is measured in any unit; and hence can be deduced series for the* sine *and* cosine *of any angle in terms of the radian measure of that angle.*

Generalizing the law indicated in the expansions of

$$\cos(\alpha+\beta+\gamma) \quad \text{and} \quad \sin(\alpha+\beta+\gamma)$$

on pp. 66, 67, assume that, if $k = \alpha_1 + \alpha_2 + \dots + \alpha_n$,

$$\cos k = \Sigma_0 - \Sigma_2 + \Sigma_4 - \dots$$

$$\sin k = \Sigma_1 - \Sigma_3 + \Sigma_5 - \dots$$

where Σ_r means *the sum of all the products formed by multiplying together the* sines *of any* r *of the angles and the* cosines *of the remaining* $(n - r)$ *angles.*

Introducing a new angle λ, we have

$$\cos(k+\lambda) = (\Sigma_0 - \Sigma_2 + \dots)\cos\lambda - (\Sigma_1 - \Sigma_3 + \dots)\sin\lambda$$

$$= \Sigma_0 \cos\lambda - (\Sigma_2\cos\lambda + \Sigma_1\sin\lambda) + \dots$$

$$= \Sigma_0' - \Sigma_2' + \Sigma_4' - \dots$$

where Σ_r' is the same for $(n+1)$ $\wedge^s$ that Σ_r is for n $\wedge^s$.

$$\sin(k+\lambda) = (\Sigma_1 - \Sigma_3 + \dots)\cos\lambda + (\Sigma_0 - \Sigma_2 + \dots)\sin\lambda$$

$$= (\Sigma_1\cos\lambda + \Sigma_0\sin\lambda) - (\Sigma_3\cos\lambda + \Sigma_2\sin\lambda) + \dots$$

$$= \Sigma_1' - \Sigma_3' + \Sigma_5' - \dots$$

∴ if the assumptions are true for any particular value of n they are also true for the next greater value.

But they are true when $n = 3$.

∴, by the *Law of Induction*, they are true generally.

From these, if $\alpha_1 = \alpha_2 = \dots = \alpha_n = x$, we get that

$$\left.\begin{aligned} \cos nx &= c^n - {}_nC_2\, c^{n-2} s^2 + {}_nC_4\, c^{n-4} s^4 - \dots \\ \text{and} \quad \sin nx &= {}_nC_1\, c^{n-1} s - {}_nC_3\, c^{n-3} s^3 + {}_nC_5\, c^{n-5} s^5 - \dots \end{aligned}\right]$$

where $\sin x = s$, $\cos x = c$

and ${}_nC_r =$ the N° of comb'ns of n things r together.

Now suppose that $n\,x = \alpha$ radians, and that this equality is maintained, when n increases, by a corresponding diminution of x, so that α remains unchanged.

Then

$$\cos\alpha = \cos^n\frac{\alpha}{n} - \left(1 - \frac{1}{n}\right)\frac{\alpha^2}{2\,!}\left[\frac{\sin\dfrac{\alpha}{n}}{\dfrac{\alpha}{n}}\right]^2\cos^{n-2}\frac{\alpha}{n}$$

$$+\left(1-\frac{1}{n}\right)\left(1-\frac{2}{n}\right)\left(1-\frac{3}{n}\right)\frac{\alpha^4}{4\,!}\left[\frac{\sin\dfrac{\alpha}{n}}{\dfrac{\alpha}{n}}\right]^4\cos^{n-4}\frac{\alpha}{n} - \ldots$$

Consider a very great number r of terms, and let n be chosen to be an integer itself very great compared with r.

Then, as $\sin\theta/\theta < 1$, and $\cos\theta < 1$,

the numerical value of each term of this series

< ,, ,, ,, the corresponding term of the series

$$1 - \frac{\alpha^2}{2\,!} + \frac{\alpha^4}{4\,!} - \frac{\alpha^6}{6\,!} + \ldots$$

But this last series is *absolutely* convergent; since, when all its terms are made pos', its test-ratio

$$\frac{(r+1)\text{th term}}{r\text{th term}} = \frac{\alpha^2}{2\,r\,(2\,r-1)}$$

and the *Limit* of this (when $r = \infty$) < 1, for all finite values of α.

∴ we may ultimately neglect all its terms after the rth.

∴, as the convergence is *absolute*, we might do the same with the 1st series, even if all its terms were pos', and *à fortiori* as it stands.

∴ $\cos\alpha$ approximates to the sum of the 1st r terms of the 1st series.

Now $\left[\dfrac{\sin\dfrac{\alpha}{n}}{\dfrac{\alpha}{n}}\right]^r$ and $\cos^{n-r}\dfrac{\alpha}{n}$, and also the coeff′, each ultimately become unity, as n increases towards ∞.

$$\therefore\quad \cos\alpha = 1 - \frac{\alpha^2}{2\,!} + \frac{\alpha^4}{4\,!} - \frac{\alpha^6}{6\,!} + \dots$$

In a precisely sim′r manner we should find that

$$\sin\alpha = \alpha - \frac{\alpha^3}{3\,!} + \frac{\alpha^5}{5\,!} - \frac{\alpha^7}{7\,!} + \dots \qquad (\textit{Newton})$$

Cor′ (1)— $$\sin x^0 = \frac{\pi x}{180} - \frac{1}{3\,!}\left(\frac{\pi x}{180}\right)^3 + \dots$$

$$\cos x^0 = 1 - \frac{1}{2\,!}\left(\frac{\pi x}{180}\right)^2 + \frac{1}{4\,!}\left(\frac{\pi x}{180}\right)^4 - \dots$$

Cor′ (2)—By actual division, we have

$$\tan\alpha = \alpha + \frac{\alpha^3}{3} + \frac{2\alpha^5}{15} + \dots$$

Applying the principle of *Reversion of Series* to this (see *Chrystal's Algebra*, vol′ ii, pp. 254 ...) we get, as successive approximations

$$\alpha = \tan\alpha$$

$$\alpha = \tan\alpha - \tfrac{1}{3}\tan^3\alpha$$

$$\alpha = \tan\alpha - \tfrac{1}{3}\tan^3\alpha + \tfrac{1}{5}\tan^5\alpha$$

From the 3rd of these, by putting x for $\tan\alpha$, we have the *approximation*

$$\tan^{-1}x = x - \frac{x^3}{3} + \frac{x^5}{5}, \quad \text{when } x \text{ is small}$$

This suggests that $\tan^{-1}x$ may be expanded in powers of x, if it is between limits which make the series conv′t.

§ **81.** *Def′*—If $\phi(x) = \phi(-x)$ then $\phi(x)$ is called an **even function** of x; and if $\phi(x) = -\phi(-x)$ then $\phi(x)$ is called an **odd function** of x.

Lemma—*If a function of* x *can be expanded in a series of ascending powers of* x, *then this series will contain*—

1°, *for an* **even** *function only* **even** *powers of* x;

and 2°, „ **odd** „ „ **odd** „ „ ;

For if $\phi(x) \equiv \alpha_0 + \alpha_1 x + \alpha_2 x^2 + \alpha_3 x^3 + \dots$

then $\phi(-x) \equiv \alpha_0 - \alpha_1 x + \alpha_2 x^2 - \alpha_3 x^3 + \dots$

$\therefore$ 1°, if $\phi(x) = \phi(-x)$ we have, by subtraction

$$\alpha_1 x + \alpha_3 x^3 + \alpha_5 x^5 + \dots \equiv 0$$

whence $\alpha_1 = 0, \quad \alpha_3 = 0, \quad \alpha_5 = 0, \dots$

and, 2°, if $\phi(x) = -\phi(-x)$ we have, by addition

$$\alpha_0 + \alpha_2 x^2 + \alpha_4 x^4 + \dots \equiv 0$$

whence $\alpha_0 = 0, \quad \alpha_2 = 0, \quad \alpha_4 = 0, \dots$

Note—Of the six T. F.s, obviously cosines and secants are *even*, but sines, tangents, cotangents, and cosecants are *odd* functions of angle; so that the expansions we have found for $\sin\alpha$ and $\cos\alpha$ are in accordance with the preceding theorem.

Now, although it is entirely illegitimate to assume that a function can be expanded in an infinite series of whose convergence we know nothing, yet the provisional assumption of such an expansion in a particular case, may be a guide towards finding a rigorous proof of its existence.

In accordance with this idea let us assume provisionally that $\tan^{-1} x$ being an *odd* function of x, can be expanded as a converging series in the form

$$\alpha_1 x + \alpha_3 x^3 + \alpha_5 x^5 + \dots$$

then, since $\tan^{-1}\dfrac{x+y}{1-xy} \equiv \tan^{-1} x + \tan^{-1} y$, *we have

* For this method I am indebted to Prof' Genese.

$$\alpha_1 \frac{x+y}{1-xy} + \alpha_3 \left(\frac{x+y}{1-xy}\right)^3 + \ldots$$
$$\equiv \alpha_1 (x+y) + \alpha_3 (x^3 + y^3) + \ldots$$

Equating coeff's of $x^{2r}y$, i. e. of $x^{2r-1}(xy)$, we get

$$(2r-1)\alpha_{2r-1} + (2r+1)\alpha_{2r+1} = 0$$

Whence $\alpha_1 = -3\alpha_3 = +5\alpha_5 = -7\alpha_7 = \ldots$

$$\therefore \quad \tan^{-1}x = \alpha_1\left(x - \frac{x^3}{3} + \frac{x^5}{5} - \ldots\right)$$

Taking x infinitesimally small, we get $\alpha_1 = 1$

$$\therefore \quad \tan^{-1}x = x - \frac{x^3}{3} + \frac{x^5}{5} - \ldots$$

If $x \not> 1$ the series is convergent.

Of this expansion the following* is a rigorous proof.

Lemma—If α, β are angles measured by the radian; and $\alpha > \beta$,

then $\tan^{-1}\alpha - \tan^{-1}\beta = \sin^{-1}\dfrac{\alpha-\beta}{\sqrt{(1+\alpha^2)(1+\beta^2)}} > \dfrac{\alpha-\beta}{\sqrt{(1+\alpha^2)(1+\beta^2)}} > \dfrac{\alpha-\beta}{1+\alpha^2}$

and $\tan^{-1}\alpha - \tan^{-1}\beta = \tan^{-1}\dfrac{\alpha-\beta}{1+\alpha\beta} < \dfrac{\alpha-\beta}{1+\alpha\beta} < \dfrac{\alpha-\beta}{1+\beta^2}$

Now suppose that $x \not> 1$, and n is a pos' integer.

Then, since $\tan^{-1}x \equiv \tan^{-1}\dfrac{x}{n} - \tan^{-1}0$

$$+ \tan^{-1}\frac{2x}{n} - \tan^{-1}\frac{x}{n}$$

$$+ \tan^{-1}\frac{3x}{n} - \tan^{-1}\frac{2x}{n}$$

---

* Due to Professor Purser of Queen's College, Belfast.

. . . . .

$$+ \tan^{-1}\frac{nx}{n} - \tan^{-1}\frac{(n-1)x}{n}$$

∴, by applying the Lemma to each of these lines, we get that

$$\tan^{-1}x > \frac{\frac{x}{n}}{1+\left(\frac{x}{n}\right)^2} + \frac{\frac{x}{n}}{1+\left(\frac{2x}{n}\right)^2} + \dots + \frac{\frac{x}{n}}{1+\left(\frac{nx}{n}\right)^2}$$

$$\text{but} < \frac{x}{n} + \frac{\frac{x}{n}}{1+\left(\frac{x}{n}\right)^2} + \frac{\frac{x}{n}}{1+\left(\frac{2x}{n}\right)^2} + \dots + \frac{\frac{x}{n}}{1+\left(\frac{n-1}{n}x\right)^2}$$

The difference of these limits is $\frac{x}{n} - \frac{\frac{x}{n}}{1+\left(\frac{nx}{n}\right)^2}$

i.e. is $\frac{1}{n}\left(\frac{x^3}{1+x^2}\right)$

∴ $\tan^{-1}x$ differs by less than $\frac{1}{n}\left(\frac{x^3}{1+x^2}\right)$ from the sum of

$$\frac{x}{n} + \frac{x}{n}\Bigg[1 - \left(\frac{x}{n}\right)^2 + \left(\frac{x}{n}\right)^4 - \left(\frac{x}{n}\right)^6 + \dots$$

$$+ 1 - \left(\frac{2x}{n}\right)^2 + \left(\frac{2x}{n}\right)^4 - \dots$$

$$+ 1 - \left(\frac{3x}{n}\right)^2 + \left(\frac{3x}{n}\right)^4 - \dots$$

. . . .

$$+ 1 - \left\{\frac{(n-1)x}{n}\right\}^2 + \left\{\frac{(n-1)x}{n}\right\}^4 - \dots\Bigg]$$

where, each series goes to infinity, but n is *at present* finite.

Collecting coeff's of like powers of x, and putting

$$S_r \text{ for } 1^r + 2^r + 3^r + \ldots + (n-1)^r$$

we get that $\tan^{-1} x$ differs from the sum of the series

$$x - \frac{S_2}{n^3} x^3 + \frac{S_4}{n^5} x^5 - \frac{S_6}{n^7} x^7 + \ldots \quad . \quad . \quad . \quad (A)$$

by a quantity which diminishes as n increases, and ultimately vanishes when $n = \infty$.

The series (A) is convergent, since it is the sum of a number of series all themselves convergent.

Now hitherto n has been kept *finite*; but it is a well-known algebraic result that

$$\text{Lim}' (n = \infty) \left[\frac{1^r + 2^r + 3^r + \ldots + n^r}{n^{r+1}} \right] = \frac{1}{r+1}$$

$$\therefore \text{ also } \quad \text{Lim}' (n = \infty) \left[\frac{S_r}{n^{r+1}} \right] = \frac{1}{r+1}$$

Hence, by increasing n indefinitely, we have, when $n = \infty$,

$$\tan^{-1} x = x - \frac{x^3}{3} + \frac{x^5}{5} - \frac{x^7}{7} + \ldots$$

where $x \ngtr 1$, numerically.

Put $\tan\theta$ for x, and we get

$$\theta = \tan\theta - \tfrac{1}{3}\tan^3\theta + \tfrac{1}{5}\tan^5\theta - \ldots$$

where θ is any angle from $\pi/4$ to $-\pi/4$, both included.

This last result is known as *Gregorie's Series.*

Note—The result $\tan^{-1} x = x - \frac{x^3}{3} + \ldots$ when used in connection with *Machin's formula* (p. 144) $4 \tan^{-1} \frac{1}{5} - \tan^{-1} \frac{1}{239} = \pi/4$ gives very rapidly converging series, from which π may be readily calculated.

For example, in order to get π correct to 7 dec'l places, it is only necessary to go as far as the 7th power of $\frac{1}{5}$, and the 1st power of $\frac{1}{239}$.

§ **82.** *To resolve* $x^n - 2\cos n\theta + x^{-n}$ *into factors.*

Since $\cos n\theta + \cos(n-2)\theta = 2\cos(n-1)\theta\cos\theta$

$$\therefore\ -2\cos n\theta = -2\cos(n-1)\theta \cdot 2\cos\theta + 2\cos(n-2)\theta$$

$$\therefore\ x^n - 2\cos n\theta + x^{-n}$$

$$\equiv (x - 2\cos\theta + x^{-1})(x^{n-1} + x^{-(n-1)})$$

$$+ \{x^{n-1} - 2\cos(n-1)\theta + x^{-(n-1)}\}\, 2\cos\theta$$

$$- \{x^{n-2} - 2\cos(n-2)\theta + x^{-(n-2)}\}$$

From which, calling the function to be factorized $\phi(n)$, we see that $\phi(n)$ is divisible by $\phi(1)$ if $\phi(n-1)$ and $\phi(n-2)$ are.

But $\phi(2) = \phi(1)\{x + 2\cos\theta + x^{-1}\}$

$\therefore\ \phi(2)$ and $\phi(1)$ are divisible by $\phi(1)$

$\therefore$ so is $\phi(3)$

Hence, by the *Law of Induction*, $\phi(n)$ is divis' by $\phi(1)$

Now $\cos n\theta = \cos(n\theta + 2r\pi)$, where r is any integer.

$$\therefore\ \phi(n) \text{ is divis' by } x - 2\cos\left(\theta + \frac{2r\pi}{n}\right) + x^{-1}$$

But, by giving r in succession the values 0, 1, 2, ..., $(n-1)$, this divisor will have n distinct values: nor has it any more; for other values of r only reproduce some of the former.

$$\therefore\ x^n - 2\cos n\theta + x^{-n}$$

$$\equiv \left(x - 2\cos\theta + \frac{1}{x}\right)\left\{x - 2\cos\left(\theta + \frac{2\pi}{n}\right) + \frac{1}{x}\right\}$$

$$\left\{x - 2\cos\left(\theta + 2\frac{2\pi}{n}\right) + \frac{1}{x}\right\}\left\{x - 2\cos\left(\theta + 3\frac{2\pi}{n}\right) + \frac{1}{x}\right\}$$

$$\cdots\cdots\left[x - 2\cos\left\{\theta + (n-1)\frac{2\pi}{n}\right\} + \frac{1}{x}\right]$$

This may be written in either of the equivalent forms

$$x^{2n} - 2x^n \cos n\theta + 1$$
$$\equiv (x^2 - 2x\cos\theta + 1)\left\{x^2 - 2x\cos\left(\theta + \frac{2\pi}{n}\right) + 1\right\}$$
$$\ldots\ldots \left[x^2 - 2x\cos\left\{\theta + (n-1)\frac{2\pi}{n}\right\} + 1\right]$$

or $x^{2n} - 2x^n y^n \cos n\theta + y^{2n}$

$$\equiv (x^2 - 2xy\cos\theta + y^2)\left\{x^2 - 2xy\cos\left(\theta + \frac{2\pi}{n}\right) + y^2\right\}$$
$$\ldots\ldots \left[x^2 - 2xy\cos\left\{\theta + (n-1)\frac{2\pi}{n}\right\} + y^2\right]^*$$

(*Cotes' Theorem*)

§ **83.** From the last §, by supposing that $x = 1$, we get

$$2(1 - \cos 2n\theta)$$
$$\equiv 2^n(1 - \cos 2\theta)\{1 - \cos 2(\theta + \pi/n)\} \ldots\ldots$$
$$\ldots\ldots [1 - \cos 2\{\theta + (n-1)\pi/n\}]$$

Taking the pos' square root of both sides gives

$$2\sin n\theta \equiv 2^n \sin\theta \sin(\theta + \pi/n)\sin(\theta + 2\pi/n) \ldots\ldots$$
$$\ldots\ldots \sin\{\theta + (n-1)\pi/n\} \quad \text{(A)}$$

Now $\sin\{\theta + (n-1)\pi/n\} = \sin(\pi/n - \theta)$

$\sin\{\theta + (n-2)\pi/n\} = \sin(2\pi/n - \theta)$

. . . .

And $\sin(\pi/n + \theta)\sin(\pi/n - \theta) = \sin^2\pi/n - \sin^2\theta$

. . . .

$$\sin n\theta$$
$$= 2^{n-1}\sin\theta(\sin^2\pi/n - \sin^2\theta)(\sin^2 2\pi/n - \sin^2\theta) \ldots\ldots$$

* Adams: *Camb' Trans'*, 1868. Ferrers: *Mess' of Math'*, 1875.

the dexter factors being all of the same form when n is odd, but having an extra factor $\cos\theta$ when n is even.

Suppose θ diminished indefinitely, then

$$n = 2^{n-1} \sin^2 \pi/n \sin^2 2\pi/n \sin^2 3\pi/n \ldots .$$

If $n\theta = \alpha$ radians, and we suppose that as n increases θ diminishes so as to keep α constant, then, from the last two results, by division,

$$\sin\alpha = n \sin\frac{\alpha}{n}\left[1 - \frac{\sin^2\frac{\alpha}{n}}{\sin^2\frac{\pi}{n}}\right]\left[1 - \frac{\sin^2\frac{\alpha}{n}}{\sin^2\frac{2\pi}{n}}\right] \ldots . \quad \text{(B)}$$

Consider a very great number r of terms, and let n be chosen to be an integer itself very great compared with r.

$$\text{Now} \quad \sin^2\frac{s\pi}{n} > \left[\frac{s\pi}{n} - \frac{1}{6}\left(\frac{s\pi}{n}\right)^3\right]^2$$

$$> \left(\frac{s\pi}{n}\right)^2\left(1 - \frac{\pi^2}{6}\right)^2 \quad \text{where } s < n$$

$$\text{say} \quad > k\left(\frac{s\pi}{n}\right)^2 \quad \text{where } k < 1$$

∴ the numerical value of each fraction in (B) < that of the corresponding fraction in the product

$$\left[1 - \frac{\sin^2\frac{\alpha}{n}}{k\left(\frac{\pi}{n}\right)^2}\right]\left[1 - \frac{\sin^2\frac{\alpha}{n}}{k\left(\frac{2\pi}{n}\right)^2}\right] \ldots . \quad \text{(C)}$$

But the sum of the fractions in (C) is *absolutely convergent*

since $\frac{1}{1^2} + \frac{1}{2^2} + \frac{1}{3^2} + \ldots$,, ,, ,, ,,

∴ the sum of the fractions in (B) ,, ,, ,, ,,

∴ the product (B) ,, ,, ,, ,,

∴, for all very great values of n, $\sin\alpha$ approximates to the product of the 1st r terms of

$$\frac{\sin\frac{\alpha}{n}}{\frac{\alpha}{n}}\cdot\alpha\left[1-\left(\frac{\sin\frac{\alpha}{n}}{\frac{\alpha}{n}}\cdot\frac{\frac{\pi}{n}}{\sin\frac{\pi}{n}}\right)^2\cdot\frac{\alpha^2}{\pi^2}\right]$$

$$\left[1-\left(\frac{\sin\frac{\alpha}{n}}{\frac{\alpha}{n}}\cdot\frac{\frac{2\pi}{n}}{\sin\frac{2\pi}{n}}\right)^2\cdot\frac{\alpha^2}{2^2\pi^2}\right]\cdots$$

Proceeding to the *Limit* when $n=\infty$, we get

$$\sin\alpha=\alpha\left(1-\frac{\alpha^2}{\pi^2}\right)\left(1-\frac{\alpha^2}{2^2\pi^2}\right)\left(1-\frac{\alpha^2}{3^2\pi^2}\right)\cdots$$

When α is between 0 and π, $\sin\alpha$ is pos′, and all the dexter factors are pos′.

When α is between π and 2π, $\sin\alpha$ is neg′, and $1-\frac{\alpha^2}{\pi^2}$ is neg′, all the other dexter factors being pos′.

Sim′ly for all other values of α.

∴ it was justifiable at the beginning to take the pos′ sign of the root only.

Cor′ (1)—Put $\pi/2$ for α, then

$$1=\frac{\pi}{2}\left(1-\frac{1}{2^2}\right)\left(1-\frac{1}{4^2}\right)\left(1-\frac{1}{6^2}\right)\cdots$$

$$\therefore\ \frac{\pi}{2}=\frac{2^2}{2^2-1}\cdot\frac{4^2}{4^2-1}\cdot\frac{6^2}{6^2-1}\cdots$$

$$=\frac{2^2}{1\cdot3}\cdot\frac{4^2}{3\cdot5}\cdot\frac{6^2}{5\cdot7}\cdots\quad(\textit{Wallis' formula})$$

Cor′ (2)—Put $\pi/6$ for α, then

$$\frac{1}{2}=\frac{\pi}{6}\left(1-\frac{1}{6^2}\right)\left(1-\frac{1}{2^2.6^2}\right)\left(1-\frac{1}{3^2.6^2}\right)\cdots$$

$$\therefore\ \pi=3\cdot\frac{36}{35}\cdot\frac{144}{143}\cdot\frac{324}{323}\cdots$$

Cor^l (3)— Since $1 - \frac{\alpha^2}{3!} + \frac{\alpha^4}{5!} \ldots = \frac{\sin \alpha}{\alpha}$

and $\left(1 - \frac{\alpha^2}{\pi^2}\right)\left(1 - \frac{\alpha^2}{2^2\pi^2}\right) \ldots = \frac{\sin \alpha}{\alpha}$

∴ taking logs and equating, we have

$$\log\left(1 - \frac{\alpha^2}{3!} + \ldots\right) = \log\left(1 - \frac{\alpha^2}{\pi^2}\right) + \log\left(1 - \frac{\alpha^2}{2^2\pi^2}\right) \ldots$$

Expand and equate coeff's of α^2, then

$$-\frac{1}{3!} = -\left(\frac{1}{\pi^2} + \frac{1}{2^2\pi^2} + \ldots\right)$$

$$\therefore \quad \frac{\pi^2}{6} = \frac{1}{1^2} + \frac{1}{2^2} + \frac{1}{3^2} + \ldots$$

Equate coeff's of α^4, then

$$\frac{1}{5!} - \frac{1}{2}\left(\frac{1}{3!}\right)^2 = -\frac{1}{2\pi^4}\left(\frac{1}{1^4} + \frac{1}{2^4} + \frac{1}{3^4} + \ldots\right)$$

$$\therefore \quad \frac{\pi^4}{90} = \frac{1}{1^4} + \frac{1}{2^4} + \frac{1}{3^4} + \ldots$$

§ 84. In eq'n (A) of last § for θ write $(\theta + \pi/2n)$ then

$$\cos n\theta = 2^{n-1} \sin(\theta + \pi/2n) \sin(\theta + 3\pi/2n) \ldots$$
$$\ldots \sin\{\theta + (2n - 1)\pi/2n\}$$

Proceeding exactly as above, there will result

$$\cos \alpha = \left(1 - \frac{2^2\alpha^2}{\pi^2}\right)\left(1 - \frac{2^2\alpha^2}{3^2\pi^2}\right)\left(1 - \frac{2^2\alpha^2}{5^2\pi^2}\right) \ldots$$

Or, since $\cos \alpha = \frac{\sin 2\alpha}{2 \sin \alpha}$, we have

$$\cos \alpha = \frac{2\alpha\left(1 - \frac{2^2\alpha^2}{\pi^2}\right)\left(1 - \frac{2^2\alpha^2}{2^2\pi^2}\right)\left(1 - \frac{2^2\alpha^2}{3^2\pi^2}\right)\left(1 - \frac{2^2\alpha^2}{4^2\pi^2}\right) \ldots}{2\alpha \qquad \left(1 - \frac{\alpha^2}{\pi^2}\right) \qquad\qquad \left(1 - \frac{\alpha^2}{2^2\pi^2}\right) \ldots}$$

$$= \left(1 - \frac{2^2\alpha^2}{\pi^2}\right)\left(1 - \frac{2^2\alpha^2}{3^2\pi^2}\right)\left(1 - \frac{2^2\alpha^2}{5^2\pi^2}\right) \ldots$$

the other factors dividing out.

Examples

1. *To find the Limit, when* m *and* n *are indefinitely increased, but* n *is greater than* m, *of the expression*

$$\left(1 - \frac{x}{m}\right)\left(1 - \frac{x}{m-1}\right)\ldots\left(1 - \frac{x}{1}\right) x \left(1 + \frac{x}{1}\right)\left(1 + \frac{x}{2}\right)\ldots\left(1 + \frac{x}{n}\right)$$

$$\text{The Limit} = \frac{\sin \pi x}{\pi} \cdot \text{Lim}'\left[\left(1 + \frac{x}{m+1}\right)\ldots\left(1 + \frac{x}{n}\right)\right]$$

$$= \frac{\sin \pi x}{\pi} \cdot \text{Lim}'\, z \text{ say}$$

$$\therefore \quad \log z = \log\left(1 + \frac{x}{m+1}\right) + \ldots + \log\left(1 + \frac{x}{n}\right)$$

$$= \frac{x}{m+1} + \ldots + \frac{x}{n} - \text{residue}$$

$$\text{Residue} < \frac{x^2}{2}\left[\frac{1}{(m+1)^2} + \ldots + \frac{1}{n^2}\right]$$

and $\therefore$ is negligible, $\because$ $\Sigma \frac{1}{r^2}$ is conv't

$$\text{Now} \quad \frac{m+1}{n+1} = \frac{m+1}{m+2} \cdot \frac{m+2}{m+3} \cdots \frac{n-1}{n} \cdot \frac{n}{n+1}$$

$$= \left(1 + \frac{1}{m+1}\right)^{-1}\left(1 + \frac{1}{m+2}\right)^{-1}\ldots\left(1 + \frac{1}{n}\right)^{-1}$$

$$\therefore \quad -\log\frac{n+1}{m+1} = -\left[\frac{1}{m+1} + \frac{1}{m+2} + \ldots + \frac{1}{n}\right] + \text{residue}$$

$$\text{Residue} < \frac{1}{(m+1)^2} + \frac{1}{(m+2)^2} + \ldots + \frac{1}{n^2}, \quad \therefore \text{ negligible}$$

$$\therefore \quad \log z = x \log\frac{n+1}{m+1} = x \log\frac{n}{m}, \quad \text{ultimately}$$

$$\therefore \quad z = \left(\frac{n}{m}\right)^x \qquad \text{,,}$$

$$\therefore \text{ Limit of original product} = \left(\frac{n}{m}\right)^x \frac{\sin \pi x}{\pi}$$

(*Proof by J. Larmor*)

2. *If* $A_1 A_2 \dots A_n$ *is a regular polygon of* n *sides*; C *the centre, and* R *the radius of its circum-circle*; P *any point in its plane*; *and if* CP *makes with any radius drawn to a corner (say* A_1*) an angle* α; *then*

$$\Pi\,(PA_r^2) = CP^{2n} - 2\,CP^n . R^n \cos n\alpha + R^{2n}$$

(*De Moivre*)

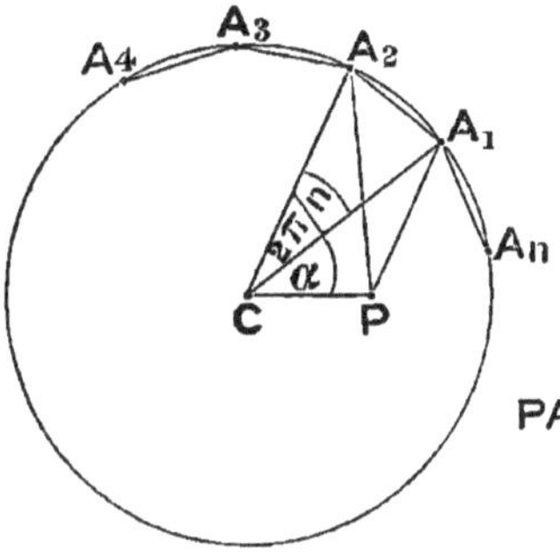

For

$$PA_1^2 = CP^2 - 2\,CP . R \cos\alpha + R^2$$

$$PA_2^2 = CP^2 - 2\,CP . R \cos\left(\alpha + \frac{2\pi}{n}\right) + R^2$$

. . . .

$$PA_n^2 = CP^2 - 2\,CP . R \cos\left\{\alpha + (n-1)\frac{2\pi}{n}\right\} + R^2$$

∴, taking products of corresponding sides, the result follows by *Cotes' Theorem* (§ 82)

Cor' (1)—Suppose P on the circumf'.

$$\text{Then} \quad \Pi\, PA_r = 2R^n \sin\frac{n\alpha}{2}$$

Cor' (2)—Suppose that CP produced meets ⊙ in A_n.

$$\text{Then} \quad \alpha = 2\pi/n$$

$$\therefore \quad \Pi\,(PA_r^2) = (CP^n - R^n)^2$$

$$\therefore \quad \Pi\, PA_r = CP^n - R^n$$

Let $m_1, m_2, \dots$ be the mid p'ts of arcs $A_1 A_2$, $A_2 A_3$, ...

$$\therefore \quad \Pi\, PA_r . \Pi\, Pm_r = CP^{2n} - R^{2n}$$

$$\therefore \quad \Pi\, Pm_r = CP^n + R^n$$

(*Cotes*)

§ **85**. By the use of the preceding results, and by other artifices, numbers of trigonometrical series can be summed.

Perhaps the most frequently useful of such artifices consists in breaking up each term of a series into two others related in some such way as the following—

Let $T_1 + T_2 + T_3 + \dots + T_n$ be a series of trig'l terms, formed according to some common law; and suppose that

$$T_r \equiv U_r - U_{r+1} \quad . \quad . \quad . \quad . \quad . \quad . \quad . \quad . \quad (\alpha)$$

where U_r, U_{r+1} are trig'l expressions.

Then the series

$$\equiv (U_1 - U_2) + (U_2 - U_3) + \dots + (U_n - U_{n+1})$$

and $\therefore$ its sum is $U_1 - U_{n+1}$

In the manipulation of series any relation, such as (α) above, by which the form of the series is modified, will be called a *formula of reduction.*

As a general rule it may be said that, where this artifice is applicable, the whole difficulty of the summation consists in finding a suitable *formula of reduction.*

Examples 1 and 2 following are done by this method.

The process of breaking up T_r may be sometimes facilitated by multiplying it by a suitable factor; see Example 3 below: this Example, as also its Corollaries, should be carefully studied.

Example 4 gives a Theorem of fundamental importance as an aid in the summation of many series by elementary methods. To it the Student is recommended to pay particular attention.

Examples

1. *To sum the series* $\operatorname{cosec} 2\alpha + \operatorname{cosec} 4\alpha + \dots + \operatorname{cosec} 2^n\alpha$.

Here $T_r \equiv \cot 2^{r-1}\alpha - \cot 2^r\alpha$

$\therefore$ the series

$$\equiv (\cot\alpha - \cot 2\alpha) + (\cot 2\alpha - \cot 4\alpha) + \dots + (\cot 2^{n-1}\alpha - \cot 2^n\alpha)$$

and $\therefore$ its sum is $\cot\alpha - \cot 2^n\alpha$

2. *To sum the series*

$$\cot^{-1}\left(\alpha + \frac{2}{\alpha}\right) + \cot^{-1}\left(3\alpha + \frac{2}{\alpha}\right) + \cot^{-1}\left(6\alpha + \frac{2}{\alpha}\right) + \ldots$$
$$\ldots + \cot^{-1}\left\{\frac{n(n+1)}{2}\alpha + \frac{2}{\alpha}\right\}.$$

Here $T_r \equiv \tan^{-1}\dfrac{2\alpha}{4 + r(r+1)\alpha^2}$

$$\equiv \tan^{-1}\frac{\dfrac{(r+1)\alpha}{2} - \dfrac{r\alpha}{2}}{1 + \dfrac{(r+1)\alpha}{2}\cdot\dfrac{r\alpha}{2}}$$

$$\equiv \tan^{-1}\frac{(r+1)\alpha}{2} - \tan^{-1}\frac{r\alpha}{2}$$

$\therefore$ the series $\equiv \left(\tan^{-1}\alpha - \tan^{-1}\dfrac{\alpha}{2}\right) + \left(\tan^{-1}\dfrac{3\alpha}{2} - \tan^{-1}\alpha\right) + \ldots$

$$\ldots + \left[\tan^{-1}\frac{(n+1)\alpha}{2} - \tan^{-1}\frac{n\alpha}{2}\right]$$

and $\therefore$ its sum is $\tan^{-1}\dfrac{(n+1)\alpha}{2} - \tan^{-1}\dfrac{\alpha}{2}$

3. *To sum the series*

$$\sin\alpha + \sin(\alpha+\beta) + \sin(\alpha+2\beta) + \ldots + \sin\{\alpha + (n-1)\beta\}.$$

Here $T_r \cdot 2\sin\dfrac{\beta}{2} \equiv 2\sin\{\alpha + (r-1)\beta\}\sin\dfrac{\beta}{2}$

$$\equiv \cos\left(\alpha + \frac{2r-3}{2}\beta\right) - \cos\left(\alpha + \frac{2r-1}{2}\beta\right)$$

$\therefore$ the series $\cdot\, 2\sin\dfrac{\beta}{2} \equiv \cos\left(\alpha - \dfrac{\beta}{2}\right) - \cos\left(\alpha + \dfrac{\beta}{2}\right)$

$$+ \cos\left(\alpha + \frac{\beta}{2}\right) - \cos\left(\alpha + \frac{3\beta}{2}\right)$$

$$\cdots\cdots\cdots\cdots\cdots$$

$$+ \cos\left(\alpha + \frac{2n-3}{2}\beta\right) - \cos\left(\alpha + \frac{2n-1}{2}\beta\right)$$

$$= \cos\left(\alpha - \frac{\beta}{2}\right) - \cos\left(\alpha + \frac{2n-1}{2}\beta\right)$$

i.e. the sum is $$\frac{\sin\left(\alpha + \frac{n-1}{2}\beta\right)\sin\frac{n\beta}{2}}{\sin\frac{\beta}{2}}$$

Cor (1)—Put $\pi + \beta$ for β, then

$\sin\alpha - \sin(\alpha+\beta) + \sin(\alpha + 2\beta) - \sin(\alpha + 3\beta) + \ldots$ to n terms

$$= \frac{\sin\left\{\alpha + \frac{n-1}{2}(\pi+\beta)\right\}\sin\frac{n(\pi+\beta)}{2}}{\cos\frac{\beta}{2}}$$

Cor (2)—If $\beta = \alpha$, we get

$$\sin\alpha + \sin 2\alpha + \sin 3\alpha + \ldots + \sin n\alpha = \frac{\sin\frac{n+1}{2}\alpha\sin\frac{n\alpha}{2}}{\sin\frac{\alpha}{2}}$$

Cor (3)—If $\beta = 2\pi/n$, then $\sin n\beta/2 = \sin\pi = 0$.

$$\therefore\ \sin\alpha + \sin(\alpha + 2\pi/n) + \sin(\alpha + 4\pi/n + \ldots$$
$$\ldots + \sin\{\alpha + (n-1)2\pi/n\} = 0$$

Cor (4)—In exactly the same way, or by putting $\pi/2 + \alpha$ for α, it will be found that

$$\cos\alpha + \cos(\alpha+\beta) + \ldots + \cos\{\alpha + (n-1)\beta\} = \frac{\cos\left(\alpha + \frac{n-1}{2}\beta\right)\sin\frac{n\beta}{2}}{\sin\frac{\beta}{2}}$$

whence $$\cos\alpha + \cos 2\alpha + \ldots + \cos n\alpha = \frac{\cos\frac{n+1}{2}\alpha\sin\frac{n\alpha}{2}}{\sin\frac{\alpha}{2}}$$

So also $\cos\alpha + \cos(\alpha + 2\pi/n) + \ldots + \cos\{\alpha + (n-1)2\pi/n\} = 0$

Cor (5)—Since $8\sin^4\alpha = \cos 4\alpha - 4\cos 2\alpha + 3$ [see § 23]

$$8\sin^4(\alpha+\beta) = \cos 4(\alpha+\beta) - 4\cos 2(\alpha+\beta) + 3$$

. . . .

$$\therefore\quad 8\left[\sin^4\alpha+\sin^4(\alpha+2\pi/n)+\sin^4(\alpha+4\pi/n)+\ldots\right.$$
$$\left.\ldots+\sin^4\{\alpha+(n-1)\,2\pi/n\}\right]$$

$$=\cos 4\alpha+\cos 4(\alpha+2\pi/n)+\ldots+\cos 4\{\alpha+(n-1)\,2\pi/n\}$$
$$-4\left[\cos 2\alpha+\cos 2(\alpha+2\pi/n)+\ldots+\cos 2\{\alpha+(n-1)\,2\pi/n\}\right]+3n$$

$$\therefore\quad \sin^4\alpha+\sin^4(\alpha+2\pi/n)+\ldots+\sin^4\{\alpha+(n-1)\,2\pi/n\}=\tfrac{3}{8}n$$

Sim'ly $\cos^4\alpha+\cos^4(\alpha+2\pi/n)+\ldots+\cos^4\{\alpha+(n-1)\,2\pi/n\}=\tfrac{3}{8}n$

The last result can be generalized thus—

By an easy inductive generalization of § 23 (which the Student should work, following exactly the method of that §) it can be shown that

$$2^{m-1}\cos^m\theta=\cos m\theta+m\cos(m-2)\theta+\ldots+{}_mC_r\cos(m-2r)\theta+\ldots$$

the last term being $\dfrac{m!}{2\left(\frac{m}{2}!\right)^2}$ when m is *even*

and $\dfrac{m!}{\frac{m+1}{2}!\;\frac{m-1}{2}!}\cos\theta$ when m is *odd*

Hence each term of the series

$$\cos^m\alpha+\cos^m(\alpha+2\pi/n)+\ldots+\cos^m\{\alpha+(n-1)\,2\pi/n\}$$

can be expressed as above in cosines of multiples of the angles, if $m<n$

$$\therefore\text{ the series }=\frac{n}{2^m}\cdot\frac{m!}{\left(\frac{m}{2}!\right)^2},\quad\text{if } m \text{ is } even$$

but $=0$, if m is *odd*.

i. e. $\cos^m\alpha+\cos^m(\alpha+2\pi/n)+\ldots+\cos^m\{\alpha+(n-1)\,2\pi/n\}$

is independent of α, *if* $m<n$

Sim'ly $\sin^m\alpha+\sin^m(\alpha+2\pi/n)+\ldots+\sin^m\{\alpha+(n-1)\,2\pi/n\}$

is independent of α *if* $m<n$.

Note—All these results should be known by heart.

4. *To sum the series*

$\sin(\alpha + n\beta) - {}_nC_1 \sin\{\alpha + (n-1)\beta\}\cos\beta + \dots$

$\dots + {}_nC_r \sin\{\alpha + (n-r)\beta\}\cos^r\beta \dots \pm \sin\alpha\cos^n\beta,$

where ${}_nC_r$ *is the number of combinations of* n *things,* r *together.*

Let the series be denoted by $\phi(n)$.

Also, for brevity, denote $\sin(\alpha + n\beta)$ by s_n, and $\cos\beta$ by c.

$$\text{Then}\quad \phi(0) = s_0 = \sin\alpha$$

$$\phi(1) = s_1 - s_0 c = \cos\alpha \sin\beta$$

Now $\sin(\alpha + n\beta) - 2\sin\{\alpha + (n-1)\beta\}\cos\beta \equiv -\sin\alpha + (n-2)\beta$

$$\therefore\quad s_n - 2s_{n-1}c = -s_{n-2} \quad (\textit{formula of reduction})$$

$$\begin{aligned}
\therefore\quad \phi(2) &= s_2 - 2s_1c + s_0c^2 \\
&= -s_0 + s_0c^2 \\
&= s_0[-(1-c^2)]
\end{aligned}$$

$$\begin{aligned}
\phi(3) &= s_3 - 3s_2c + 3s_1c^2 - s_0c^3 \\
&= (s_3 - 2s_2c) - (s_2 - 2s_1c)c + s_1c^2 - s_0c^3 \\
&= -s_1 + s_0c + s_1c^2 - s_0c^3 \\
&= (s_1 - s_0c)[-(1-c^2)]
\end{aligned}$$

$$\begin{aligned}
\phi(4) &= s_4 - 4s_3c + 6s_1c^2 - 4s_1c^3 + s_0c^4 \\
&= (s_4 - 2s_3c) - 2(s_3 - 2s_2c)c + 2(s_2 - 2s_1c)c^2 + s_0c^4 \\
&= -s_2 + 2s_1c - 2s_0c^2 + s_0c^4 \\
&= s_0 - 2s_0c^2 + s_0c^4 \\
&= s_0[-(1-c^2)]^2
\end{aligned}$$

The law indicated (of which a generalized inductive proof is added) is that for the *even* numbers (beginning with zero) $\phi(n)$ is a G. P. whose 1st term is $\sin\alpha$, for the *odd* numbers $\phi(n)$ is a G. P. whose 1st term is $\cos\alpha\sin\beta$; and the common ratio for each G. P. is $-\sin^2\beta$:

$$\text{i.e. that}\quad \phi(2m) = (-1)^m \sin\alpha \sin^{2m}\beta$$

$$\text{and}\quad \phi(2m+1) = (-1)^m \cos\alpha \sin^{2m+1}\beta^*$$

---

* Due to Mr. R. Chartres, of Manchester, by whom it, and several original Exercises depending on it, were kindly sent to me.—R. C. J. N.

Assume, when n is *even*, that $\phi(n)$, or the series

$$\sin(\alpha+n\beta)-n\sin\{\alpha+(n-1)\beta\}\cos\beta+\ldots$$

$$\pm{}_{n}C_{r}\sin\{\alpha+(n-r)\beta\}\cos^{r}\beta\ \ldots\pm\sin\alpha\cos^{n}\beta$$

$$=(-1)^{\frac{n}{2}}\sin\alpha\sin^{n}\beta$$

Writing $\alpha+\beta$ for α we get that

$$\sin\{\alpha+(n+1)\beta\}-n\sin(\alpha+n\beta)\cos\beta+\ldots$$

$$\pm{}_{n}C_{r}\sin\{\alpha+(n+1-r)\beta\}\cos^{r}\beta\ \ldots\pm\sin(\alpha+\beta)\cos^{n}\beta$$

$$=(-1)^{\frac{n}{2}}\sin(\alpha+\beta)\sin^{n}\beta$$

Also multiplying $\phi(n)$ by $-\cos\beta$ gives

$$-\sin(\alpha+n\beta)\cos\beta+n\sin\{\alpha+(n-1)\beta\}\cos^{2}\beta-\ldots$$

$$\pm{}_{n}C_{r-1}\sin\{\alpha+(n+1-r)\beta\}\cos^{r}\beta\ \ldots$$

$$\pm n\sin(\alpha+\beta)\cos^{n}\beta\mp\sin\alpha\cos^{n+1}\beta$$

$$=-(-1)^{\frac{n}{2}}\sin\alpha\cos\beta\sin^{n}\beta$$

Adding the last two results we get that

$$\sin\{\alpha+(n+1)\beta\}-(n+1)\sin(\alpha+n\beta)\cos\beta+\ldots$$

$$\pm{}_{n+1}C_{r}\sin\{\alpha+(n+1-r)\beta\}\cos^{r}\beta\ \ldots$$

$$\pm(n+1)\sin(\alpha+\beta)\cos^{n}\beta\mp\sin\alpha\cos^{n+1}\beta$$

$$=(-1)^{\frac{n}{2}}\cos\alpha\sin^{n+1}\beta$$

$$\therefore\ \text{if}\quad\phi(2m)=(-1)^{m}\sin\alpha\sin^{2m}\beta$$

$$\text{then}\quad\phi(2m+1)=(-1)^{m}\cos\alpha\sin^{2m+1}\beta$$

$$\text{Sim'ly if}\quad\phi(2m+1)=(-1)^{m}\cos\alpha\sin^{2m+1}\beta$$

$$\text{then}\quad\phi(2m+2)=(-1)^{m+1}\sin\alpha\sin^{2m+2}\beta$$

But we have seen that the theorem is true for $\phi(3)$ and $\phi(4)$.

$\therefore$, by the *Law of Induction*, it is true for all pos′ integral values of n.

5. *To sum, both to infinity and to* n *terms, the series*

$$\sin\alpha + x\sin(\alpha+\beta) + x^2\sin(\alpha+2\beta) + \dots$$

where x *is less than unity.*

The series . $(1 - 2x\cos\beta + x^2)$

$$\begin{array}{r|l|l}
\equiv \sin\alpha + \sin(\alpha+\beta) & x + \sin(\alpha+2\beta) & x^2 + \dots \\
-2\sin\alpha\cos\beta & \quad -2\sin(\alpha+\beta)\cos\beta & \quad - \dots \\
 & \quad +\sin\alpha & \quad + \dots
\end{array}$$

Whence we see that (using the notation of the last Example) the rth column, if $r > 2$,

$$\text{is} \quad (s_r - 2s_{r-1}c + s_{r-2})\,x^r$$

and ∴ vanishes by using the *formula of reduction* in Example 4.

$$\therefore \text{ the sum } ad\ inf' = \frac{\sin\alpha - x\sin(\alpha-\beta)}{1 - 2x\cos\beta + x^2}$$

But the sum to n terms will have two extra terms, owing to the last two columns not vanishing.

Thus the sum to n terms

$$= \frac{\sin\alpha - x\sin(\alpha-\beta) - x^n\sin(\alpha+n\beta) + x^{n+1}\sin\{\alpha+(n-1)\beta\}}{1 - 2x\cos\beta + x^2}$$

6. *To find the sum to infinity of the series*

$$\sin\alpha + x\sin(\alpha+\beta) + \frac{x^2}{2!}\sin(\alpha+2\beta) + \frac{x^3}{3!}\sin(\alpha+3\beta) + \dots$$

Let the sum be denoted by S : then since

$$e^{-x\cos\beta} = 1 - x\cos\beta + \frac{x^2}{2!}\cos^2\beta - \frac{x^3}{3!}\cos^3\beta + \dots$$

we have, by multiplication of the two series, $S \,.\, e^{-x\cos\beta}$

$$\begin{array}{r|l|l|l}
= \sin\alpha + \sin(\alpha+\beta) & x + \sin(\alpha+2\beta) & \dfrac{x^2}{2!} + \sin(\alpha+3\beta) & \dfrac{x^3}{3!} \\
-\sin\alpha\cos\beta & \quad -2\sin(\alpha+\beta)\cos\beta & \quad -3\sin(\alpha+2\beta)\cos\beta & \\
 & \quad +\sin\alpha\cos^2\beta & \quad +3\sin(\alpha+\beta)\cos^2\beta & \\
 & & \quad -\sin\alpha\cos^3\beta &
\end{array}$$

. . . .

i. e., using the notation and results of Example 4, $S.e^{-x\cos\beta}$

$$= \phi(0) + x\phi(1) + \frac{x^2}{2!}\phi(2) + \frac{x^3}{3!}\phi(3) + \dots$$

$$= \sin\alpha\left[1 - \frac{x^2}{2!}\sin^2\beta + \frac{x^4}{4!}\sin^4\beta - \dots\right]$$

$$+ \cos\alpha\left[x\sin\beta - \frac{x^3}{3!}\sin^3\beta + \frac{x^5}{5!}\sin^5\beta - \dots\right]$$

$$= \sin\alpha\cos(x\sin\beta) + \cos\alpha\sin(x\sin\beta)$$

$$\therefore \quad S = e^{x\cos\beta}\sin(\alpha + x\sin\beta)$$

Exercises

1. If $17\alpha = \pi$, prove that

$$\cos\alpha + \cos 3\alpha + \cos 5\alpha + \dots + \cos 15\alpha = \tfrac{1}{2}$$

2. Prove that—

(1) $\tan\alpha + 2\tan 2\alpha + \dots + 2^{n-1}\tan 2^{n-1}\alpha = \cot\alpha - 2^n\cot 2^n\alpha$

(2) $\cos\alpha + \dfrac{\cos^2\alpha}{2} + \dfrac{\cos^3\alpha}{3} + \dots = \log_e \dfrac{\cot\frac{\alpha}{2}}{\sin\alpha}$

(3) $1 + \dfrac{\cos\theta}{\cos\theta} + \dfrac{\cos 2\theta}{\cos^2\theta} + \dfrac{\cos 3\theta}{\cos^3\theta} + \dots$ to n terms $= 0$

where $n\theta = \pi$

3. If $\tan\alpha$ is less than unity, prove that—

(1) $\tan^2\alpha - \tfrac{1}{2}\tan^4\alpha + \tfrac{1}{3}\tan^6\alpha - \dots$

$= \sin^2\alpha + \tfrac{1}{2}\sin^4\alpha + \tfrac{1}{3}\sin^6\alpha + \dots$

(2) $\left.\begin{array}{l}(1 - 2\tan^2\alpha + 3\tan^4\alpha - 4\tan^6\alpha + \dots)\\ \times(1 + 2\cos^2\alpha + 3\cos^4\alpha + 4\cos^6\alpha + \dots)\end{array}\right] = \cot^4\alpha$

4. Prove that $\sin(n+1)\alpha\operatorname{cosec}\alpha$

$$= \cos^n\alpha + \cos^{n-1}\alpha\cos\alpha + \cos^{n-2}\alpha\cos 2\alpha + \dots + \cos n\alpha$$

NOTE—*Comes, by repeated substitution, from*

$$\sin(n+1)\alpha\operatorname{cosec}\alpha = \cos n\alpha + \cos\alpha\sin n\alpha\operatorname{cosec}\alpha$$

5. Sum to n terms each of the following—

(1) $\tan\frac{\alpha}{2}\sec\alpha + \tan\frac{\alpha}{4}\sec\frac{\alpha}{2} + \tan\frac{\alpha}{8}\sec\frac{\alpha}{4} + \dots$

(2) $\operatorname{cosec}\alpha + \operatorname{cosec}\frac{\alpha}{2} + \operatorname{cosec}\frac{\alpha}{4} + \operatorname{cosec}\frac{\alpha}{8} + \dots$

(3) $\cot\alpha\operatorname{cosec}\alpha + 2\cot 2\alpha\operatorname{cosec} 2\alpha + 2^2\cot 2^2\alpha\operatorname{cosec} 2^2\alpha + \dots$

(4) $\sec\alpha\sec(\alpha+\beta) + \sec(\alpha+\beta)\sec(\alpha+2\beta) + \dots$

(5) $\cos\alpha\cos\beta + \cos 2\alpha\cos 2\beta + \cos 3\alpha\cos 3\beta + \dots$

(6) $\frac{\sec\alpha\sec 3\alpha}{\operatorname{cosec} 2\alpha} + \frac{\sec 3\alpha\sec 5\alpha}{\operatorname{cosec} 4\alpha} + \frac{\sec 5\alpha\sec 7\alpha}{\operatorname{cosec} 6\alpha} + \dots$

(7) $\sin^2\alpha + \sin^2 2\alpha + \sin^2 3\alpha + \dots$

(8) $\sin^3\alpha + \sin^3 2\alpha + \sin^3 3\alpha + \dots$

(9) $\cos 2\alpha\operatorname{cosec}^2 2\alpha + 2\cos 4\alpha\operatorname{cosec}^2 4\alpha + 4\cos 8\alpha\operatorname{cosec}^2 8\alpha + \dots$

(10) $\sin\frac{2r\pi}{n} + \sin\frac{4r\pi}{n} + \sin\frac{6r\pi}{n} + \dots$

(11) $\sin\alpha\sin 2\alpha\sin 3\alpha + \sin 2\alpha\sin 3\alpha\sin 4\alpha + \dots$

Pet' Camb': '90

(12) $\cos\alpha + 2\cos\frac{\alpha}{2} + 4\cos\frac{\alpha}{2}\cos\frac{\alpha}{4} + 8\cos\frac{\alpha}{2}\cos\frac{\alpha}{4}\cos\frac{\alpha}{8} + \dots$

Dickson: E. T. lv

(13) $\frac{\sin\alpha}{\cos^2\alpha} + \frac{\sin 3\alpha}{\cos^2 2\alpha\cos^2\alpha} + \frac{\sin 5\alpha}{\cos^2 3\alpha\cos^2 2\alpha} + \dots$

(14) $\log(1 + 2\cos\alpha) + \log(1 + 2\cos 3\alpha) + \log(1 + 2\cos 9\alpha) + \dots$

(15) $\tan\pi/2^{n+1} + 2\tan\pi/2^n + \dots + 2^{n-2}\tan\pi/2^3 + 2^{n-1}$

(16) $\tan^{-1}2 + \tan^{-1}\frac{4}{1+3\cdot 4} + \tan^{-1}\frac{6}{1+8\cdot 9} + \tan^{-1}\frac{8}{1+15\cdot 16} + \dots$

Joh' Camb': '79

(17) $\tan^{-1}\frac{1}{3} + \tan^{-1}\frac{1}{7} + \tan^{-1}\frac{1}{13} + \tan^{-1}\frac{1}{21} + \dots$

(18) $\tan^{-1}\alpha + \tan^{-1}\frac{\alpha}{1+2\alpha^2} + \tan^{-1}\frac{\alpha}{1+6\alpha^2} + \tan^{-1}\frac{\alpha}{1+12\alpha^2} + \dots$

NOTE—*All the foregoing can be done by finding, in each case, a suitable* **formula of reduction.**

6. Sum to n terms each of the series whose nth terms are—

(1) $\sin^3 3^{n-1}\theta / 3^{n-1}$

(2) $(-1)^{n-1} \cos^3 3^{n-1}\theta / 3^{n-1}$

(3) $\cos 3^{n-1}\theta / \sin 3^n \theta$

(4) $3^{n-1} \sin 2 \times 3^{n-1}\theta / (1 + 2\cos 2 \times 3^{n-1}\theta)$

(5) $(2^n \cos 3^{n-1}\theta - 2^{n-1}\cos 3^n \theta)/\sin 3^n \theta$

(6) $(1 + 2\cos 2^{2n-1}\theta)/\sin 2^{2n}\theta$

(7) $3^{n-1}(1 - 2\cos 2^n \theta)/\sin 2^n \theta$

(8) $4^{n-1}(5\sin 3\times 4^{n-1}\theta - 3\sin 5\times 4^{n-1}\theta)/(\cos 3\times 4^{n-1}\theta - \cos 5\times 4^{n-1}\theta)$

(9) $3^{n-1}(1 + 4\sin 4^{n-1}\theta \sin 3 \times 4^{n-1}\theta)/\sin 4^n \theta$

(10) $(3\sin 3^{n-1}\theta - \sin 3^n \theta)/(3^{n-1}\cos 3^n \theta)$

Wolstenholme: 361

NOTE—*In each case find a* **formula of reduction.**

7. Prove that $\cot\theta + \frac{1}{2}\tan\frac{\theta}{2} + \frac{1}{4}\tan\frac{\theta}{4} + \dots = \frac{1}{\theta}$

NOTE—*This can be done by a* **formula of reduction**; *but if the Student knows how to differentiate, he will see that it comes at once by taking logs of* $\sin\theta/\theta = \cos\frac{\theta}{2}\cos\frac{\theta}{4}\cos\frac{\theta}{8}\dots$ (see p. 170) *and then differentiating.*

8. Sum to infinity each of the following—

(1) $\sin\alpha + \frac{\sin\alpha}{1}\sin 2\alpha + \frac{\sin^2\alpha}{2!}\sin 3\alpha + \frac{\sin^3\alpha}{3!}\sin 4\alpha + \dots$

(2) $1 + e^{-c\theta}\cos n\theta + e^{-2c\theta}\cos 2n\theta + e^{-3c\theta}\cos 3n\theta + \dots$

NOTE—*Deduce from Example* 5.

(3) $(1 - 3^{-\frac{1}{2}}) - \frac{1}{3}(1 - 3^{-\frac{3}{2}}) + \frac{1}{5}(1 - 3^{-\frac{5}{2}}) - \dots$

NOTE—*Use the expansion of* $\tan^{-1} x$.

(4) $\tan^{-1}\frac{1}{2} + \tan^{-1}\frac{1}{8} + \tan^{-1}\frac{1}{18} + \dots + \tan^{-1}\frac{1}{2r^2} + \dots$

NOTE—*Show that sum of* r *terms is* $\tan^{-1} r/(r+1)$ *and deduce.*

9. Show that $\sin^3 \alpha$ expanded in terms of α is

$$\frac{3}{4}\left\{\frac{3^2-1}{3!}\alpha^3 - \frac{3^4-1}{5!}\alpha^5 + \ldots \pm \frac{3^{2n}-1}{(2n+1)!}\alpha^{2n+1} \mp \ldots\right\}$$

Pet' Camb': '60

NOTE—*Recollect that* $\sin^3 \alpha = \frac{3}{4}\sin\alpha - \frac{1}{4}\sin 3\alpha$

10. Show that, for all integral values of n, $n \cot n\alpha$

$$= \cot\alpha + \cot(\alpha + \pi/n) + \cot(\alpha + 2\pi/n) + \ldots + \cot\{\alpha + (n-1)\pi/n\}$$

Math' Tri': '66

11. If $S_n = \frac{1}{1^n} + \frac{1}{2^n} + \frac{1}{3^n} + \ldots$

$$\Sigma_n = \frac{1}{1^n} + \frac{1}{3^n} + \frac{1}{5^n} + \ldots$$

prove that $\Sigma_n = \frac{2^n - 1}{2^n} S_n$; and deduce Σ_2 and Σ_4 from § 83 *Cor'* (3).

12. Show that

$$\tan^2\theta = \frac{\sin^2\theta}{1+\sin\alpha} + \frac{\sin^2\theta}{1+\sin\alpha}\cdot\frac{\sin^2\theta+\sin\alpha}{1+\sin\beta}$$

$$+ \frac{\sin^2\theta}{1+\sin\alpha}\cdot\frac{\sin^2\theta+\sin\alpha}{1+\sin\beta}\cdot\frac{\sin^2\theta+\sin\beta}{1+\sin\gamma} + \ldots \textit{ ad inf'}.$$

Joh' Camb': '43

13. Deduce from *Gregorie's Series* that

$$\frac{\theta}{\tan\theta} = 1 - \frac{1}{3}\sin^2\theta - \frac{2}{3\cdot5}\sin^4\theta - \frac{2\cdot4}{3\cdot5\cdot7}\sin^6\theta - \ldots$$

14. Prove that, if θ is small, the expression

$$\frac{1}{2}\sqrt{1+\sin\theta}\,\log(1-\theta) + \tan^{-1}\theta\,\sin\left(\frac{\pi}{3}+\theta\right) = \frac{1}{2}(\sqrt{3}-1)\,\theta$$

to the second order of small quantities.

Joh' Camb': '90

15. In Example 4, writing $\pi/2 + \alpha$ for α, and letting $f(n)$ represent the resulting series, prove that

$$\phi(n+1) = f(n)\sin\beta \quad \text{and} \quad f(n+1) = -\phi(n)\sin\beta$$

16. By means of the last Exercise sum to infinity

$$\cos\alpha + x\cos(\alpha+\beta) + \frac{x^2}{2!}\cos(\alpha+2\beta) + \frac{x^3}{3!}\cos(\alpha+3\beta) + \ldots$$

17. By means of *Chartres' Theorem* (Example 4 preceding) prove that—

(1) $\sin n\alpha - {}_nC_1 \sin(n-1)\alpha\cos\alpha + {}_nC_2\sin(n-2)\alpha\cos^2\alpha - \ldots$

$\pm {}_nC_r \sin(n-r)\alpha\cos^r\alpha \ldots - n\sin\alpha\cos^{n-1}\alpha$

$= 0$, if n is *even*; but $= (-1)^{\frac{n-1}{2}}\sin^n\alpha$, if n is *odd*

(2) $\sin(n+1)\alpha - {}_nC_1\sin n\alpha\cos\alpha + {}_nC_2\sin(n-1)\alpha\cos^2\alpha - \ldots$

$\pm {}_nC_r \sin(n+1-r)\alpha\cos^r\alpha \ldots + \sin\alpha\cos^n\alpha$

$= (-1)^{\frac{n}{2}}\sin^{n+1}\alpha$, if n is *even*; but $= (-1)^{\frac{n-1}{2}}\cos\alpha\sin^n\alpha$ if n is *odd*.

18. Also by the same Theorem sum to infinity—

(1) $\sin\alpha + x\sin 2\alpha + x^2\sin 3\alpha + \ldots$

(2) $\sin\alpha + \sin(\alpha+\beta)\sin\theta + \sin(\alpha+2\beta)\sin^2\theta + \ldots$

(3) $\sin\alpha + \sin(\alpha+\beta)\cos\beta + \sin(\alpha+2\beta)\cos^2\beta + \ldots$

(4) $\sin\alpha + \sin 2\alpha\cos\alpha + \sin 3\alpha\cos^2\alpha + \ldots$

19. Sum to infinity, by treating as the sum of two series, each of—

(1) $1 - \frac{\cos 2\alpha}{2!} + \frac{\cos 4\alpha}{4!} - \ldots$

NOTE—*In Exercise* 16, *if* $x = 1$, $\alpha = 0$, *and* β *is first* $\pi/2 - \theta$, *then* $\pi/2 + \theta$, *two series result: add these, and then write* α *for* θ.

(2) $2\cos\alpha + \frac{3}{2}\cos^2\alpha + \frac{4}{3}\cos^3\alpha + \ldots$

20. Following a process analogous to that in Corollary (5) of Example 3, sum to n terms

$$\sin^4\alpha + \sin^4(\alpha+\beta) + \sin^4(\alpha+2\beta) + \ldots$$

NOTE—*See* p. 71.

21. Show that each of the following is a particular case of Example 6—

(1) $\sin\alpha - \frac{\sin 2\alpha}{2!} + \frac{\sin 3\alpha}{3!} - \ldots$

(2) $\sin\alpha\cos\alpha + \frac{\sin 2\alpha\cos^2\alpha}{2!} + \frac{\sin 3\alpha\cos^3\alpha}{3!} + \ldots$

22. Show that each of the following is a particular case of Exercise 16—

(1) $\cos\alpha + \frac{\cos\alpha}{1}\cos 2\alpha + \frac{\cos^2\alpha}{2!}\cos 3\alpha + \frac{\cos^3\alpha}{3!}\cos 4\alpha + \ldots$

(2) $\cos\alpha + \frac{\sin\alpha}{1}\cos 2\alpha + \frac{\sin^2\alpha}{2!}\cos 3\alpha + \frac{\sin^3\alpha}{3!}\cos 4\alpha + \ldots$

23. If $S = x\sin\theta + \frac{x^2}{2!}\sin 2\theta + \frac{x^3}{3!}\sin 3\theta + \dots$

and $C = 1 + x\cos\theta + \frac{x^2}{2!}\cos 2\theta + \frac{x^3}{3!}\cos 3\theta + \dots$

prove that $x^2 = \left(\tan^{-1}\frac{S}{C}\right)^2 + \left\{\frac{1}{2}\log(S^2 + C^2)\right\}^2$

and $\tan\theta = 2\tan^{-1}\frac{S}{C}\Big/\log(S^2 + C^2)$

24. With the notation, and by means of Example 4 (*Chartres' Theorem*) sum each of the following—

(1) $\phi(0) + \phi(2) + \phi(4) + \dots$ to infinity, and to n terms

(2) $\phi(0) - \phi(2) + \phi(4) - \dots$,, ,, ,,

(3) $\phi(1) + \phi(3) + \phi(5) + \dots$,, ,, ,,

(4) $\phi(1) - \phi(3) + \phi(5) - \dots$,, ,, ,,

(5) $\phi(0) + x\phi(2) + x^2\phi(4) + \dots$,, ,, ,,

(6) $\phi(0) + \frac{x^2}{2!}\phi(2) + \frac{x^4}{4!}\phi(4) + \dots$,,

(7) $x\phi(1) + \frac{x^3}{3!}\phi(3) + \frac{x^5}{5!}\phi(5) + \dots$,,

(8) $\phi(2) - \frac{1}{2}\phi(4) + \frac{1}{3}\phi(6) - \dots$,,

25. With the notation of Exercise 15 for ϕ write f, in each of the foregoing, and sum the eight series produced.

26. Following exactly the process on pages 316 to 318 show that

$$\sin^{-1}x = x + \frac{1}{2}\cdot\frac{x^3}{3} + \frac{1\cdot 3}{2\cdot 4}\cdot\frac{x^5}{5} + \dots$$

Deduce $\tan^{-1}x = \frac{1}{\sqrt{1+x^2}}\left\{1 + \frac{1}{2}\cdot\frac{1}{3}\cdot\frac{1}{1+x^2} + \frac{1\cdot 3}{2\cdot 4}\cdot\frac{1}{5}\cdot\frac{1}{(1+x^2)^2} + \dots\right\}$

27. Show that

$$(1 + \sec\alpha)(1 + \sec 2\alpha)(1 + \sec 4\alpha)\dots \text{ to } n \text{ factors} = \frac{\tan 2^{n-1}\alpha}{\tan\frac{\alpha}{2}}$$

28. Find the Limits of each of the following infinite products—

(1) $\left(1 - 3\tan^2\frac{\alpha}{3}\right)\left(1 - 3\tan^2\frac{\alpha}{9}\right)^3\left(1 - 3\tan^2\frac{\alpha}{27}\right)^9\ldots$

(2) $\left(3\cot^2\frac{\alpha}{3} - 1\right)\left(3\cot^2\frac{\alpha}{9} - 1\right)^3\left(3\cot^2\frac{\alpha}{27} - 1\right)^9\ldots$

(3) $\left(1 - 4\sin^2\frac{\alpha}{3}\right)\left(1 - 4\sin^2\frac{\alpha}{9}\right)\left(1 - 4\sin^2\frac{\alpha}{27}\right)\ldots$

(4) $\left(1 - \frac{4}{3}\sin^2\frac{\alpha}{3}\right)\left(1 - \frac{4}{3}\sin^2\frac{\alpha}{9}\right)\left(1 - \frac{4}{3}\sin^2\frac{\alpha}{27}\right)\ldots$

29. By substituting the series for $\sin\theta$, $\cos\theta$, $\sin 2\theta$, $\cos 2\theta$ in the formulæ—

(1) $\sin 2\theta = \sin\theta\cos\theta$

(2) $\cos 2\theta = 2\cos^2\theta - 1$

(3) $\cos 2\theta = 1 - 2\sin^2\theta$

and equating coefficients of θ^{2n+1} in (1), of θ^{4n} in (2), and of θ^{4n+2} in (3), three algebraic series can be summed: find them.

Purkiss: E. T. iv

30. Show that—

(1) $\tan x = \frac{2}{\pi - 2x} - \frac{2}{\pi + 2x} + \frac{2}{3\pi - 2x} - \frac{2}{3\pi + 2x} + \ldots$

(2) $\frac{\sec x}{4\pi} = \frac{1}{\pi^2 - 4x^2} - \frac{3}{3^2\pi^2 - 4x^2} + \frac{5}{5^2\pi^2 - 4x^2} - \ldots$

Math' Tri': '80

NOTE—*Since* $\cos(\theta - x)/\cos x \equiv \cos\theta + \sin\theta\tan x$; *by expanding each term of the sinister in factors, dividing and re-arranging*; *and expanding* $\sin\theta$, $\cos\theta$ *in the dexter, in powers of* θ; *we get two series in powers of* θ: *equate coeff's of* θ. *This will give* (1), *then apply* (1) *to*

$$\sec x \equiv \tan\left(\frac{\pi}{4} + \frac{x}{2}\right) - \tan x.$$

31. Similarly to the last find series for $\cot x$ and $\operatorname{cosec} x$.

NOTE—*The expressions for* $\tan x$ *and* $\cot x$, *in this and the preceding, come very easily by taking* logs *of* $\cos x$ *and* $\sin x$ *expressed in factors, and differentiating.*

32. Resolve $\sin x + \cos x$ into quadratic factors.

Joh' Camb': '45

33. Prove that $1 + \sin\theta$

$$= 2\left(\frac{\pi+2\theta}{4}\right)^2\left(\frac{3\pi-2\theta}{4\pi}\right)^2\left(\frac{5\pi+2\theta}{4\pi}\right)^2\left(\frac{7\pi-2\theta}{8\pi}\right)^2\left(\frac{9\pi+2\theta}{8\pi}\right)^2\left(\frac{11\pi-2\theta}{12\pi}\right)^2\cdots$$

King's Camb': '90

34. If n is a positive integral power of 2, prove that

$$\sqrt{n} = 2^{\frac{n-1}{2}} \sin\frac{\pi}{n}\sin\frac{2\pi}{n}\ldots\sin\frac{(n-2)\pi}{2n}$$

NOTE—*Deduce from* $\sin x = 2\sin\frac{x}{2}\sin\frac{x+\pi}{2}$

$$= 2^3\sin\frac{x}{4}\sin\frac{x+\pi}{4}\sin\frac{x+2\pi}{4}\sin\frac{x+3\pi}{4} = \&c$$

35. From $\sin n\phi = 2^{n-1}\sin\phi\sin\left(\phi+\frac{\pi}{n}\right)\ldots\sin\left(\phi+\frac{n-1}{n}\pi\right)$ deduce—

(1) $1 = 2^{n-1}\sin\frac{\pi}{2n}\sin\frac{3\pi}{2n}\sin\frac{5\pi}{2n}\ldots\sin\frac{(2n-1)\pi}{2n}$

(2) $n = 2^{n-1}\sin\frac{\pi}{n}\sin\frac{2\pi}{n}\sin\frac{3\pi}{n}\ldots\sin\frac{(n-1)\pi}{n}$

(3) $\tan\phi\tan\left(\phi+\frac{\pi}{n}\right)\tan\left(\phi+\frac{2\pi}{n}\right)\ldots\tan\left(\phi+\frac{n-1}{n}\pi\right)$

$= (-1)^{\frac{n}{2}}$ when n is *even*; but $= (-1)^{\frac{n-1}{2}}\tan n\phi$ when n is *odd.*

NOTE—*Let the Student pay attention to these results: they will be useful in some of the Exercises following.*

36. Prove Huygens' approximation for the length of a circular arc—

$$\text{Length of arc} = 2\beta + \tfrac{1}{3}(2\beta - \alpha)$$

where α = chord of arc, β = chord of half the arc.

NOTE—*If* r = *rad', x* = *arc, then* $\alpha = 2r\sin\frac{x}{2r}$, $\beta = 2r\sin\frac{x}{4r}$: *expand the* sines *in powers of the* $\wedge^s$, *and neglect* x^5 &c.

37. The circumference of the inner of two concentric circles (radii R, r) is divided into n equal parts by $P_1, P_2, \ldots P_n$: if A is a fixed point on the outer circumference, prove that

$$\Sigma PA^2 = n(R^2 + r^2)$$

38. If P is a point within a square ABCD, such that $\Pi\, PA = XA^4$, where X is the cross of the diagonals; and if α denote the angle XPA; prove that (a being a side of the square)

$$2\, XP^4 = a^4 \cos 4\alpha$$

T. C. D.

NOTE—*Use De Moivre's Prop' given as Example* 2.

39. The circumference of a circle of radius R is divided into n parts by points $P_1, P_2, \ldots P_n$, each of which subtends the same angle at a point A within the circle: if CA is a, and $r_1, r_2, \ldots r_n$ are the lengths of $AP_1, AP_2, \ldots AP_n$, prove that

$$\Sigma\, r = (R^2 - a^2)\, \Sigma \frac{1}{r}$$

40. $A_1, A_2, \ldots A_n$ is a regular polygon; C is the centre and R is the radius of its circum-circle; P is any point, and CP is a; $PN_1, PN_2, \ldots$ are perpendiculars on $A_1A_2, A_2A_3, \ldots$, prove that

$$\Sigma\, (N_1 N_2) = n\,(R^2 + a^2) \sin^2 \frac{2\pi}{n}$$

41. $A_1 A_2 \ldots A_{2n+1}$ is a regular polygon, and P is any point in the arc $A_1 A_{2n+1}$: prove that

$$PA_1 + PA_3 + \ldots + PA_{2n+1} = PA_2 + PA_4 + \ldots + PA_{2n}$$

42. If $p_1, p_2, \ldots p_{2n}$ are perpendiculars from any point in the circumference of a circle of radius r, on the sides of a regular circumscribing polygon of 2 n sides, prove that

$$p_1 p_3 \ldots p_{2n-1} + p_2 p_4 \ldots p_{2n} = r^n / 2^{n-1}$$

Math' Tri': '42

NOTE—*Use the formulæ*

$$\sin n\phi = 2^{n-1} \sin\phi \sin\left(\phi + \frac{\pi}{n}\right) \ldots \sin\left(\phi + \frac{n-1}{n}\pi\right)$$

$$\cos n\phi = 2^{n-1} \sin\left(\phi + \frac{\pi}{2n}\right) \sin\left(\phi + \frac{3\pi}{2n}\right) \ldots \sin\left(\phi + \frac{2n-1}{2n}\pi\right)$$

43. One corner of a regular polygon of n sides, each of which is a, is joined to all the rest: prove that—

(1) the *sum* of the joins $= \dfrac{a}{2} \operatorname{cosec}^2 \dfrac{\pi}{2n}$

(2) the *product* of the joins $= n\left(\dfrac{a}{2} \operatorname{cosec} \dfrac{\pi}{n}\right)^{n-1}$

44. ABC is a triangle, right-angled at C, and CP_1, CP_2, ... CP_n are drawn dividing the right angle into $n + 1$ equal parts, where P_1, ... P_n are in AB : prove that

$$\Sigma \frac{1}{CP^2} = \frac{2}{AB^2}\left\{n + \cot\frac{\pi}{2(n+1)}\right\}$$

45. C is the centre of a circle radius r; and A is a point at which the circle subtends a right angle: from A are drawn $2n-1$ secants (of which APQ is a type) and tangents are drawn at P, Q: if perpendiculars p, q are dropped from A on these tangents, prove that

$$\Pi(pq) = n\left\{r^{2n-1}/2^{n-1}\right\}^2$$

NOTE—*Show that* $\Pi(pq) = r^{2(2n-1)} \sin\frac{\pi}{2n} \ldots \sin\frac{2n-1}{2n}\pi$ *and write* $2n$ *for* n *in Exercise* 35, (2).

46. n points are taken at equal distances along the circumference of a circle of radius a—the first being fixed and the other varying: if ρ is the distance of their mean centre (for equal multiples) from the centre of the circle; and ϕ the angle this distance makes with the radius through the fixed point, prove that

$$n\rho\sin\frac{\phi}{n} = a\sin\left(\frac{n-1}{n}\right)\phi$$

NOTE—*If* $\bar{x}$, $\bar{y}$ *are rect'r coord's of mean centre*; x, y *those of any point*, $n\bar{x} = \Sigma x$, $n\bar{y} = \Sigma y$: *then change to polar coord's* (see p. 21). *The result is the polar eq'n to the Locus of the* Centre of Gravity *of* n *equal particles arranged as the points.*

CHAPTER XIX

Disjecta Membra

This chapter consists of various matters, more or less important, inadvertently omitted in preceding chapters.

§ 86. *The trigonometrical solution of a cubic equation.*

It is of course well known that a cubic eq'n can be at once thrown into a form lacking the *square* of the variable.

$$\therefore \quad x^3 - qx - r = 0$$

may be taken to represent *any* cubic eq'n.

Writing y/n for x, we get

$$y^3 - qn^2y - rn^3 = 0 \quad \ldots\ldots\ldots\ldots \quad (1)$$

Compare this with

$$\cos^3\phi - \tfrac{3}{4}\cos\phi - \tfrac{1}{4}\cos 3\phi = 0 \ldots\ldots\ldots \quad (2)$$

The two will be identical if $\cos\phi = y$, provided that we can choose n so that

$$\left.\begin{aligned} qn^2 &= \tfrac{3}{4} \\ \text{and} \quad rn^3 &= \tfrac{1}{4}\cos 3\phi \end{aligned}\right\}$$

$$\left.\begin{aligned} \text{i.e. if} \quad n &= \frac{1}{2}\sqrt{\frac{3}{q}} \\ \text{and} \therefore \quad \cos 3\phi &= \frac{r}{2}\sqrt{\frac{27}{q^3}} \end{aligned}\right\}$$

It is $\therefore$ a necessary condition for the identity of (1) and (2) that

$$\frac{r^2}{4} < \frac{q^3}{27}$$

The method will $\therefore$ apply to the irreducible case of *Cardan's Algebraic Solution*, i.e. to the case when all three roots are real and unequal.

Let α be the least value of 3ϕ satisfying

$$\cos 3\phi = \frac{r}{2}\sqrt{\frac{27}{q^3}}$$

Then one value of y is $\cos\frac{\alpha}{3}$

The other values are $\cos\frac{2\pi \pm \alpha}{3}$ (See p. 132)

∴ the roots are

$$2\sqrt{\frac{q}{3}}\cos\frac{\alpha}{3}, \quad 2\sqrt{\frac{q}{3}}\cos\frac{2\pi+\alpha}{3}, \quad 2\sqrt{\frac{q}{3}}\cos\frac{2\pi-\alpha}{3}$$

Example

Solve $8x^3 - 36x^2 + 42x - 13 = 0$ *by trigonometry.*

Put $y + \frac{3}{2}$ for x and the eq'n becomes

$$y^3 - \tfrac{3}{2}y - \tfrac{1}{2} = 0$$

Writing z/n for y, we get

$$z^3 - \tfrac{3}{2}n^2 z - \tfrac{1}{2}n^3 = 0$$

$$\text{Since } \tfrac{1}{4}\left(-\tfrac{1}{2}n^3\right)^2 < \tfrac{1}{27}\left(-\tfrac{3}{2}n^2\right)^3$$

all the roots are real and unequal, and the method applies.

∴ comparing with $\cos^3\phi - \frac{3}{4}\cos\phi - \frac{1}{4}\cos 3\phi \equiv 0$

we have $z = \cos\phi$, $n = \frac{1}{\sqrt{2}}$, $2n^3 = \cos 3\phi$

whence $\cos 3\phi = \frac{1}{\sqrt{2}} = \cos\frac{\pi}{4}$, or $\cos\left(2\pi \pm \frac{\pi}{4}\right)$

$$\therefore \quad \phi = \frac{1}{3}\left(\frac{\pi}{4}\right), \quad \text{or} \quad \frac{1}{3}\left(2\pi \pm \frac{\pi}{4}\right)$$

$$z = \frac{+\sqrt{3}+1}{+2\sqrt{2}}, \quad \text{or} \quad \frac{-\sqrt{3}+1}{+2\sqrt{2}}, \quad \text{or} \quad \frac{1}{-\sqrt{2}}$$

$$\therefore \quad y = \frac{\pm\sqrt{3}+1}{2}, \quad \text{or} \quad -1$$

$$\therefore \quad x = \frac{\pm\sqrt{3}+4}{2}, \quad \text{or} \quad \tfrac{1}{2}$$

Exercises

1. Solve the equation $x^3 - 6x - 4 = 0$ by trigonometry.

2. In the equations $x^3 \pm qx - r = 0$, if $27r^2 > 4q^3$, so that *Cardan's method* applies, the values of x are given by

$$x = \sqrt[3]{\left\{\frac{r}{2} + \sqrt{\frac{r^2}{4} \pm \frac{q^3}{27}}\right\}} + \sqrt[3]{\left\{\frac{r}{2} - \sqrt{\frac{r^2}{4} \pm \frac{q^3}{27}}\right\}}$$

upper signs going together, and lower together: show that these values can be put in the forms

$$\sqrt[3]{r \sec\theta}\left(\cos^{\frac{2}{3}}\frac{\theta}{2} - \sin^{\frac{2}{3}}\frac{\theta}{2}\right)$$

$$\sqrt[3]{r}\left(\cos^{\frac{2}{3}}\frac{\theta}{2} + \sin^{\frac{2}{3}}\frac{\theta}{2}\right)$$

3. Solve by trigonometry (using a table-book) the equation

$$x^3 - 18x^2 + 87x - 70 = 0$$

4. Show that, similarly to the case of a cubic, the quintic equation

$$x^5 + px^3 + qx + r = 0$$

may be solved trigonometrically, by comparing it with

$$\cos 5\theta \equiv 16\cos^5\theta - 20\cos^3\theta + 5\cos\theta$$

provided that $p^2 = 5q$, and that p and $4pq^2 + 125r^2$ are both negative.

Apply the method to solve

$$x^5 + 5x^4 - 20x^2 - 5x + 3 = 0.$$

5. AB, BC, CD are three consecutive chords of a semi-circle, whose respective lengths are as the numbers 1, 2, 3; find a cubic equation to give the length of the diameter AD; and solve it by trigonometry.

6. A circle, centre C, cuts in A, B, the arms of an angle ACB: it is an easily proved geometrical fact that, *if* P can be found in AC produced so that, PB cutting the circle in Q, PQ is equal to CA, then the parallel through C to PQB is a trisector of ACB: show that, taking the radius as unity, and denoting CP and ACB by x and α respectively, the equation giving x is

$$x^3 - 3x - 2\cos\alpha = 0$$

And solve it by trigonometry, in the particular case when α is 60°, interpreting the meaning of each of the three values of x.

§ **87**. *The geometrical division of a given angle into parts whose* T. F.s *are in a given ratio.*

Let AOB be the given $\wedge$, and m : n the given ratio.

Case 1—*for* sines *and* cosecants.

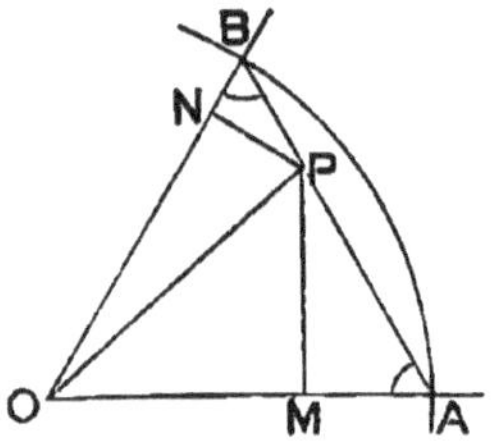

With O as centre, and any radius, describe a $\odot$ cutting the arms of the $\wedge$ in A, B.

Divide AB in P, so that PA : PB = m : n
Then OP divides AOB as req'd.
For dropping PM, PN $\perp^s$ on OA OB, we have

$$\sin POA : \sin POB = PM : PN$$
$$= PA : PB \quad \text{by sim'r } \Delta^s$$
$$= m : n$$

And $\operatorname{cosec} POB : \operatorname{cosec} POA = m : n$

Case 2—*for* cosines and secants.

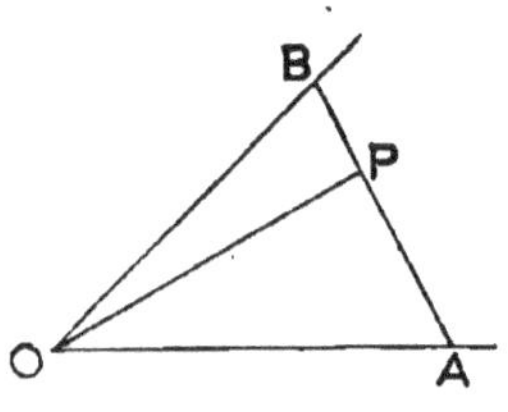

Take any length OB in one arm, and OA in the other so that

$$m : n = OB : OA$$

Draw OP $\perp$ to AB; then OP divides as req'd.

For $\cos POA : \cos POB = OP/OA : OP/OB$
$$= m : n$$

And $\sec POB : \sec POA = m : n$

Case 3—*for* tangents *and* cotangents.

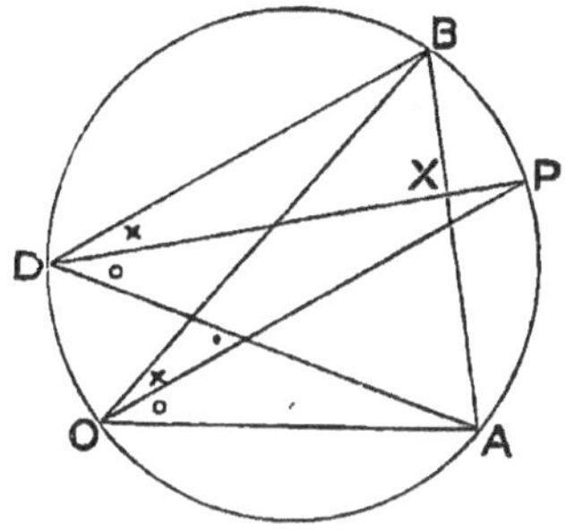

Describe any ⊙ thro′ O cutting the arms in A, B.

Divide AB in X, so that

$$XA : XB = m : n$$

Draw XD ⊥ to AB to meet arc AOB in D; and let DX meet ⊙ again in P.

Then OP divides as req′d.

For $\tan POA : \tan POB = \tan PDA : \tan PDB$

$$= XA/XD : XB/XD$$

$$= m : n$$

And $\cot POB : \cot POA = m : n$

§ 88. *Geometrical construction of an angle of which a* T.F. *is given.*

Example

If $\cos\theta = \sqrt{\frac{2}{3}}$, *construct* θ *geometrically.*

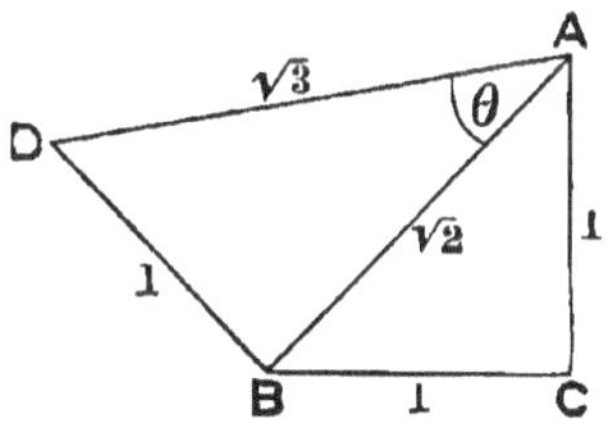

Take CA, CB of unit length and ⊥ to each other.

Then $AB = \sqrt{2}$

Draw BD ⊥ to BA, and of unit length.

Then $AD = \sqrt{3}$

$$\therefore \cos BAD \left(\text{which} = \frac{AB}{AD}\right) = \sqrt{\frac{2}{3}}$$

i.e. $B\hat{A}D = \theta$

Exercises

Construct geometrically $\sec^{-1} 3$, $\sin^{-1} \frac{1}{2\sqrt{3}}$, $\operatorname{cosec}^{-1} 5$, $\tan^{-1} \sqrt{5}$.

See also Exercise 44, p. 150.

§ 89. *Some of the results of §§ 32 and 33 can be easily found by geometry.*

Examples

1. *To find* $\sin 18°$ *geometrically.*

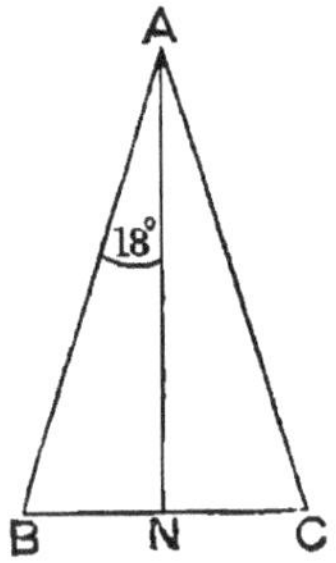

Let ABC be a $\triangle$, described by *Euclid* iv. 10, such that

$$\hat{B} = 2\,B\hat{A}C = \hat{C}$$

Drop AN $\perp$ on BC

Then $B\hat{A}N = 18°$

For AB put r, and for BC put x.

Then $\sin 18° = \dfrac{x}{2r}$

But $x^2 = r(r - x)$ (by *Euc'* iv. 10)

$$\therefore \left(\frac{x}{r}\right)^2 + \frac{x}{r} = 1$$

$$\therefore 2\frac{x}{r} + 1 = \sqrt{5}$$

$$\therefore \sin 18° = \frac{\sqrt{5} - 1}{4}$$

where $+\sqrt{5}$ must be taken, $\because$ $\sin 18°$ is pos'.

2. *To find* $\tan 15°$ *geometrically.*

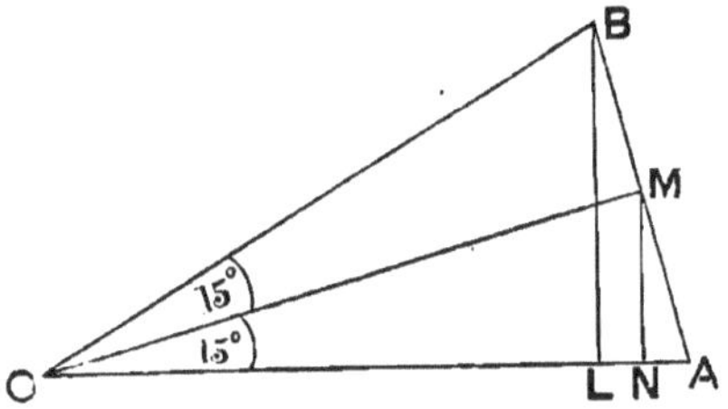

Let $A\hat{O}B$ be 30°.

Bisect AOB by OM; and draw AMB $\perp$ to OM.

Drop BL, MN $\perp^s$ on OA.

Then (by § 31) if BL is taken as unity, OB will be 2, and MN will be $\frac{1}{2}$. Put x for BM, and y for OM.

$$\text{Then} \quad \tan 15^\circ = \frac{x}{y}$$

$$\text{But} \quad x : 2 = \tfrac{1}{2} : y \quad (\text{Sim'r } \Delta^s \text{ OMB, ONM})$$

$$\therefore \quad xy = 1$$

$$\text{Also} \quad x^2 + y^2 = 4$$

$$\therefore \quad \frac{x}{y} + \frac{y}{x} = 4$$

$$\therefore \quad \left(\frac{x}{y}\right)^2 - 4\left(\frac{x}{y}\right) = -1$$

$$\therefore \quad \tan 15^\circ = 2 - \sqrt{3}$$

where $-\sqrt{3}$ must be taken, $\because$ $\tan 15^\circ < 1$, obviously.

Exercises

1. Find geometrically $\sin 15^\circ$, $\tan 22^\circ\frac{1}{2}$, $\cos 36^\circ$, $\sec 72^\circ$.

2. Prove geometrically that

$$\cot 11^\circ\tfrac{1}{4} = 1 + \sqrt{2} + \sqrt{2}\sqrt{2 + \sqrt{2}}$$

§ 90. *On so-called 'geometrical proofs' of some of the fundamental formulæ of Chapter IV.*

As in §§ 18, 19 and 20. we saw that the geometrical parts of the proofs of the formulæ therein treated of only applied to special cases, and required additional reasoning to make them general; so with other fundamental formulæ geometrical constructions may be used to prove them *in particular cases*, but such constructions do not (as a rule) prove them for other cases. Excepting in the few instances when we can exhaust the possible cases, the only way in which purely geometrical proofs can be made universal is by the generalizing principles of *projection*. But these principles are difficult to apply thoroughly, or to see the full force of, until considerable familiarity with the algebraic treatment of geometry by coordinates has been attained; and unless a proof by projection is carefully and thoroughly elaborated, it amounts to a mere 'begging of the question.'

Here follow some geometrical proofs in special cases.

1. T. F.s of 2α *in terms of* T. F.s of α, when 2α *is less than a right angle.*

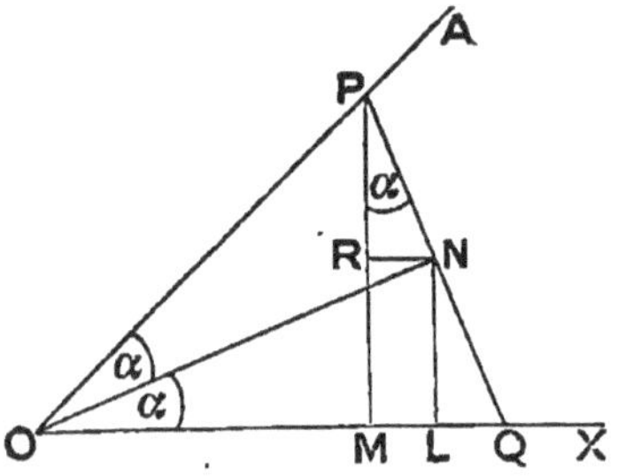

Let OA, starting from coincidence with OX, revolve thro′ 2α, where $2\alpha < 90°$.

Take equal lengths OP, OQ in OA, OX respectively.
Let bisector of 2α meet PQ in N.
Drop ⊥s PM, NL on OX; and NR on PM.

Obviously $\triangle ONP \equiv \triangle ONQ$

so that N is mid p′t of PQ, and ∴ PM = 2 LN, and PR = NL.

$$\therefore \quad \sin 2\alpha = \frac{MP}{OP} = \frac{2\,LN}{ON}\cdot\frac{ON}{OP} = 2\sin\alpha\cos\alpha$$

$$\cos 2\alpha = \frac{OM}{OP} = \frac{OL-RN}{OP} = \frac{OL}{ON}\cdot\frac{ON}{OP} - \frac{RN}{PN}\cdot\frac{PN}{OP} = \cos^2\alpha - \sin^2\alpha$$

$$\tan 2\alpha = \frac{MP}{OM} = \frac{2\,LN}{OL-RN} = \frac{\dfrac{2\,LN}{OL}}{1-\dfrac{RN}{PR}\cdot\dfrac{NL}{OL}} = \frac{2\tan\alpha}{1-\tan^2\alpha}$$

2. *Expansion of* $\tan(\alpha-\beta)$ *when* α *is within* $5°$ *of* $220°$, *and* β *within* $5°$ *of* $40°$.

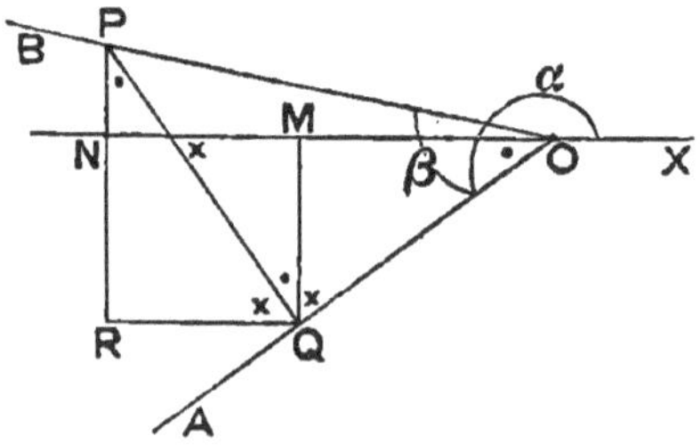

Let OA, initially along OX, revolve in pos′ direc′ thro′ α; and then back in neg′ direc′ thro′ β to OB; where α, β have *about* the values assigned.

In OB take any p't P; and drop $\perp^s$, PQ on OA, PN, QM on XO produced, and QR on PN.

Then the sim'ly marked $\wedge^s$ are clearly equal.

$$\text{Now} \quad \frac{NP}{ON} = -\frac{NP}{NO} = -\frac{RP - MQ}{MO + RQ} = \frac{\frac{MQ}{MO} - \frac{RP}{MO}}{1 + \frac{MQ}{MO} \cdot \frac{RQ}{MQ}}$$

$$\text{And} \quad \frac{RP}{MO} = \frac{RQ}{MQ} = \frac{QP}{QO}, \quad \text{by sim'r } \Delta^s$$

$$\therefore \quad \tan(\alpha - \beta) = \frac{\tan\alpha - \tan\beta}{1 + \tan\alpha \tan\beta}$$

3. *To show that the expansion of* sin $(\alpha - \beta)$ *can be deduced immediately from* **Ptolemy's Theorem.**

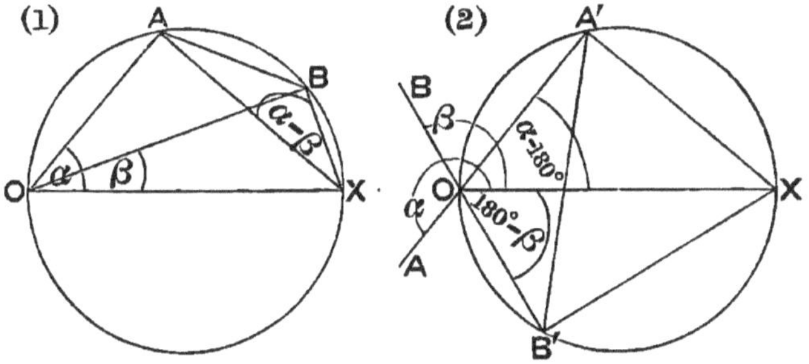

With any diam' OX (which = D) let a ⊙ be described; and let OA, OB, initially along OX, revolve, in pos' direc' thro' α, β respectively.

We will consider two cases—

In fig' (1) each of α and $\beta < 90°$.

,, (2) α is between 180° and 270°, and β is between 90° and 180°.

In fig' (1) OA, OB meet ⊙ in A, B.

,, (2) AO, BO ,, A′, B′.

Then, making joins as in the fig's, we have by *Ptolemy's Theorem*

$$AB.OX = AX.OB - OA.BX \quad \text{in fig}' (1)$$

$$A'B'.OX = A'X.OB' + OA'.B'X \quad ,, \ (2)$$

$$\therefore \quad D^2 \sin\overline{\alpha - \beta} = D^2 \sin\alpha\cos\beta - D^2\cos\alpha\sin\beta, \quad \text{in both}$$

$$\text{whence} \quad \sin(\alpha - \beta) = \sin\alpha\cos\beta - \cos\alpha\sin\beta$$

Sim'ly other cases can be done.

4. *To prove geometrically the second group of formulæ on page 65, when α is about 250° and β about 150°.*

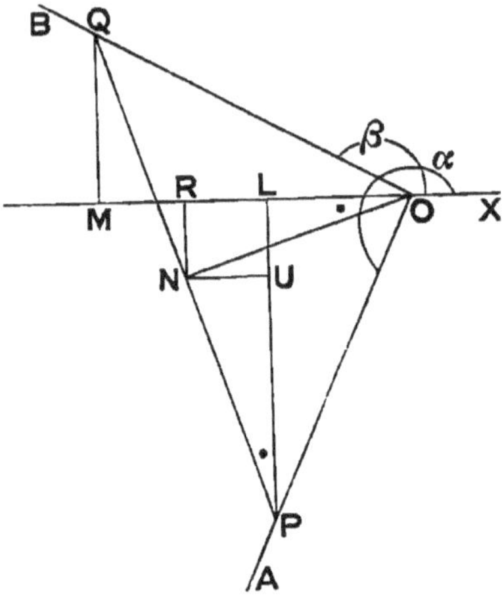

Let OA, OB, initially along OX, revolve respectively, in pos' direc', thro' α, β, where these $\wedge^s$ have *about* the values assigned.

Along OA, OB take OP, OQ, any equal lengths.

Drop ON $\perp$ on PQ.

Then Δ ONP $\equiv$ Δ ONQ.

Drop PL, NR, QM $\perp^s$ on XO produced; and NU $\perp$ on PL.

$$\text{Then}\quad X\hat{O}N = \frac{\alpha+\beta}{2}, \quad \text{and}\quad P\hat{O}N = \frac{\alpha-\beta}{2} = Q\hat{O}N$$

$$\therefore\quad \sin\alpha + \sin\beta = \frac{LP}{OP} + \frac{MQ}{OQ}$$

$$= \frac{2\,RN}{ON}\cdot\frac{ON}{OP}$$

$$= 2\sin\frac{\alpha+\beta}{2}\cos\frac{\alpha-\beta}{2}$$

$$\text{And}\quad \cos\alpha - \cos\beta = \frac{OL}{OP} - \frac{OM}{OQ}$$

$$= -\frac{LM}{OP}$$

$$= -\frac{2\,UN}{PN}\cdot\frac{PN}{OP}, \quad \text{since } RL = RM$$

$$= -2\frac{RN}{ON}\cdot\frac{PN}{OP}, \quad \text{by sim'r } \Delta^s$$

$$= -2\sin\frac{\alpha+\beta}{2}\sin\frac{\alpha-\beta}{2}$$

The other two can be sim'ly proved.

5. *To prove geometrically the formulæ at the end of page* 68, *and top of page* 69, *taking the simple case when* α *and* β *are acute angles.*

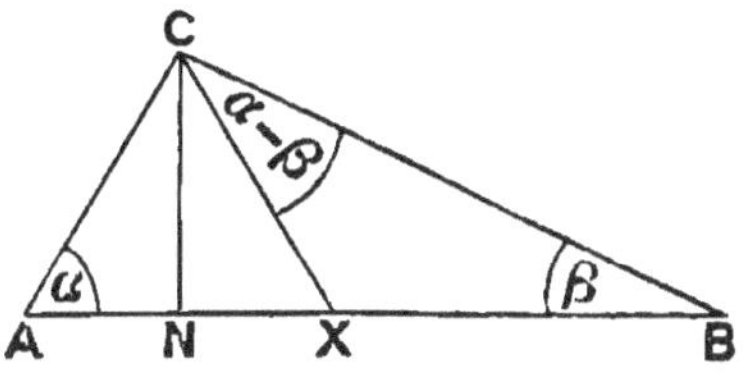

Arrange α, β so as to be the $\wedge$ˢ A, B respectively of $\triangle$ ABC. Then if $\alpha > \beta$, CB > CA.

Draw CX equal to CA, so that X is in AB; and drop CN ⊥ on AB.

$$\text{Since} \quad \frac{CA}{\sin B} = \frac{CX}{\sin B}$$

∴ the ⊙ˢ round CAB, CXB, have diam's of same length, D say.

$$\text{Then} \quad CB^2 - CA^2 = BN^2 - AN^2 = AB \, . \, BX$$

$$\therefore \quad \left(\frac{CB}{D}\right)^2 - \left(\frac{CA}{D}\right)^2 = \frac{AB}{D} \cdot \frac{BX}{D}$$

$$\text{whence} \quad \sin^2\alpha - \sin^2\beta = \sin(\alpha+\beta)\sin(\alpha-\beta)$$

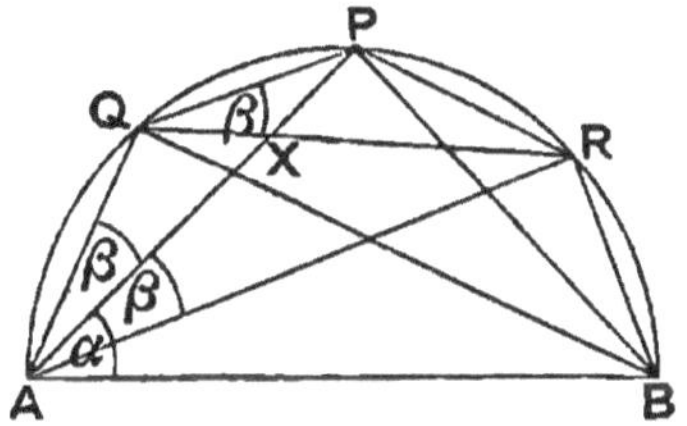

Next, on any line AB, equal to D, as diam' describe a semi-⊙; and let BAP, PAQ be α, β respectively, where $\alpha > \beta$, and P, Q are on the arc.

Draw AR to meet arc, so that $\widehat{PAR} = \beta$.

Let QR cut AP in X; and make joins as in fig'.

Then AQ.AR = AP.AX (See *Euclid Revised*, p. 287)

= AP (AP – PX)

= AP² – PQ², since PQ touches ⊙ AQX.

$$\therefore \quad \frac{AQ}{D} \cdot \frac{AR}{D} = \left(\frac{AP}{D}\right)^2 - \left(\frac{PQ}{D}\right)^2$$

$$\text{whence} \quad \cos(\alpha+\beta)\cos(\alpha-\beta) = \cos^2\alpha - \sin^2\beta$$

6. *Geometrical proof of the formulæ of* § 68.

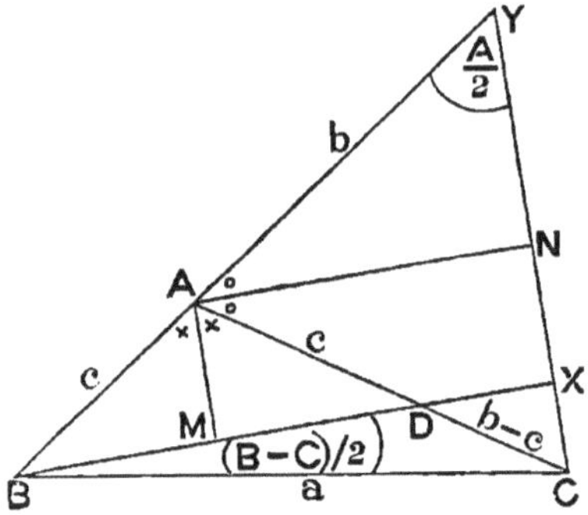

AM, AN are intern′ and extern′ bisectors of $\hat{A}$.

Rest of fig′ will, for brevity, be considered to explain itself.

Then $$\frac{\tan\frac{B-C}{2}}{\cot\frac{A}{2}} = \frac{XC}{XY} = \frac{CD}{YB} = \frac{b-c}{b+c}$$

Again $$\frac{\sin\frac{A}{2}}{\cos\frac{B-C}{2}} = \frac{BX}{b+c}\cdot\frac{a}{BX}$$

and $$\frac{\cos\frac{A}{2}}{\sin\frac{B-C}{2}} = \frac{CX}{b-c}\cdot\frac{a}{CX}$$

whence $$(b+c)\frac{\sin\frac{A}{2}}{\cos\frac{B-C}{2}} = a = (b-c)\frac{\cos\frac{A}{2}}{\sin\frac{B-C}{2}}$$

Same fig′ gives $\sin^2\frac{A}{2} = \frac{NA}{b}\cdot\frac{MD}{c} = \frac{4\,MX\,.\,MD}{4\,bc}$

But, by *Euc′* ii. 8, $4\,MX\,.\,MD + DX^2 = BX^2$

$\therefore\ 4\,MX\,.\,MD = BX^2 - DX^2 = a^2 - (b-c)^2 = 4\,s_2 s_3$

$$\therefore\ \sin\frac{A}{2} = \sqrt{\frac{s_2 s_3}{bc}}$$

7. *A circle touches the arms of an angle* AOB *in* A, B, *and* OXY, *any line drawn to cut the circle in* X, Y, *divides* AOB *into parts* θ, ϕ : *to show that* $\sin\theta \sin\phi$ *varies as* XY^2

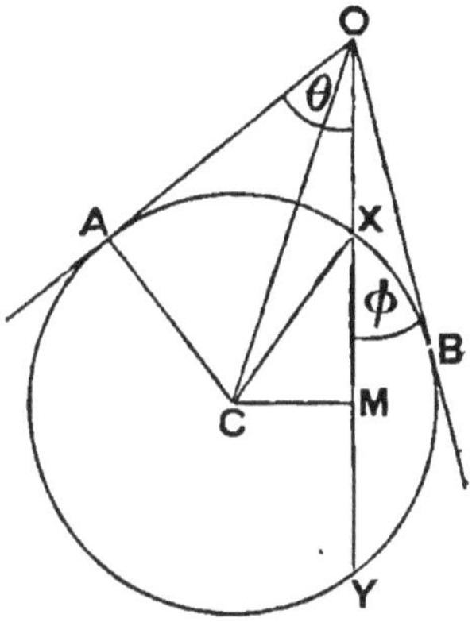

Let C be centre of ⊙: drop CM ⊥ on XY; and join CO, CA, CX.

$$C\hat{O}A = \frac{\theta + \phi}{2}$$

$$C\hat{O}M = \frac{\theta - \phi}{2}$$

Then $\sin\theta \sin\phi = \sin^2\frac{\theta+\phi}{2} - \sin^2\frac{\theta-\phi}{2}$ (See p. 69)

$$= \left(\frac{CA}{CO}\right)^2 - \left(\frac{CM}{CO}\right)^2$$

$$= \frac{CX^2 - CM^2}{CO^2}$$

$$= \left(\frac{MX}{CO}\right)^2 = \left(\frac{XY}{2\,CO}\right)^2$$

and ∴ $\propto XY^2$

Cor'— $\dfrac{\text{Product } \perp^s \text{ from A, B on OXY}}{XY^2}$ is const'

The Student can now easily prove for himself other cases of the preceding results; and may try to find out similar proofs of other formulæ. Enough of this kind of work—entirely unimportant as it is—has been now given here.

§ 91. *Some generalizing methods.*

1. From any known relation between sines and cosines of $\wedge^s$ A, B, C of a Δ, endless other relations can be deduced thus—

Let l, m, n be integers connected by the relation

$$l + m + n = 6k - 1$$

where k may be *any* pos′ or neg′ integer.

$$\text{Put } \alpha \text{ for } lA + mB + nC$$
$$\beta \quad ,, \quad mA + nB + lC$$
$$\gamma \quad ,, \quad nA + lB + mC$$

Then, since $\overline{2k\pi - \alpha} + \overline{2k\pi - \beta} + \overline{2k\pi - \gamma} = \pi$,

$$\therefore \quad \overline{2k\pi - \alpha}, \quad \overline{2k\pi - \beta}, \quad \overline{2k\pi - \gamma} \text{ are } \wedge^s \text{ of a } \Delta.$$

$$\text{Now} \quad \sin(2k\pi - \alpha) = -\sin\alpha$$
$$\text{and} \quad \cos(2k\pi - \alpha) = \cos\alpha$$

Hence it is at once obvious that all relations between sines and cosines of A, B, C, will hold for α, β, γ, *if* the signs of all sines are changed.

2. The plan, given on page 245, by which all formulæ involving sides and $\wedge^s$ of a Δ can be extended through the relations connecting the sides and $\wedge^s$ of the pedal Δ with those of the original Δ, may be still further extended, by using the pedal Δ of the pedal Δ, and then again the pedal Δ of this, and so on.

Thus, denoting corresponding parts of the successive pedal Δ^s by dotted letters, we have

$$A' = \pi - 2A, \quad A'' = \pi - 2A' = 4A - \pi, \quad A''' = 3\pi - 8A \ldots$$

$$a' = a\cos A, \quad a'' = -a\cos A\cos 2A, \quad a''' = a\cos A\cos 2A\cos 4A$$

$$\text{Now} \quad \Delta' = \Delta \,.\, 2\Pi\cos A \quad \text{(p. 245)}$$

$$\therefore \quad \Delta'' = -\Delta' \,.\, 2\Pi\cos 2A = -\Delta \,.\, 2^2\Pi(\cos A\cos 2A)$$

$$\Delta''' = -\Delta'' \,.\, 2\Pi\cos 4A = \Delta \,.\, 2^3\Pi(\cos A\cos 2A\cos 4A)$$

. . . .

And the area of the nth pedal Δ can be expressed in terms of the area and $\wedge^s$ of ABC.

MISCELLANEOUS EXERCISES

1. Given that—

$$\log \tan 76^\circ\, 11'\, 15'' = \cdot 6093 \qquad \log 761875 = 5\cdot 8819$$
$$\log \tan 53^\circ\, 3' = \cdot 1237 \qquad \log 233050 = 5\cdot 3675$$
$$\log \pi = \cdot 4971 \qquad \log 180000 = 5\cdot 2553$$

show that, if $76^\circ\, 11'\, 15'' = \theta$ radians, then $x = \theta$ is a solution of the equation $\tan x = \tan^{-1} x$.

Ox' 2nd Public Exam': '77

2. In a triangle given a, b and Δ, show that the product of the possible values of c is

$$\sqrt{16\,\Delta^2 + (a^2 - b^2)^2}$$

Draw a figure showing the ambiguity.

3. If $\left.\begin{aligned} \sin\alpha + \sin\beta + \sin\gamma &= 0 \\ \text{and } \cos\alpha + \cos\beta + \cos\gamma &= 0 \end{aligned}\right\}$

find the value of $\sin\dfrac{\alpha-\beta}{2}\sin\dfrac{\beta-\gamma}{2}\sin\dfrac{\gamma-\alpha}{2}$.

Ox' 1st Public Exam': '77

4. Show that the Limit, when $x = \pi/2$, of

$$\frac{\sqrt{3 + \cos 2x} - 2\sin x}{x \sin 2x + 2x\cos x} \quad \text{is} \quad \frac{\sqrt{3}}{2\pi}$$

5. Reduce to a single term the expression

$$\frac{\tan\alpha}{\tan(\alpha-\beta)\tan(\alpha-\gamma)} + \frac{\tan\beta}{\tan(\beta-\gamma)\tan(\beta-\alpha)} + \frac{\tan\gamma}{\tan(\gamma-\alpha)\tan(\gamma-\beta)}$$

R. U. Schol': '83

6. If $x\cos(\alpha_1 + \alpha_2) + y\sin(\alpha_1 + \alpha_2) - \cos(\alpha_1 - \alpha_2)$ is denoted by A_{12}, prove that $A_{12}\,.\,A_{34}\,.\,A_{56} - A_{23}\,.\,A_{45}\,.\,A_{61}$ is divisible by $x^2 + y^2 - 1$

Q. C. B. '73

7. Find a relation between the lengths of the six joins of four co-planar points.

NOTE—*Take any three joins which form a* Δ, *and let* α, β, γ *be the* $\wedge^s$ *they subtend at the 4th p't: then (Ex' 1, p. 79)* $\Sigma \cos^2 \alpha = 1 + 2 \Pi \cos \alpha$; *and each* cosine *is expressible in terms of 3 of the joins.*

8. If XYZ is the pedal triangle of ABC, prove that the mean centre of the six points A, B, C, X, Y, Z, for respective multiples

$$\sin^2 A,\ \sin^2 B,\ \sin^2 C,\ \cos^2 A,\ \cos^2 B,\ \cos^2 C,$$

coincides with the mean centre of X, Y, Z for any equal multiples.

W. H. H. Hudson: E. T. xx

9. Draw graphs to show the relative magnitudes of $\cos 5\theta$ and $\cos \theta$ for all values of θ.

Prove that the ratio of these quantities always lies between $-1\frac{1}{4}$ and $+5$; and mark on the diagram points corresponding to these limiting values.

R. U. Schol': '91

10. Show that the equation

$$x \tan \frac{\alpha}{2} = (\sqrt{1+x} - 1)(\sqrt{1-x} - 1)$$

is satisfied by $x = -\sin 2\alpha$

R. U. Schol': '82

11. Three arcs α, β, γ are taken consecutively along the circumference of a circle whose radius is unity: prove that the area of the quadrilateral formed by joining the four points so determined is

$$2 \sin \frac{\alpha+\beta}{2} \sin \frac{\beta+\gamma}{2} \sin \frac{\gamma+\alpha}{2}.$$

R. U. Schol': '82

12. Solve the equation in x

$$x^2 \sin 2\alpha + 2x(\sin \alpha + \cos \alpha) + 2 = 0$$

And find the equation whose roots are the squares of its roots.

13. Two ships are sailing uniformly with velocities u, v, along straight lines inclined at an angle α: if a, b are their distances at one time from the cross of their courses, show that their *minimum* distance apart is

$$(av - bu) \sin \alpha / \sqrt{u^2 + v^2 - 2uv \cos \alpha}$$

Math' Tri': '43

14. In any triangle prove that

$$\Sigma\left(\frac{\cos A}{c \sin B}\right) \times \Sigma\left(\frac{1}{a \sin C}\right) = 2\Sigma\left(\frac{1}{ab}\right)$$

15. Solve by trigonometry the equations

$$\left.\begin{aligned} x + 2y - xy^2 + \sqrt{3}(1 - 2xy - y^2) &= 0 \\ y + 2x - x^2y + (2 + \sqrt{3})(1 - 2xy - x^2) &= 0 \end{aligned}\right\}$$

16. Reduce to a single term

$$\sin nA + \sin nB + \sin nC$$

where A, B, C are angles of a triangle; and n is any positive integer.

17. If D, E, F are the mid points of the sides of a triangle ABC, and G its centroid; and if $D\alpha$, $E\beta$, $F\gamma$ are perpendicular and proportional to the sides —all being drawn outwards, or all inwards—prove that the mean centre of α, β, γ, for equal multiples, is G.

18. If α, β, γ, δ are four values of θ satisfying the equation

$$\sin 2\theta - p \cos\theta - q \sin\theta + r = 0$$

prove that $\alpha + \beta + \gamma + \delta = n\pi$

Wolstenholme: 414

19. If $a \sin(x + \alpha) + b \sin(y + \beta) \equiv \rho \sin(x + \alpha + y + \beta + \phi)$ determine ρ and ϕ in terms of the other letters.

20. If $f(\theta) = \sqrt{m^2 - n^2 \sin^2\theta}$, show that the area of the triangle whose sides are

$$f(\theta + \pi/4), \quad \sqrt{2}\, f(\theta), \quad f(\theta - \pi/4) \quad \text{is} \quad m^2(m^2 - n^2).$$

21. ABCD is a cyclic quadrilateral, circumscribing a circle: AB, DC meet in P: BC, AD meet in Q: if ρ_1 is the in-radius of triangle PBC; ρ_2 the in-radius of QCD; ρ_3 the ex-radius of PAD within the angle P; ρ_4 the ex-radius of QAB, within the angle Q; prove that

$$\text{in-radius of ABCD} = \sqrt[4]{\rho_1 \rho_2 \rho_3 \rho_4}.$$

22. If x, y, z are connected with the sides of a triangle ABC by the relation $ax + by + cz = 0$, prove that

$$\frac{y^2 + z^2 + 2yz \cos A}{a^2} = \frac{z^2 + x^2 + 2zx \cos B}{b^2} = \frac{x^2 + y^2 + 2xy \cos C}{c^2}$$

23. If P, Q, R are the respective mid points of the arcs BC, CA, AB of the circum-circle of a triangle ABC, prove that

4 area $\triangle$ PQR = 2 area hexagon PCQARB = area principal ex-central $\triangle$

24. Two circles are drawn through the circum-centre of a triangle to touch the same two sides; if ρ_1, ρ_2 are their radii, prove that

$$\frac{1}{\rho_1} + \frac{1}{\rho_2} - \frac{r}{\rho_1 \rho_2} = \frac{2R + r}{R^2}$$

Tucker: E. T. xvi

25. Prove that $2\Sigma \sin^2(A-B) + 2\Sigma \sin(A-B)\sin(C-A)\cos A$

$$= (1 - 8\Pi \cos A)\,\Sigma \sin^2 A$$

26. If $x\cot\alpha + y\cot\beta + z\cot\gamma = (x+y+z)\cot\alpha\cot\beta\cot\gamma$
and $(x+y)\cot\alpha\cot\beta + (y+z)\cot\beta\cot\gamma + (z+x)\cot\gamma\cot\alpha = 0$

prove that $\Sigma x \sin 2\alpha = 0$
and $\Sigma x \cos^2\alpha = 0$

Queen's Camb': '48

27. If $x_1 = \dfrac{\sin(\theta-\alpha)}{\sin(\alpha-\gamma)\sin(\alpha-\beta)}$, $x_2 = \dfrac{\sin(\theta-\beta)}{\sin(\beta-\alpha)\sin(\beta-\gamma)}$,

$x_3 = \dfrac{\sin(\theta-\gamma)}{\sin(\gamma-\beta)\sin(\gamma-\alpha)}$, and $S_r = x_1^r + x_2^r + x_3^r$,

prove that $\dfrac{S_7}{7} = \dfrac{S_2}{2} \cdot \dfrac{S_5}{5}$

NOTE—*Show that* x_1, x_2, x_3 *are roots of* $x^3 + p_2 x + p_3 = 0$; *and deduce from Newton's relations between roots and coeff's.*

28. C is the centre of a circle; AC a radius produced through C to B, so that CB is one-third CA; P is any point on the circumference; PB produced meets the circle in Q; QC produced meets PA in R: if θ is the angle CRA, and ϕ the angle CAR, prove that

$$\frac{1 - \sin\theta}{1 + \sin\theta} = \left(\frac{1 - \sin\phi}{1 + \sin\phi}\right)^3$$

W. Roberts: E. T.

29. If $m\sin(\theta + \phi) = \cos(\theta - \phi)$ prove that

$$\frac{1}{1 - m\sin 2\theta} + \frac{1}{1 - m\sin 2\phi} = \frac{2}{1 - m^2}$$

30. If $(1 + 2\cos^{\frac{2}{3}}\alpha)(1 + 2\cos^{\frac{2}{3}}\beta) = 3$ prove that

$$\frac{(1 + 8\cos^2\alpha)^{\frac{3}{2}}}{\sin^3\alpha\cos\alpha} = \frac{(1 + 8\cos^2\beta)^{\frac{3}{2}}}{\sin^3\beta\cos\beta}$$

Wolstenholme: E. T. xxix

31. Given that

$$\left.\begin{aligned} A^2\cos^2(\theta - \alpha) + B\cos(\theta + \alpha)\cos(\theta - \alpha) &= \frac{A^2 - B^2}{4A} \\ \text{and}\quad A^2\cos^2(\theta - \beta) + B\cos(\theta + \beta)\cos(\theta - \beta) &= \frac{A^2 - B^2}{4A} \end{aligned}\right]$$

find $A^2\cos^2(\alpha-\beta) + B\cos(\alpha+\beta)\cos(\alpha-\beta)$ in terms of A and B.

R. U. Schol': '86

32. Show that the roots of $x^3 - 3x - 1 = 0$ are

$$2\cos 20^\circ, \quad -2\sin 10^\circ, \quad -2\cos 40^\circ.$$

R. U. Schol': '87

33. If $\mu\tan\phi = \tan\phi'$, where μ is given, find $\tan\phi$, $\tan\phi'$ so that $\phi' - \phi$ is *maximum*.

34. If
$$\frac{\cos\theta\cos\frac{\phi}{2}}{\cos\left(\theta - \frac{\phi}{2}\right)} + \frac{\cos\phi\cos\frac{\theta}{2}}{\cos\left(\phi - \frac{\theta}{2}\right)} = 1,$$

prove that $\cos\theta + \cos\phi = 1$

35. If α, β, γ are the three values of θ, unequal, positive and each less than two right angles, satisfying the equation

$$a\tan 3\theta + b\tan\theta + c = 0$$

$$\left.\begin{aligned} \text{prove that}\quad \sin(\alpha + \beta + \gamma) &= 2\sin\alpha\sin\beta\sin\gamma \\ \text{and}\quad 2(a + b)\tan(\alpha + \beta + \gamma) + c &= 0 \end{aligned}\right\}$$

Jes', Christ's and Emm', Camb': '90

36. P is any point in the plane of a triangle ABC: PX, PY, PZ are isoclinals to the sides, making with them the same angle ϕ measured one way round: prove that

$$\frac{\text{area XYZ}}{\text{area ABC}} = \pm\frac{R^2 - SP^2}{4R^2}\operatorname{cosec}^2\phi$$

according as P is inside or outside the triangle; where S is the centre, and R the radius of the circum-circle of ABC.

NOTE—*Compare with Exercise* 25 *on p.* 261.

37. Given that

$$\tan\theta\tan\phi = \left\{\frac{(m-1)(3+m)}{(m+1)(3-m)}\right\}^2 \quad \text{and} \quad \frac{\tan\theta}{\tan\phi} = \frac{(m+1)(3+m)}{(m-1)(3-m)}$$

find the value of $\sqrt{\sin\theta\sin\phi} + \sqrt{\cos\theta\cos\phi}$

38. Find, in its simplest form, the fourth root of

$$4\sin^2\theta\sin^2\phi\cos^2(\theta+\phi) + 2(\sin^2\theta+\sin^2\phi)\sin^2(\theta+\phi) - (\sin^2\theta+\sin^2\phi)^2$$

39. If $\tan(\theta-\phi) = \cos\alpha\tan\phi$, show that

$$(1-\sin^2\alpha\sin^2\phi)\left(1-\tan^4\frac{\alpha}{2}\sin^2\theta\right)$$

may be expressed as a perfect square.

40. If $\cos\theta = \dfrac{\cos\phi(4-\sin^2\phi)}{4+5\sin^2\phi}$ show that the value of

$$\frac{\tan\left(\frac{\theta}{2}+\phi\right)}{\tan\phi}$$ is independent of the angles.

41. If $\sin\phi = \dfrac{(1+m)\sin\theta}{1+m\sin^2\theta}$ and $\sin\chi = \dfrac{(1-m)\sin\theta}{1-m\sin^2\theta}$, express $\tan\dfrac{\phi-\chi}{2}$ in terms of θ.

42. If $\tan\phi_1 = \dfrac{(1+m)\tan\phi}{1-m\tan^2\phi}$ and $\tan\phi_2 = \dfrac{(1-m)\tan\phi}{1+m\tan^2\phi}$, find m in terms of ϕ_1 and ϕ_2.

NOTE—*The last six Exercises are taken from the T. C. D. Schol' papers of* '74, '75.

43. If $\dfrac{\sin(\alpha+\beta-\theta)}{\sin(\alpha+\beta)\cos^2\gamma} = \dfrac{\sin(\alpha+\gamma-\theta)}{\sin(\alpha+\gamma)\cos^2\beta}$, prove that each

$$= \frac{\sin(\beta+\gamma-\theta)}{\sin(\beta+\gamma)\cos^2\alpha};$$ and that

$$\cot\theta = \frac{\sin^2(\alpha+\beta+\gamma) + \cos(\beta+\gamma)\cos(\gamma+\alpha)\cos(\alpha+\beta)}{\sin(\beta+\gamma)\sin(\gamma+\alpha)\sin(\alpha+\beta)}$$

Wolstenholme: E. T. xxi. *Math' Tri'*: '78

44. If PQRS, pqrs are the quadrilaterals formed respectively by the internal and external bisectors of the angles of a convex quadrilateral; then it is a well-known geometrical result that PQRS and pqrs are cyclic: prove that their respective circum-radii are

$$\frac{AC}{4}\left[\frac{\cos\frac{\alpha+\gamma}{2}\cos\frac{\beta-\gamma}{2}\sim\cos\frac{\beta+\delta}{2}\cos\frac{\alpha-\gamma}{2}}{\sin\frac{\alpha+\beta}{2}\sin\frac{\gamma+\delta}{2}\cos\frac{\alpha-\gamma}{2}\cos\frac{\beta-\delta}{2}}\right]$$

$$\frac{AC}{4}\left[\frac{\cos\frac{\alpha+\gamma}{2}\cos\frac{\beta-\gamma}{2}+\cos\frac{\beta+\delta}{2}\cos\frac{\alpha-\gamma}{2}}{\cos\frac{\alpha+\beta}{2}\cos\frac{\gamma+\delta}{2}\cos\frac{\alpha-\gamma}{2}\cos\frac{\beta-\delta}{2}}\right]$$

where $\alpha, \beta, \gamma, \delta$ are respectively the internal angles made by AC with AB, BC, CD, DA.

Heppel: E. T. lvi

45. In a triangle ABC, prove that

$$\begin{vmatrix} a & b & c \\ \sin^2\frac{A}{2} & \sin^2\frac{B}{2} & \sin^2\frac{C}{2} \\ \cos^2\frac{A}{2} & \cos^2\frac{B}{2} & \cos^2\frac{C}{2} \end{vmatrix} = \frac{(a+b+c)(a-b)(b-c)(c-a)}{2abc}$$

Pet' Camb': '90

46. If $\theta_1 + \theta_2 + \theta_3 + \alpha_1 + \alpha_2 + \alpha_3 = 0$, prove that

$$\begin{vmatrix} \tan(\theta_1+\alpha_1) & \tan(\theta_2+\alpha_1) & \tan(\theta_3+\alpha_1) \\ \tan(\theta_1+\alpha_2) & \tan(\theta_2+\alpha_2) & \tan(\theta_3+\alpha_2) \\ \tan(\theta_1+\alpha_3) & \tan(\theta_2+\alpha_3) & \tan(\theta_3+\alpha_3) \end{vmatrix} = 0$$

Show also that if sin is written for tan the result holds true.

47. If $P \equiv \begin{vmatrix} ac^2 & ba^2 & cb^2 \\ ab^2 & bc^2 & ca^2 \\ \cos A & \cos B & \cos C \end{vmatrix}$ $\quad Q \equiv \begin{vmatrix} ac & a^2 & bc \\ ab & bc & a^2 \\ \frac{1}{2} & \cos B & \cos C \end{vmatrix}$

where the letters are parts of a triangle ABC, prove that

$$P = Q\,\Sigma a^2$$

48. Prove that the evaluation of each of the following determinants leads to the given result—

(1) $$\begin{vmatrix} 1 & 1 & 1 & 1 \\ 1 & 1 & \cos\gamma & \cos\beta \\ 1 & \cos\gamma & 1 & \cos\alpha \\ 1 & \cos\beta & \cos\alpha & 1 \end{vmatrix} = -16\,\Pi \sin^2\frac{\alpha}{2}$$

(2) $$\begin{vmatrix} 1 & 1 & 1 \\ \sin\alpha & \sin\beta & \sin\gamma \\ \cos\alpha & \cos\beta & \cos\gamma \end{vmatrix} = \Sigma \sin(\alpha-\beta)$$

(3) Same determinant $= -4\,\Pi \sin\dfrac{\beta-\gamma}{2}$

(4) $$2\begin{vmatrix} \cos\frac{\beta-\gamma}{2} & \cos\frac{\gamma-\alpha}{2} & \cos\frac{\alpha-\beta}{2} \\ \cos\frac{\beta+\gamma}{2} & \cos\frac{\gamma+\alpha}{2} & \cos\frac{\alpha+\beta}{2} \\ \sin\frac{\beta+\gamma}{2} & \sin\frac{\gamma+\alpha}{2} & \sin\frac{\alpha+\beta}{2} \end{vmatrix} = \text{either of last results}$$

NOTE—*Multiply the results of* (2) *and* (3) *giving the square of the determinant.*

(5) $$8\begin{vmatrix} 0 & \sin^2\frac{C}{2} & \sin^2\frac{B}{2} \\ \sin^2\frac{C}{2} & 0 & \sin^2\frac{A}{2} \\ \sin^2\frac{B}{2} & \sin^2\frac{A}{2} & 0 \end{vmatrix} = (\Sigma \sin A)^2$$

where A, B, C are angles of a triangle.

Baltzer

ANSWERS

Note—*Many of the results following are approximate: unless otherwise stated,* 3·1416 *is taken as the numerical value of* π.

Page 15

1. $13^\circ\cdot\dot{8}4615\dot{3}$ **2.** $^r(\cdot 22689)$ **3.** $(1\cdot 6534)$

4. $\pi/12$ **5.** $180\,n\,\alpha/\pi\,\beta$ **6.** 1 ft $6\frac{1}{2}$ in

7. $143^\circ\ 14'\ 20''\cdot 809$ **8.** 6875·4 yds

9. 10 in, 14·142 in, 17·328 in

10. 30°, 25°, 125°; or 30°, $15 \times 100'$, $45 \times 10000''$

11. 1·15192 miles

Page 16

12. $57'\cdot 29578$ **13.** 57·29578 ft

15. $180/(180 - \pi) = 1^\circ\cdot 019$ **16.** 3·15 **17.** ·3071

20. 3·1416 **23.** ·162 sq ft

Page 17

24. ·0905 **25.** 185575 miles per sec

26. 2276 miles per hr **27.** 1108000000 grs

28. 26590439766400000 using 3·1415926

26590502400000000 using 3·1416

error in excess 62633600000 sq miles

29. N° sides are $\left.\begin{matrix}5\\8\end{matrix}\right]$ or $\left.\begin{matrix}6\\12\end{matrix}\right]$ or $\left.\begin{matrix}8\\32\end{matrix}\right]$ or $\left.\begin{matrix}9\\72\end{matrix}\right]$

30. $\dfrac{10\,\pi}{1819}, \ \dfrac{9\,\pi}{1819}, \ \dfrac{1800\,\pi}{1819}$

31. $\dfrac{2250\,\pi}{6289}, \ \dfrac{2500\,\pi}{6289}, \ \dfrac{1539\,\pi}{6289}$

Page 18 **35.** Same as **29.**

36. $\frac{20\,mn - 18}{10\,n - 9}$, $\frac{20\,mn - 18}{m\,(10\,n - 9)}$; where m = 1st ratio, n = 2nd „

37. $3^\circ\,2'\,27''$, $2^\circ\,44'\,12''$, $174^\circ\,13'\,21''$

38. 4°, 16° **39.** $d^2(5\pi - 6\sqrt{3})/24$

Page 31 **1.** $\frac{2x(x+1)}{2x^2+2x+1}$, $\frac{2x+1}{2x^2+2x+1}$

2. $\sin\alpha = \frac{1}{\sqrt{2}} = \cos\alpha$

3. $\pm q/\sqrt{p^2+q^2}$, $\pm p/\sqrt{p^2+q^2}$ **4.** 1

5. $\cos\alpha = \cdot 999$, $\tan\alpha = \cdot 012$, $\cot\alpha = 83\cdot 321$,
$\sec\alpha = 1\cdot 001$, $\operatorname{cosec}\alpha = 83\cdot 333$

6. $3\sqrt{2} + 2$ **8.** $\sin\alpha = \sqrt{2\,\text{vers}\,\alpha - \text{vers}^2\alpha}$

Page 51 **1.** 0 **2.** 1 **3.** $(1 - \sin\alpha)(1 + \cos\alpha)$ **4.** 0

Page 55 **1.** 1 **2.** $-\sin 3^\circ - \cos 13^\circ$

Page 74 **1.** + + **2.** − − **3.** $2n\pi \pm \pi/4$

4. $2n\pi + 3\pi/4$, $2n\pi + 5\pi/4$

Page 85 **1.** (1) ·3535, ·169 (2) ·612, 1·959, − 1·379

(3) $\frac{527}{625}$, $\frac{11753}{10296}$, -7 or $\frac{1}{7}$

(4) $\frac{5}{12}$ or $\frac{12}{5}$, $\pm\frac{2035}{2197}$ or $\pm\frac{828}{2197}$

Page 86 (5) $\pm\frac{20\sqrt{6}}{49}$, $\pm\frac{\sqrt{6}\pm 1}{\sqrt{14}}$ (6) $\pm\frac{2mn}{m^2+n^2}$

Page 90 **15.** $\sin x = \frac{a^2+b^2-c^2}{2ab}$, $\sin y = \frac{c^2+a^2-b^2}{2ca}$

$$\sin\theta = \frac{c^2 - b^2}{a\sqrt{2b^2 + 2c^2 - a^2}}$$

Page 91 **21.** $\tan(\alpha + \beta)$ **22.** $\sqrt{\dfrac{1 \pm \sqrt{1 - m^2}}{m^2 - 1 \mp \sqrt{1 - m^2}}}$

23. $-\frac{4}{5}, -\frac{3}{5}$

28. (1) $\dfrac{(a^2 + b^2)^2 - 4b^2}{4(a^2 + b^2)}$ (2) $\dfrac{(a^2 + b^2)^2 - 4a^2}{4(a^2 + b^2)}$

(3) $\dfrac{4ab}{(a^2 + b^2)^2 - 4a^2}$ (4) $\dfrac{4a}{a^2 + b^2 + 2b}$

(5) $\dfrac{(b^2 - a^2)(a^2 + b^2 - 2)}{a^2 + b^2}$

(6) $b\left\{4b^2 - 3 - 3\dfrac{(a^2 + b^2)^2 - 4a^2}{a^2 + b^2}\right\}$

Page 100 **1.** $\pm\frac{1}{2}, \pm\frac{\sqrt{3}}{2}$ **2.** $\pm 1, \pm 2$ **3.** $1 \pm \frac{1}{\sqrt{2}}$

Page 102 **1.** 0 **3.** $\sqrt{2}\sqrt{2 + \sqrt{2}} - \sqrt{2} - 1$

5. $\sqrt{6} + \sqrt{3} - \sqrt{2} - 2, \sqrt{6} - \sqrt{3} - \sqrt{2} + 2$

Page 103 **10.** $P\hat{X}C = 67°\frac{1}{2}, P\hat{C}X = 22°\frac{1}{2}$

Page 107 **2.** $32x^5 - 16x^4 - 32x^3 + 12x^2 + 6x - 1 = 0$

3. $8x^3 + 4x^2 - 4x - 1 = 0$

4. $128x^7 - 64x^6 - 192x^5 + 80x^4 + 80x^3 - 24x^2 - 8x + 1 = 0$

Page 126 **4.** $n\pi \pm \pi/4$ **5.** $2n\pi + \alpha$

6. $2n\pi + \alpha$ **7.** No

Page 127 **8.** (1) $2n\pi \pm \pi/3$ (2) $2n\pi \pm \pi/3$ (3) $n\pi \pm \pi/3$

(4) $3\theta - \pi/4 = 2n\pi \pm \pi/3$ (5) $n\pi/2, 2n\pi \pm 2\pi/3$

(6) $(4n \pm 1)\pi/2, (6n \pm 1)\pi/3(n - 1)$ (7) $n\pi$

(8) $n\pi,\ n\pi/2+(-1)^n\pi/4$ (9) $n\pi \pm \pi/6$

(10) $n\pi,\ 2n\pi \pm 2\pi/3$ (11) $n\pi + \pi/4$

(12) $n\pi + \pi/4$ (13) $2n\pi \pm \pi/5,\ 2n\pi \pm 3\pi/5$

(14) $2n\pi \pm \pi/6$ (15) $(4n+3)\pi/4,\ (4n+3)\pi/8$

(16) $(4n+1)\pi/8,\ (4n-1)\pi/4$

(17) $(4n+1)\pi/8,\ -(4n-1)\pi/4,\ \pi/4$

(18) $2n\pi \pm \pi/6$ (19) $(2n+1)\pi/5,\ (2n+1)\pi/7$

(20) $n\pi/2 + \alpha,\ n\pi/2$ (21) $(4n-1)\pi/6 \pm \alpha$

Page 128 (22) $n\pi \pm \pi/3$ (23) $n\pi + 3\pi/4$ (24) $2n\pi + \pi/4$

(25) $2n\pi \pm \pi/2$ or $\cos\theta = \dfrac{1 \pm \sqrt{2353}}{56}$

(26) $(2n+1)\pi/4 \pm \alpha$ (27) $\theta = n\pi$ or $\tan\theta = \pm\sqrt{\frac{5}{3}}$

12. $3, \frac{1}{3}$; $m\pi,\ m\pi \pm \pi/4$ **15.** $(6n \pm 1)\pi/3,\ (2n+1)\pi$

Page 129 **16.** 150° **17.** 70°, 110°, 190°, 230°, 310°, 350°

20. $n\pi \pm \pi/4,\ n\pi \pm \pi/6$ **21.** $\pm\sqrt{3},\ \pm\dfrac{1}{\sqrt{3}}$

Page 130 **25.** (1) $n\pi + (-1)^n\pi/6$ (2) $n\pi \pm \pi/3$

Page 134 **2.** $\tan\alpha,\ \tan(\alpha \pm \pi/3)$; $x^3 - 3x^2\tan 3\alpha - 3x + \tan 3\alpha = 0$

5. $\cot\alpha + \cot(\alpha + \pi/3) + \cot(\alpha - \pi/3) = 3\cot 3\alpha$

$$\cot\alpha\cot(\alpha+\pi/3) + \cot(\alpha+\pi/3)\cot(\alpha-\pi/3) + \cot\alpha\cot(\alpha-\pi/3) = -3$$

$\cot\alpha\cot(\alpha+\pi/3)\cot(\alpha+2\pi/3) = \cot 3\alpha$

6. 6

Page 135 **8.** $\pm\tan\alpha,\ \tan(\pi/4 \pm \alpha)$ **9.** 12

10. $x^6 \tan 6\alpha + 6x^5 - 15x^4 \tan 6\alpha - 20x^3$

$$+ 15x^2 \tan 6\alpha + 6x - \tan 6\alpha = 0$$

15. $x^5 - 55x^4 + 330x^3 - 462x^2 + 165x - 11 = 0$

Page 146 **32.** Out of place here: compare the fig′ on p. 279, with the results at end of p. 277.

Page 147 **37.** (1) $\pm \frac{1}{3}$ (2) $\sqrt{2 \cot \pi/12}$ (3) $1 \pm \sqrt{2\alpha/(1+\alpha)}$

(4) $\pm 1, \frac{1}{2}$ (5) $\sqrt{3}$ (6) $\pm(\sqrt{3} + 1)$

38. (1) $n\pi \pm \frac{\pi}{4}$ or $n\pi \pm \frac{1}{2}\cos^{-1}\frac{1 \pm \sqrt{17}}{8}$

(2) $2n\pi \pm \cos^{-1}\frac{1 \pm \sqrt{17}}{8}$

(3) $n\pi + \tan^{-1}\frac{\sin(\alpha + \beta)}{\sin\alpha \sin\beta}$

(4) $2n\pi \pm \cos^{-1}\frac{1}{\sqrt{5}}$ or $2n\pi \pm \cos^{-1}\frac{1}{5\sqrt{13}}$

(5) $2\theta = n\pi - \tan^{-1}\frac{2\sin\alpha\sin\beta}{\sin(\alpha + \beta)}$

(6) $n\pi - \alpha - (-1)^n \sin^{-1}\frac{\sin 2\alpha}{2}$

(7) $2\theta - \alpha = \tan^{-1}\left(\frac{m - n}{m + n}\tan\alpha\right)$

(8) $(-1)^n \frac{\pi}{4} + \cos^{-1}\left\{\frac{2n + (-1)^n}{2\sqrt{2}}\right\}$

Page 148 (9) $2\theta = n\pi + \tan^{-1}\frac{2\sin\alpha\sin\beta}{\sin(\alpha + \beta)}$

(10) $\cos\theta = \frac{\sqrt{q}}{\sqrt[4]{pq - p^2}}$, $\cos\phi = \sqrt[4]{\frac{p}{q - p}}$

(11) Add, subtract, and multiply results; this gives $\cos\theta/\cos\phi$: then by division $\tan\theta/\tan\phi$ is found. The rest is obvious.

(12) $\theta + \phi = m\pi + (-1)^m \sin^{-1}(a + b)$,

$\theta - \phi = n\pi + (-1)^n \sin^{-1}(a - b)$

(13) $\theta = r\pi + \tan^{-1}\left\{\frac{m}{n^2-4m^2}\left(\pm 2\sqrt{n^2-3m^2} - n\right)\right\}$

(14) $\left.\begin{array}{l} 2\theta = \sin^{-1}\frac{3}{10} + \sin^{-1}\frac{1}{10} \\ 2\phi = \sin^{-1}\frac{3}{10} - \sin^{-1}\frac{1}{10} \end{array}\right\}$

or $\left.\begin{array}{l} 2\theta = \sin^{-1}\frac{1}{30} + \sin^{-1}\frac{1}{10} \\ 2\phi = \sin^{-1}\frac{1}{30} - \sin^{-1}\frac{1}{10} \end{array}\right\}$

(15) $\theta + \phi = \sin^{-1}\frac{\alpha^2 + \beta^2 - 2}{2}$, $\theta - \phi = \sin^{-1}\frac{\alpha^2 - \beta^2}{\alpha^2 + \beta^2}$

(16) $n\pi + \pi/2$, $2n\pi \pm \cos^{-1}\sqrt{\frac{7 \pm \sqrt{-7}}{8}}$

(17) $n\pi$, $\cos\theta = \pm 2\sin^2\frac{\alpha}{2}$ or $\pm 2\cos^2\frac{\alpha}{2}$,

as $\alpha <$ or $> \frac{\pi}{2}$

(18) $2n\pi \pm \sec^{-1}(\tan\alpha\tan\beta \pm \sec\alpha\sec\beta)$

(19) $n\pi \pm \pi/12$, $\theta = n\pi + \frac{1}{2}\cos^{-1}\left(\frac{1}{5 - 4\sqrt{3}}\right)$

(20) $2n\pi \pm \pi/3$, $2n\pi \pm \cos^{-1}\frac{1}{3}$, $n\pi + \tan^{-1}2$

Page 150 **46.** $2Rr\sin\alpha/\sqrt{R^2 + r^2 + 2Rr\cos\alpha}$

Page 158 **1.** (1) $(a^2 - b^2)^2 = 16b$ (2) $a + b + 1 = ab$

(3) $(ab)^{\frac{2}{3}}(a^{\frac{2}{3}} + b^{\frac{2}{3}}) = 1$

(4) $(a^2 + b^2)^2 - 3(a^2 + b^2) - 2b = 0$

(5) $a^{\frac{2}{3}} - b^{\frac{2}{3}} = 2^{\frac{2}{3}}$

(6) $\cos(\alpha - \beta) = (aa' + bb')/(ab' + a'b)$

Page 159

(7) $2(a^2 + b^2)^3 = (a^2 - b^2)^2$

(8) $\cos(\alpha - \beta) + \sin\alpha \sin\beta = 0$

(9) $\sin\alpha / \sin\beta = (1 + m)/(1 - m)$

(10) $\sin\alpha \tan z + \sin\beta \tan x = \sin(\alpha + \beta) \tan y$

(11) $(a^2 - b^2 - c^2)\sin\alpha = 2bc$

(12) $(a^2 + b^2 - c^2)^3 = 4a^2b^2(a^2 + b^2 - c^2) + a^2b^2c^2$

(13) $\{2(x^2 + y^2)^2 - 3ax + a^2\}^2 = a^2\{(x - a)^2 + y^2\}$

(14) $a/(b - c) + b/(c - a) + d/(a - b) = 0$

(15) $\cos\dfrac{\alpha - \beta}{2} = \pm k \sin(\alpha + \beta)$

(16) $2m = \cos\alpha \pm \cos 2\beta$

(17) $(a^2 + b^2)b^2 = 4c^2(a^2 + b^2 - c^2)$

(18) $$\frac{(2a^2\cos\omega - 2xy)^2}{(2a^2\cos\omega - 2xy)^2 + (x^2 - y^2)^2} = \frac{\{x^2 + y^2 - a^2(1 + \cos^2\omega)\}^2}{(2a^2 - x^2 - y^2)^2\cos^2\omega}$$

2. (1) $\left(\frac{x}{a}\right)^{\frac{2}{m}} + \left(\frac{y}{b}\right)^{\frac{2}{m}} + \left(\frac{z}{c}\right)^{\frac{2}{m}} = 1$

Page 160

(2) $\dfrac{\tan^2\alpha}{\tan^2\gamma} = \dfrac{\cos\gamma(\cos\beta - \cos\alpha)}{\cos\alpha(\cos\beta - \cos\gamma)}$

(3) $(a + b)(m + n) = 2mn$ (4) $a^2 - c^2 = m(b^2 - c^2)$

(5) $(a - b)\{(a + b)^2 - c^2\} + 4abcm = 0$

(6) $2ab = c(b^2 - 1)$ (7) $(a^2 - b^2)^2 = 2c^2(a^2 + b^2)$

(8) $a + b + 2cx + 2dy = bx^2 + ay^2$ (9) $m^4 + n^4 = \frac{1}{2}$

Page 161 **3.** $(x/a)^2 + (y/b)^2 = 1$ **4.** $(ax)^{\frac{2}{3}} + (by)^{\frac{2}{3}} = (a^2 - b^2)^{\frac{2}{3}}$

5. $A^{\frac{4}{3}} x^{\frac{2}{3}} / a^{\frac{2}{3}} + B^{\frac{4}{3}} y^{\frac{2}{3}} / b^{\frac{2}{3}} = (A^2 - B^2)^{\frac{2}{3}}$

Page 162 **9.** $y^2 = ax$

10. $b^2 = (a + c)^2, \ \cos\theta = -b/(a+c) = \pm 1$

12. $(a - b)\, x^2 = b^2 (a + b - 2c)$

13. $(x/a)^2 \pm 2xy\sqrt{\mu^2 - 1} / (ab\mu) + (y/b)^2 = 2$

Page 163 **17.** $\left(A - \frac{1}{c^2}\right)\left(B - \frac{1}{a^2}\right)\left(B - \frac{1}{b^2}\right) + C\left(\frac{1}{a^2} + \frac{1}{b^2} - \frac{1}{c^2}\right)$

$= BC^2$

Page 164 **20.** $\{a^2 b^2 - (a^2 - b^2) c^2\} \sin x \sin y$

$+ \{a^2 b^2 + (a^2 - b^2) c^2\} \cos x \cos y = (a^2 + b^2) c^2 - a^2 b^2$

22. $(x + y)^{\frac{2}{3}} + (x - y)^{\frac{2}{3}} = 2a^{\frac{2}{3}}$

23. $(x/a + y/b)^{\frac{2}{3}} + (x/a - y/b)^{\frac{2}{3}} = 2$

24. $a^2 - b^2 - c^2 + 2ac + 4bc = 0$

Page 165 **25.** $abm^3 + b(a + 1)m^2 + (b + c)m + bc - ad = 0$

26. $(x\cos\alpha + y\sin\alpha)^{\frac{2}{3}} + (x\sin\alpha - y\cos\alpha)^{\frac{2}{3}} = (2a)^{\frac{2}{3}}$

28. $\rho^2 - a\rho\cos\alpha - 2a^2 = 0$

30. $p^2 - 2pq\cos\omega + q^2 = \sin^2\omega$

where $\omega = 3\alpha - 2\beta$, $p = 4\left(\frac{x}{a}\right)^3 - 3\left(\frac{x}{a}\right)$, $q = 1 - 2\left(\frac{y}{b}\right)^2$

Page 166 **31.** $\left(\frac{x}{\cos B}\right)^2 + \left(\frac{y}{\cos C}\right)^2 - \frac{2xy\cos 2A}{\cos B\cos C}$

$- 4R\sin A\left(\frac{x}{\cos B} + \frac{y}{\cos C}\right) + 4R^2\sin^2 A = 0$

34. $\{c^3 + c^2 - 2(a^2+b^2)\}^2 \{(c+1)^2 - 4(a^2+b^2)\}$
$= 16a^2b^2(2c+1)^3$

35. By addition and subtraction, we get

$$\sin 3\theta = k\cos\theta \text{ and } \cos 3\theta = k' \sin\theta$$

where $\theta = \phi + \pi/12$, $k = \alpha(\sqrt{3}-1)/(x+y)$,

$$k' = \alpha(\sqrt{3}-1)/(x-y)$$

Page 180 **2.** (1) 0 (2) 1

Page 181 **3.** (1) $\pi/648000$ (2) $61\pi/648000$

(3) α/β (4) $(\alpha/\beta)^2$

4. 2 **8.** 1·0355 **12.** $-\alpha^2/(2\beta^2)$

Page 182 **15.** (1) $\frac{1}{2}$ (2) $\frac{1}{6}$ (3) $\frac{1}{2}$ (4) n

(5) $-\frac{7}{52}$ (6) $\frac{7}{26}$ (7) $\frac{8}{125}$

(8) $8\left(\frac{29}{3}\right)^2$ (9) $\frac{1}{12}$ (10) $\sqrt[3]{e}$

Page 190 **1.** .6073257 **2.** 34° 29′ 45″·7

3. ·5003233 **4.** 60° 1′ 22″.6

Page 193 **1.** 9·7299388 **2.** 9·5906363

3. 9·5050321 **4.** 22° 28′ 16″·43

5. 20° 35′ 15″·95 **6.** 44° 59′ 30″·7

Page 199 **3.** Within $\left(\frac{1}{200}\right)''$

Page 216 **13.** 28 feet

Page 219 **41.** 3, 4, 5, 6

Page 224 **1.** (1) $2\sec\alpha\sec\beta\sin\frac{\alpha+\beta}{2}\sin\frac{\alpha-\beta}{2}$

(2) $-4\sin\frac{\alpha}{2}\sin\frac{\beta}{2}\cos\left(A+\frac{\alpha+\beta}{2}\right)$

(3) 1°, $\left.\begin{array}{l}2\tan\theta\sin\frac{\phi+\chi}{2}\sin\frac{\phi-\chi}{2}=\sin\phi \\ \text{where } b=a\cos\chi\end{array}\right\}$

2°, $a\sin(\theta+\phi)=b\sin\theta$

Page 226 **10.** $2\sec\sin^{-1}\frac{b}{a}$

12. $\cos x=-\frac{\cos A\cos(\phi+B)}{\cos\phi}$,

where $\tan\phi=\cos y\tan A$

15. $\left.\begin{array}{l}4\sec^2\phi\cos^4\frac{x+y}{2}\sin^2\frac{x-y}{2} \\ \text{where } \tan\phi=\tan^2\frac{x+y}{2}\cot\frac{x-y}{2}\end{array}\right\}$

Note that if the $-$ is changed to $+$ the expression

$=2\sin(x+y)\cos\frac{x+y}{2}\cos\frac{x-y}{2}$

Page 235 **1.** 75° 57′ 49″·5 **2.** 9.6733937

Page 236 **3** No, right-angled. **4.** $b=79$, $c=77$

5. $\frac{1}{2\sqrt{2}}\{\sqrt{3c^2-4m^2}\pm\sqrt{12m^2-c^2}\}$, $\frac{\sqrt{3}}{8}(c^2-4m^2)$

6. 70° 53′ 36″·2, 60°, 49° 6′ 23″·8

7. 10·0835683 **8.** $\log c=\cdot9271876$

9. 71° 5′ 45″, 48° 54′ 15″

10. 87° 58′, 50° 46′ **11.** 83° 58′ 28″

13. 18°, 36°, 126°; $4(\sqrt{5}\pm1)/\sqrt{10-2\sqrt{5}}$, 4,

Page 282 **12.** 6

Page 284 **29.** $a^2 \left\{\pi + n\left(\frac{\pi}{2} + \cot\frac{\pi}{n}\right)\right\}$

Page 294 **3.** $a^2/2b$ **6.** 29·719 feet **7.** 11·715728 miles

Page 295 **10.** $(k^2 + hk\cot\beta)/(2k + h\cot\beta - h\cot\alpha)$

13. $h\left\{\sqrt{\cot^2\gamma - \cot^2\alpha} - \sqrt{\cot^2\beta - \cot^2\alpha}\right\}$

14. $2\sqrt{[mn(m+n)/\{m\operatorname{cosec}^2\gamma + n\operatorname{cosec}^2\alpha - (m+n)\operatorname{cosec}^2\beta\}]}$

where 2α, 2β, 2γ are $\wedge^s$ subtended at P, Q, R; and $PQ = m$, $QR = n$

Page 296 **16.** 4·142, 10·824 miles **17.** 15·2c25 miles

19. $(\cot^2\gamma - \cot^2\beta)h^2 + 2tv\sin\alpha\{\cot^2\gamma + \cot\alpha\cot\beta\cos(\delta+\epsilon)\}h$

$= (tv\sin\alpha)^2(\cot^2\alpha - \cot^2\gamma)$

ABC Balloon's path; BB′, CC′ two heights; D place of observation; $BAB' = \alpha$, $BDB' = \beta$, $CDC' = \gamma$, $B'DA = \delta$, $DAB' = \epsilon$; t = time, v = velocity, h = height

22. $\frac{1}{4}(\sqrt{30} + \sqrt{6})$

Page 298 **30.** $\tan^{-1}\dfrac{\tan\alpha\tan\beta + \tan\beta\tan\gamma - 2\tan\alpha\tan\gamma}{2\tan\beta - \tan\alpha - \tan\gamma}$

Page 304 **3.** (1) $\frac{3\sqrt{3}}{2}$ (2) $\frac{3}{2}$ (3) $\frac{1}{8}$ (4) $\frac{3\sqrt{3}}{8}$

5. (1) $\frac{\alpha}{3}$ (2) $m\alpha/(m+n)$ (3) $-45°$

Page 334 **5.** (1) $\tan\alpha - \tan\frac{\alpha}{2^n}$ (2) $\cot\frac{\alpha}{2^n} - \cot\alpha$

(3) $\frac{1}{2}\operatorname{cosec}^2\frac{\alpha}{2} - 2^{n-1}\operatorname{cosec}^2 2^{n-1}\alpha$

(4) $\sec\alpha\sec(\alpha + n\beta)\sin n\beta/\sin\beta$

(5) $$\frac{\cos\frac{n+1}{2}(\alpha+\beta)\sin\frac{n}{2}(\alpha+\beta)}{2\sin\frac{\alpha+\beta}{2}} + \frac{\cos\frac{n+1}{2}(\alpha-\beta)\sin\frac{n}{2}(\alpha-\beta)}{2\sin\frac{\alpha-\beta}{2}}$$

(6) $$\frac{\sec(2n+1)\alpha - \sec\alpha}{2\sin\alpha}$$

(7) $$\frac{n\sin\alpha - \sin n\alpha\cos(n+1)\alpha}{2\sin\alpha}$$

(8) $$\frac{3\sin\frac{n\alpha}{2}\sin\frac{n+1}{2}\alpha}{4\sin\frac{\alpha}{2}} - \frac{\sin\frac{3n\alpha}{2}\sin\frac{3(n+1)}{2}\alpha}{4\sin\frac{3\alpha}{2}}$$

(9) $\frac{1}{2}\operatorname{cosec}^2\alpha - 2^{n-1}\operatorname{cosec}^2 2^n\alpha$ (10) 0

(11) $$(2\cos 2\alpha + 1)\frac{\sin\frac{n+3}{4}\alpha\sin\frac{n\alpha}{2}}{4\sin\frac{\alpha}{2}} - \frac{\sin\frac{n+1}{2}3\alpha\sin\frac{3n\alpha}{2}}{4\sin\frac{3\alpha}{2}}$$

(12) $\sin\alpha\cot\frac{\alpha}{2^n} - 1$ (13) $\operatorname{cosec}\alpha\tan^2 n\alpha$

(14) $\log\sin\frac{3^n\alpha}{2} - \log\sin\frac{\alpha}{2}$ (15) $\cot\pi/2^{n+1}$

(16) $\tan^{-1} n(n+1)$ (17) $\frac{\pi}{4} - \tan^{-1}\frac{1}{n+1}$

(18) $\tan^{-1} n\alpha$

Page 335 **6.** (1) $\frac{3}{4}(\sin\theta - 3^{-n}\sin 3^n\theta)$

(2) $\frac{3}{4}\{\cos\theta + (-3)^n\cos 3^n\theta\}$

(3) $\frac{1}{2}(\cot\theta - \cot 3^n\theta)$ (4) $\frac{1}{4}(\cot\theta - 3^n\cot 3^n\theta)$

(5) $\cot\theta - 2^n\cot 3^n\theta$ (6) $\cot\theta - \cot 4^n\theta$

(7) $\cot\theta - 3^n\cot 2^n\theta$ (8) $\cot\theta - 4^n\cot 4^n\theta$

(9) $\cot\theta - 3^n\cot 4^n\theta$ (10) $\frac{3}{2}(3^{-n}\tan 3^n\theta - \tan\theta)$

8. (1) $e^{\sin\theta\cos\theta}\sin(\theta + \sin^2\theta)$

(2) $\frac{1 - e^{-c\theta}\cos n\theta}{1 - 2e^{-c\theta}\cos n\theta + e^{-2c\theta}}$

(3) $\pi/12$ (4) $\pi/4$

Page 336 **11.** $\pi^2/8$, $\pi^4/96$ **16.** $e^{x\cos\beta}\cos(\alpha + x\sin\beta)$

Page 337 **18.** (1) $\frac{\sin\alpha}{1 - 2x\cos\alpha + x^2}$ (2) $\frac{\sin\alpha - \sin(\alpha-\beta)\sin\theta}{1 - 2\cos\beta\sin\theta + \sin^2\theta}$

(3) $\cos(\alpha - \beta)/\sin\beta$ (4) $\operatorname{cosec}\alpha$

19. (1) $\frac{1}{2}e^{\sin\theta}\cos(\cos\theta) + \frac{1}{2}e^{-\sin\theta}\cos(\cos\theta)$

(2) $\cos\theta/(1 - \cos\theta) - \log(1 - \cos\theta)$

20. $\frac{\cos\{4\alpha + (n-1)2\beta\}\sin 2n\beta}{8\sin 2\beta}$

$- \frac{\cos\{2\alpha + (n-1)\beta\}\sin n\beta}{2\sin\beta} + \frac{3n}{}$

21. (1) $\frac{\sin(\sin\alpha)}{e^{\cos\alpha}}$ (2) $e^{\cos^2\alpha}\sin(\cos\alpha\sin\alpha)$

22. (1) $e^{\cos\alpha\cos\beta}\cos(\alpha+\cos\alpha\sin\beta)$

(2) $e^{\sin\alpha\cos\beta}\cos(\alpha+\sin\alpha\sin\beta)$

Page 338 **24.** (1) $\frac{\sin\alpha}{1+\sin^2\beta}$, $\frac{\sin\alpha\{1-(-1)^n\sin^{2n}\beta\}}{1+\sin^2\beta}$

(2) $\frac{\sin\alpha}{\cos^2\beta}$, $\frac{\sin\alpha(1-\sin^{2n}\beta)}{\cos^2\beta}$

(3) $\frac{\cos\alpha\sin\beta}{1+\sin^2\beta}$, $\frac{\cos\alpha\sin\beta\{1-(-1)^n\sin^{2n}\beta\}}{1+\sin^2\beta}$

(4) $\frac{\cos\alpha\sin\beta}{\cos^2\beta}$, $\frac{\cos\alpha\sin\beta(1-\sin^{2n}\beta)}{\cos^2\beta}$

(5) $\frac{\sin\alpha}{1+x\sin^2\beta}$, $\frac{\sin\alpha\{1-(-1)^n x^n\sin^{2n}\beta\}}{1+x\sin^2\beta}$

(6) $\sin\alpha\cos(x\sin\beta)$ (7) $\cos\alpha\sin(x\sin\beta)$

(8) $2\sin\alpha\log\cos\beta$

25. (1) $\frac{\cos\alpha}{1+\sin^2\beta}$, $\frac{\cos\alpha\{1-(-1)^n\sin^{2n}\beta\}}{1+\sin^2\beta}$

(2) $\frac{\cos\alpha}{\cos^2\beta}$, $\frac{\cos\alpha(1-\sin^{2n}\beta)}{\cos^2\beta}$

(3) $-\frac{\sin\alpha\sin\beta}{1+\sin^2\beta}$, $-\frac{\sin\alpha\sin\beta\{1-(-1)^n\sin^{2n}\beta\}}{1+\sin^2\beta}$

(4) $-\frac{\sin\alpha\sin\beta}{\cos^2\beta}$, $-\frac{\sin\alpha\sin\beta(1-\sin^{2n}\beta\}}{\cos^2\beta}$

(5) $\frac{\cos\alpha}{1+x\sin^2\beta}$, $\frac{\cos\alpha\{1-(-1)^n x^n\sin^{2n}\beta\}}{1+x\sin^2\beta}$

(6) $\cos\alpha\cos(x\sin\beta)$ (7) $-\sin\alpha\sin(x\sin\beta)$

(8) $2\cos\alpha\log\cos\beta$

Page 339 **28.** (1) $\cos\alpha$ (2) $\sin\alpha$ (3) $\cos\alpha$ (4) $\sin\alpha/\alpha$

29. (1) $$\frac{2^{2n}}{(2n+1)!} = \frac{1}{(2n+1)!} + \frac{1}{(2n-1)!\,2!} + \frac{1}{(2n-3)!\,4!} + \dots + \frac{1}{(2n)!}$$

(2) $$\frac{2^{4n-2}}{(4n)!} = \frac{1}{(2n-1)!\,(2n+1)!} + \frac{1}{(2n-3)!\,(2n+3)!} + \dots + \frac{1}{(4n-1)}$$

(3) $$\frac{2^{4n}}{(4n+2)!} = \frac{1}{(2n)!\,(2n+2)!} + \frac{1}{(2n-2)!\,(2n+4)!} + \frac{1}{(2n-4)!\,(2n+6)!} + \dots + \frac{1}{(4n+2)!}$$

31. $$\cot x = \frac{1}{x} - \frac{1}{\pi - x} + \frac{1}{\pi + x} - \frac{1}{2\pi - x} + \frac{1}{2\pi + x} - \dots$$

$$\operatorname{cosec} x = \frac{1}{x} + \frac{2x}{\pi^2 - x^2} - \frac{2x}{2^2\pi^2 - x^2} + \frac{2x}{3^2\pi^2 - x^2} - \dots$$

32. $$\sqrt{2}\left(x + \frac{\pi}{4}\right)\left\{1 - \left(\frac{4x+\pi}{4\pi}\right)^2\right\}\left\{1 - \left(\frac{4x+\pi}{8\pi}\right)^2\right\}\dots$$

Page 345 **1.** $-2,\ 1 \pm \sqrt{3}$ **3.** 1, 7, 10

4. -3, 1·7936, ·2841, $-$·55754, $-$3·5201

5. $x^3 - 14x - 12 = 0$, 4·1133

6. 1·87949 is the solution; but $-$·3474 and $-$1·53209 can be geometrically interpreted

Page 357 **3.** $\frac{\sqrt{3}}{8}$ **5.** $\tan\alpha \tan\beta \tan\gamma$

Page 358 **7.** $$\Sigma\{a^2 l^2 (a^2 - b^2 - c^2 + l^2 - m^2 - n^2)\} + \Sigma(a^2 m^2 n^2) + a^2 b^2 c^2 = 0$$

12. $-\sec\alpha$, $-\operatorname{cosec}\alpha$; $x^2 \sin^2 2\alpha - 4x + 4 = 0$

Page 359 **15.** $x = \tan\left\{(2n - m)\frac{\pi}{3} + \frac{\pi}{6}\right\}$

$$y = \tan\left\{(2m - n)\frac{\pi}{3} + \frac{\pi}{4}\right\}$$

16. $-4\cos\frac{n\pi}{2}\sin\frac{nA}{2}\sin\frac{nB}{2}\sin\frac{nC}{2}$,

when n is $4m$, or $4m + 2$

$4\sin\frac{n\pi}{2}\cos\frac{nA}{2}\cos\frac{nB}{2}\cos\frac{nC}{2}$,

when n is $4m + 1$ or $4m + 3$

19. $\rho^2 = \frac{1}{2}(a^2 + b^2)$, $\tan\frac{\phi}{2} = \tan\frac{x - y + \alpha - \beta}{2}$

Page 361 **31.** $(A^2 - B^2)/4A$ **33.** $\tan\phi = \sqrt{\mu}$, $\tan\phi' = \frac{1}{\sqrt{\mu}}$

Page 362 **37.** ± 1, $\pm(m^2 - 3)/(2m)$ **38.** $\pm\sin(\theta + \phi)$

39. $\{\cos(2\phi - \theta)\cos\phi/\cos(\theta - \phi)\}^2$ **40.** $\frac{3}{2}$

41. $\pm m\sin\theta\cos\theta/\sqrt{1 - m^2\sin^2\theta}$

42. $(\sin\phi_1 \pm \sin\phi_2)^2/(\sin^2\phi_1 - \sin^2\phi_2)$

THE END

www.ingramcontent.com/pod-product-compliance
Lightning Source LLC
Chambersburg PA
CBHW021352210326
41599CB00011B/845
9783744738941